ONE OF TEN BILLION EARTHS

How We Learn about Our Planet's Past
and Future from Distant Exoplanets

Karel Schrijver

OXFORD
UNIVERSITY PRESS

OXFORD
UNIVERSITY PRESS

Great Clarendon Street, Oxford, OX2 6DP,
United Kingdom

Oxford University Press is a department of the University of Oxford.
It furthers the University's objective of excellence in research, scholarship,
and education by publishing worldwide. Oxford is a registered trade mark of
Oxford University Press in the UK and in certain other countries

First published 2018
First published in paperback 2021

Impression: 1

Published in the United States of America by Oxford University Press
198 Madison Avenue, New York, NY 10016, United States of America

British Library Cataloguing in Publication Data

Data available

Library of Congress Cataloging in Publication Data

Data available

ISBN 978–0–19–879989–4 (Hbk.)
ISBN 978–0–19–284533–7 (Pbk.)

Printed and bound by
CPI Group (UK) Ltd, Croydon, CR0 4YY

Preface

This book is a journey to planets near and far, to explore their habitability, their past and their future, and in particular it is about the Earth-like planets and, above all, Earth itself. This is not only a trek through the Solar System and some of the exoplanet systems, but also an expedition through what astronomers have learned about them and how they learned it all.

For millennia humans have forged fairly complex societies, with cities full of specialized citizens, from farmers to sailors, from architects to soldiers. As far as our understanding of the heavens was concerned, however, we did not progress much beyond looking at the stars and wondering what they were, including the few wandering stars that were later identified as planets, while still often considering the sky thronged with divine beings and filled with heralds of things to come. Knowledge about the Universe increased very slowly, beginning to accelerate in the late-medieval Renaissance, the "rebirth" of inquisitive, critical thinking. But it was not until around 1920 that we began to understand what a star really is and how it can shine for so long and so brightly, because for that we needed to develop, among other things, the science of nuclear physics.

Disciplines of science initially move slowly, then can accelerate to an astonishing pace as two things happen. First, we see that as the network of observable or understood phenomena grows— through the development of instruments, methodologies, theories, and in recent decades computers—more connections can be made to create additional new insights. The bigger that network is, the faster it can add new links of understanding. Second, a rapidly advancing field attracts new generations of curious scientists with fresh ideas and perspectives, as well as funding to support them.

Thus it came to be that within a few decades from the discovery of radioactivity at the very end of the nineteenth century we mostly understood the lives and deaths of all the stars in all their various forms. Another such leap occurred around the transition into the twenty-first century. It was not until 1995, with the use of newly devised, sensitive instruments, that the very first solid evidence was found of a distant planetary system like our own. Over the following twenty years, we came to know that planets outside the Solar System, known as exoplanets, are more common than the stars themselves that they orbit.

The story of the Solar System and that of other planetary systems illustrates how the science of astrophysics works and why the world around us looks the way it does. This story rests on facts of nature—as scientific stories should—but a long list of facts alone does not make for a good story. So, these facts are brought to life by how they were discovered and by what they teach us about the Universe. I hope this exploration will reveal to you as it did to me some of the beauty and diversity of the Universe in which we live, the unfolding story of astrophysical approaches, and the way our imagination pushes back the frontier of knowledge about our planetary system and the billions of others that surround us.

Astrophysics, like all physics, is about interpreting what we observe in the context of other things, whether they are occasional similarities or are rules so frequently encountered and tested that they are viewed as "laws": gravity, electromagnetism, relativity, quantum mechanics, thermodynamics, etc. The story of our Solar System and of other planetary systems is about understanding phenomena based on the combination of all things relevant to that understanding. It is, obviously, hard to know what is relevant before it is understood, and it is impossible (and boring) to provide all the facts without any context. I have had to choose what to present when, and in what context. But I did want to preserve some of the excitement of the hunt for understanding that all astronomers share. I hope to have found a comprehensible balance.

The technical details and quotes found in the book are supported by a "further reading" section for each chapter at the end of the book. Many of the references in these sections are published in the professional literature, written by and for astrophysicists. Often access is restricted to those with library subscriptions or granted only after payment. However, many modern-day authors submit copies of their work to preprint servers where they can be accessed for free, although generally in a form that predates the final editing. The one most commonly used in astrophysics is https://arxiv.org; where available, the reading list entry includes the link to the arχiv entry. The astrophysical literature can be searched most readily with the SAO/NASA Astrophysics Data System (http://www.adsabs.harvard.edu), which enables a look through the papers that it references and papers that cite it. As added benefit, it provides one-click access to any papers that were submitted to the arχiv preprint server.

Contents

1

From One to Astronomical

Just how many planetary systems exist in the Universe, and how these evolve in a bewildering variety of ways, has been uncovered over the course of just two decades in surprising leaps forward in the science of extra-solar planets and in the appreciation of how special Earth and its solar-system environment are. All this was enabled by the curiosity of a growing population of searching astrophysicists and probing planetary scientists, guided by the modern scientific methods, and propelled by rapidly advancing technologies from micro-electronics to space travel. This book illustrates key discoveries that teach us about the past and likely future of Earth and about the habitability or lack thereof of billions upon billions of planets orbiting other stars, which are now known as exoplanets.

The account of planetary systems told in the following chapters leads from unhurried, cloaked formation to ultimate brilliant if not cataclysmic demise: Stars are born in batches in dense hatcheries, then leave their nest for a largely segregated existence as they separate from their siblings to move about their galaxy. Eventually they expire in puffs or blasts, ejecting newly created elements to partake in another cycle of celestial birth and death. Throughout their existence, most stars do maintain company: many are born with one or more stellar sisters, and most have at least one full-sized planet orbiting them, along with swarms of smaller heavenly bodies with sizes like entire continents down to pebbles and grains of dust. Planetary systems change over time, with phases in which planets resettle closer to or further away from their central star. In doing so, they subject their fellow

One of Ten Billion Earths. Karel Schrijver, Oxford University Press (2018). © Karel Schrijver.
DOI: 10.1093/oso/9780198799894.001.0001

planets to bombardments by asteroids if not catastrophically shepherd them into their star or expel them into interstellar space, along the way causing planetary collisions and supplying or removing atmospheric gases and water. Planetary habitability comes and goes with the aging of the central star, with the changes in the magma motions inside the planet, and—at least on Earth, the only world for which we know unambiguously that life exists—with the evolution of life on its surface.

Galaxies, stars, planets, and more

The observable universe may contain of order two trillion[A] galaxies. That is a number with twelve zeros: 2,000,000,000,000 galaxies! Each of these contains at least a million stars, but many—like our own Milky Way Galaxy (or just Galaxy for short)—contain over 100 billion (100,000,000,000) stars, and some as many as 100 trillion (100,000,000,000,000) stars. Until just a decade ago, we did not know that it is likely that most of these stars have a planetary system, typically with multiple planets. Each planetary system is thought to have roughly 100,000 objects somewhere in its vicinity that are larger than 100 kilometers (or 60 miles) in diameter. Combining these numbers, there must be—very, very roughly—10,000,000,000,000,000,000,000,000,000 (10 octillion) large clumps of some mixture of gas, ice, rock, and metal in the visible universe, moving in and between 100,000,000,000,000,000,000,000 (100 sextillion) planetary systems. There could be ten or a hundred times more or less than that, but whatever the numbers turn out to be, there are many large objects floating about in the observable Universe, without even including the multitudes of

A Throughout this volume large numbers are given in the so-called "short scale" common in the U.S. and U.K.: each subsequent multiplication by 1,000 starting from a million (six zeros) yields a billion (nine zeros), a trillion (twelve zeros), a quadrillion (fifteen zeros), a quintillion (eighteen zeros), a sextillion (twenty-one zeros), a septillion (twenty-four zeros), to the largest used here: an octillion (twenty-seven zeros).

smaller objects, all the way down to particles of dust and loose fragments of atoms.

Earth is estimated to be one of several billion Earth-like planets in the Galaxy orbiting within a distance from their stars at which water, if it is present, could exist as a liquid on the planetary surface. Although one of many in terms of size and orbit, Earth remains the only planet that we know to be able to support human life, and in fact the only one that we know supports any life at all. Whereas for many centuries Earth was, in the human mind, at the center of the Universe (see Figure 1.1), it has now attained the status as the sole known hospitable island within reach in the vast archipelago of remote exoplanets that are revealing to us why our world is the way it is.

All these galaxies, with their stars and planetary systems, are contained within the part of the Universe that we can see, which stretches for 13.8 billion light years[B] in all directions from Earth. We know of no reason to assume that the Universe stops there or, in fact, that it stops anywhere. We just cannot see any further than that because no light from beyond that distance has been able to reach Earth since the beginning of time!

These are all large numbers. Astrophysicists are used to working with such numbers. They are trained to compare such large numbers to other large numbers in order to deduce something about the Universe. As an astrophysicist, I can write down these estimates about what surrounds the Earth in the Galaxy and beyond, but I cannot grasp their meaning with any correspondence to reality. People just cannot visualize what numbers that large mean, but we can work with them as computational results of the combination of observations with logical arguments.

Although we cannot visualize the overall contents of the Universe, these numbers, and the observations that they are based on,

---

[B] A light year is the distance that light travels in one year going at 299,792 kilometers or 186,282 statute miles per second. For comparison: the full width of the orbit of Neptune, the most distant planet from the Sun and a good size measure for the Solar System, is almost 0.001 light years.

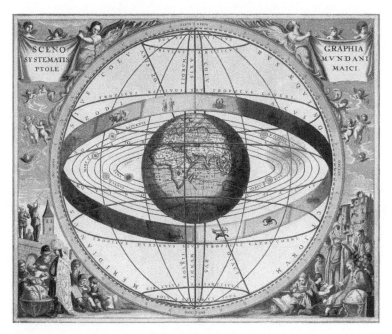

Figure 1.1 A depiction of the Ptolemaic geocentric system, a world view from the Middle Ages and Renaissance. In this view, the Earth is at the center of the Universe, orbited by, in order of increasing distance, the Moon, Mercury, Venus, Mars, Jupiter and Saturn, all fitting inside a sphere that holds the "fixed stars." This version, which places the Sun between Mars and Jupiter, is from "Harmonia Macrocosmica" by Andreas Cellarius (1596–1665, a cartographer born in Neuhausen, Germany, died in Hoorn, the Netherlands). It was first published in 1660, 117 years after the publication of the heliocentric view by Copernicus, still twenty-seven years before Isaac Newton's publication of his theories of gravity and motion pushed the Ptolemaic world view out of favor so that the Copernican heliocentric world view could take center stage with the planets moving with varying speeds on elliptical orbits.[1]

do tell us a lot. What is clear from them, for example, is that there are plenty of other planetary systems around to compare to our own. Many are too far away to study in any useful detail; most are indeed too distant to observe at all. But those that are sufficiently nearby to reveal some detail about themselves—still many

thousands—offer a great deal of diversity in their properties: the number and orbits of their planets, the masses and rough chemical make-up of these planets, some of their past and future.... The information we take from them helps us understand more about the Universe, but also about the formation, lifetime, and demise of our own planetary system, the one we call the Solar System. The speed at which we learn about this is as amazing as the number of planetary systems we know of: most of what we know of planetary systems other than our own was put together within just two decades.

The first cool-star exoplanet, in Florence

In the fall of 1995, I was participating in a conference of astrophysicists entitled *Cool stars, Stellar Systems, and the Sun*. The meeting was one in a series that was held every second year, alternatingly somewhere either in Europe or in the USA. That year, we were in Florence, Italy. The conference venue was a rather modern building, but it was just a few minutes' walk north of the medieval city center full of art treasures and architectural marvels. The speakers at the conference addressed a few hundred colleagues from around the world. They were discussing magnetism of stars very much like our own Sun, including that of the Sun itself.

The work that I was presenting there was about the properties of the million-degree outermost atmospheres of stars like the Sun caused by the magnetism generated deep inside stars. But this conference had topics ranging from newly discovered tiny deepred stars, called brown dwarfs, to the largest giant stars. And more: among the long list of entries on the agenda was a presentation by Dr. Michel Mayor of the Geneva Observatory. He reported on the observation of signatures of a planet orbiting one of those Sun-like stars, work done with a graduate student, Didier Queloz (and if it had not been for a fatal car accident early on, it would also have included their colleague Antoine Duquennoy). This was not just a discovery of one more of something; it was the first of

its kind. The presentation was followed by deserved applause, but only mildly more than the typical acknowledgement by the scientific audience at a conference. In hindsight, it is extraordinary how little fanfare there was at what turned out to be, in fact, the beginning of a radical change in how we view our place in the Universe: *the planetary system that is our home was proven not to be the sole such system.*

It was interesting that this finding was announced in such a fitting place: centuries earlier, Florence, then the capital of the Grand Duchy of Tuscany, saw the beginning of the Renaissance in which arts and sciences flourished anew after stalling for many centuries in the intellectual freeze of the Middle Ages. Part of that was the discovery of moons orbiting Jupiter, and the understanding that the other planets do not orbit Earth but revolve around a Sun that itself rotates and displays surface features in the form of sunspots.

Mayor's exoplanet discovery was the first of many that, just a few years later, would reveal our Solar System to be one of many similar planetary systems. It was fitting that the announcement happened in a city where you can sit in a quiet church in its medieval center, the Basilica di Santa Croce, surrounded by tombs and funerary monuments debossed with some of the best-known names of that earlier Renaissance. Among them is Galileo Galilei (1564–1642) who added fuel to another revolution by advocating the concept launched by Nicolaus Copernicus (1473–1543) that the planets orbit the Sun rather than the Earth, and that the Earth is one of them, and thus itself also moves as part of the heavens rather than being underneath all that. This earlier insight led humanity to realize that Earth (depicted with the Moon in Figure 1.2) is not at the center of the Solar System and—as one then thought—of the Universe[2]; it occurred not long after the first successful circumnavigation of the Earth in 1522 showed beyond doubt that it was—as many had already concluded—spherical. That endeavor, by the way, was initiated by Fernão de Magalhães (1480–1521), but completed by the skeleton remains of his expedition: Magalhães himself was killed in the Philippines

Figure 1.2 Once considered to be at the center of the Universe by its inhabitants, Earth is but one of many planets. This image shows Earth and its Moon as seen from NOAA's DSCOVR spacecraft, which hovers one million miles above the Earth in the direction of the Sun. The Moon's far side can be seen brightly lit by a Sun overhead; the Moon's dark side at the time of this picture is facing Earth. The far side of the Moon was first observed only in 1959 from the Soviet Union's Luna 3 spacecraft. This image, colored as the human eye would see things, is a combination of three exposures using red, green, and blue filters. The narrow sliver of color around the edges of the Moon is an artifact caused by the Moon's orbital motion during the roughly 30 seconds separating the first and last of the combined exposures (See Plate 1 for a color version).

more than a year before eighteen surviving members of the original crew of about 240 managed to return to Spain.

The more recent exoplanet revolution makes us realize what we might have anticipated but could not take to be true until

then: our Solar System is just one of an unimaginable number of similar systems that exist throughout the Universe. Until 1995 we knew of one single planetary system associated with a common type of star, namely our own Solar System. Two decades after that, thousands of such systems had been observed. Statistical estimates based upon these observations suggest that most stars in the Galaxy have at least one planet. With well over 100 billion stars in the Milky Way Galaxy, that makes for a lot of planetary systems in just one galaxy. With billions of similar galaxies within the segment of the Universe that we can observe, the number of planetary systems out there is expected to be truly astronomical.

Transformation and the principles of uniformity and mediocrity

This transformation in our knowledge about planets in the Universe is astonishing, but it was not unexpected. Among the fundamental assumptions that astrophysicists make in their work, two have always guided our thinking. First, we assume that—unless there is firm evidence to the contrary—the laws of physics are universal: the same chemical elements are assumed to exist everywhere, and the same laws of quantum physics, electromagnetism, and gravity are taken to apply in all places and at all times, not only where we happen to live. This assumption is referred to as the principle of uniformity. Second, we assume that—again, unless there is firm evidence to the contrary—our place in the Universe is not special. Now, of course, Earth is special in the sense that we live on this particular planet that happens to be located within the particular planetary family—which we call the Solar System—that revolves around "our" star, which we call the Sun. But we should not assume it to be special in the sense that it has an extraordinary history, is in a very distinctive place in the Universe, is temporarily in a special state in some way, or has any characteristic that is well out of the common range of what surrounds us in the Galaxy, and that even the Galaxy as a whole

would not be particularly unusual in any way. This assumption is referred to as the principle of mediocrity.

Unless one or both of these two fundamental assumptions about physical laws and circumstances were invalid, we may conjecture that our solar system is not exceptional, and therefore other planetary systems should be expected to exist. Granted, they need not look identical. In fact, they would likely differ in the number and properties of the constituent bodies—including their central stars and orbiting planets—but quite a few should exist given the billions upon billions of stars around us. And that is why astrophysicists have been looking for what we now call exoplanets for a long time.

What we did not know was whether planetary systems would be common or would form under some set of restrictive particular circumstances that we were not aware of. In other words, we did not know whether we should expect there to be a handful in the Galaxy or many billions. The expected number makes quite a difference to how many scientists would be willing to invest their careers in the hunt for extrasolar planetary systems. The answer also makes a big difference in how much funding governments, industry, or private donors would make available to pursue the hunt. .

Although our minds were prepared to find other planetary systems, our technology was not able to pull it off for most of civilization's history. Planets, including those as large as Jupiter, are small compared to the stars that they orbit. Moreover, they are very, very faint compared to stars, because they themselves do not emit light other than the generally feeble infrared glow associated with their atmospheric or surface temperature; most of their brightness comes from the reflection of part of the tiny fraction of the light from their neighboring star (or stars) that hits the planet. And any star around which they might orbit is very, very far away from Earth, so that seeing a much fainter object right next to them is not technologically feasible in almost all cases for traditional imaging optical telescopes. Each of these

aspects makes it a challenge to spot planets outside the Solar System. It takes very particular types of measurements (which we will get to later on) with very sensitive instruments, and patient, repeated observations of many stars, before planets outside our own planetary systems can be detected. But astrophysicists have a lot of imagination, and engineers working with them take pride in their innovation skills. All of them are tenacious in getting to their goals. So, in 1995, technology crossed the threshold at which a new field of science became possible: the study of planets orbiting ordinary stars other than our Sun.

Once one such exoplanet was found, astrophysicists knew what to look for. As a result, within just a few years of the first discovery dozens of other detections were reported. The push for yet more suitable and sensitive techniques that followed these first discoveries led to dedicated telescopes on the ground and in space. This yielded several thousand known exoplanetary systems within two decades of the first sighting. By knowing exactly under what conditions exoplanets could and could not be detected for any of the methods now in use, astrophysicists estimated how many exoplanetary systems should be out there, even if these were—at least for now—undetectable or at least undetected. The result was astounding: perhaps almost every star has such a system, although statistics allow that only every third star has one. Regardless, that results in more planetary systems than we can visualize. Just imagine:

With the naked eye, we can see about 9,000 stars. Not all at once, because no matter where we are on Earth, half the sky is hidden. Even those that are in principle visible at any time and place cannot be seen at one glance because our eyes cannot capture the width of the entire night sky. But if we look at the sky on a clear night, well away from city lights, the several thousand stars that we see certainly look like a large number.

Astrophysicists have put forward solid arguments that for each of these 9,000 stars which the human eye can see, there are very likely more than 10 million others in our Galaxy, and that most

of those have a planetary system. If we assigned a dozen planetary systems to each person on Earth, there would possibly still be billions of planetary systems left over in the Galaxy.

We do not yet know how many planets there typically are in exoplanet systems. Perhaps our solar system is not atypical with its eight planets: Mercury, Venus, Earth, Mars, Jupiter, Saturn, Uranus, and Neptune. But there is a lot more orbiting the Sun than these large spheres: a few dwarf planets (including Pluto) and many asteroids, comets, moons, and more. As we will discuss later on, however, there is more in the exoplanet zoo: there are what are called the "rogue planets" or "nomads". These are planets that do not orbit any star in particular but that wander about the Galaxy, moving out there either entirely alone or accompanied by their moons and perhaps ring systems such as our giant planets have. By some estimates, there may be some 100,000 objects per star that are larger than a hundred kilometers (60 miles) in diameter floating unbound between the stars. Plus many, many smaller objects.

Planet detectives

Humankind went from knowing about one planetary system to knowing it was surrounded in all directions by alien worlds in the span of two decades. How did that come to be? What do these worlds look like? What are we learning from all this? What does this mean for finding life on other planets? And how do we know what we know?

That last question—"How do we know what we know?"—is an important one to tackle. In many ways, we can think of astrophysicists as detectives looking at a crime scene: everything could be a clue, so every detail deserves attention, and every object, its place, and any trace on it needs to be reviewed as to whether it provides information about the what, how, and why of what occurred in that place before, during, and after the crime.

Astrophysicists, like detectives, cannot perform experiments of the kind physicists can in their laboratories; they have to take their clues only from what they observe. Such a forensic science is challenging, particularly if you have seen only a single crime scene, or rather, if you have seen only a single planetary system, as was the case until the last few years of the twentieth century. Of course, much can be, and indeed must be, learned from this one "case study" available. Within the Solar System, we find evidence for a diversity of bodies, for their evolution over time, and of their interaction, all from pebbles left over from meteors up to the largest of the giant planets. But now, although we cannot see them in any detail, we have access to thousands of detected planetary systems, and many, many more we know are out there but have not been able to investigate as yet. With new methods we now use the wobble of stars to weigh their planets; we use gravitational lensing of light rays to "see" interstellar planets that were ejected from their birth systems; we can perform population studies to learn about planetary systems from birth to death despite the fact that none of them changes structurally in a human life span; we use deceased stars to sniff out the chemical makeup of decayed planets using the atmospheres of white dwarfs that were "polluted" by planetary debris crashing into them; we use ancient, long-gone radioactivity to know about the setting in which the Sun and its planets initially formed; and—among the most transformative of changes compared to earlier times—we use powerful supercomputers to test, in ever more detail, our ideas by experimenting with virtual worlds through carefully tunable histories subject to the laws of the Universe, although bounded in their fidelity by the capabilities of our computers.

Astrophysicists realize full well that what they infer from observations and experiments is tentative at first, always subject to scrutiny and change, and possibly doomed to miserable rejection although hopeful of favorable adoption at any time in the future. Their goal is to advance our knowledge, with two tools of

logic: hypothesis and falsification. They formulate a hypothesis, knowing that they can never prove it to be the actual truth. All they have at their disposal is the methodology to prove something is wrong, or rather that it is inconsistent with other knowledge. One could look at the works of Karl Popper (1902–1994), philosopher of science, to see the details of that process, and the logic behind it. One quote from him on this is: "Bold ideas, unjustified anticipations, and speculative thought, are our only means for interpreting nature: our only organon, our only instrument, for grasping her. And we must hazard them to win our prize. Those among us who are unwilling to expose their ideas to the hazard of refutation do not take part in the scientific game." Or one might read words by the German playwright and poet Bertolt Brecht (1898–1956). He condensed the essence of science into this observation: "Es ist nicht das Ziel der Wissenschaft, der unendlichen Weisheit eine Tür zu öffnen, sondern eine Grenze zu setzen dem unendlichen Irrtum"—"The goal of science is not to open a door to infinite wisdom, but rather to set bounds to infinite mistake."

This formulation by Brecht avoids a common impression that scientists prove theories to be true by their efforts at falsification. Another perspective, rather more positive, optimistic, and constructive, is that they point out the parts of hypotheses and theories that may need revision or refinement. This is often the case when new developments enable us to look where we could not look before: further away, at larger densities, lower temperatures, long ago in time, at higher energies, or on subatomic scales. This was eloquently articulated, as often in his writings, in an essay by Isaac Asimov (1919–1992) on "The Relativity of Wrong". He points out that "The basic trouble, you see, is that people think that 'right' and 'wrong' are absolute; that everything that isn't perfectly and completely right is totally and equally wrong. However, I don't think that's so. It seems to me that right and wrong are fuzzy concepts," and then explains what he means. He notes that long ago, people thought the Earth was

flat. For all the traveling they could do, this seemed a perfectly acceptable approximation. Only when they started traveling many hundreds of miles from home did it become evident that this was not an acceptable description: the way constellations in the heavens changed, or how the elevation of the Sun shifted, required the next approximation, namely that the Earth was, in fact, a sphere. But Isaac Newton already realized that this was not right either: a spinning, self-gravitating body would settle into a somewhat oblate, ellipsoid shape because of how gravitational and centrifugal forces compete. Nowadays this no longer suffices either: orbiting spacecraft respond to distortions in the Earth's pull depending on positions of the continents, the oceans, and the polar ice caps and orbit calculations need to take that all into consideration.

Similarly, as Asimov points out, the shift from a geocentric to a heliocentric solar system was in some way only a slight improvement: when, after Nicolaus Copernicus and Johannes Kepler, the planets were taken to move in ellipses around the Sun, this yielded a much better approximation to the practical process of predicting planetary positions. Similarly, one can argue that Newton described gravity just fine for everyday applications for a long time; the correction introduced by Albert Einstein by his general theory of relativity is minute, except, that is, in some astrophysical conditions and for very fast particles. And also whenever timing is of the essence: without the Einsteinian formulas, the Newtonian approximation would make satellite-based navigation (GNSS, or GPS) fail miserably: high-altitude spacecraft clocks drift by tens of microseconds per day compared to ground-based clocks, and although that sounds rather ignorable a GPS system without a correction for this effect of general relativity would see position errors accumulate at a rate of some 10 kilometers per day.

So, the process often is one in which a better approximation is developed for the problem at hand. And yet, from a perspective far enough away, things do look fundamentally different: a flat Earth does not compare well with a somewhat misshapen

ellipsoid; simple geometry is different from an Einsteinian universe in which time, space, and gravity influence each other; and living in the only solar system in the Universe is experienced as completely distinct from living in just one of a hundred billion such systems in a galaxy. These are tremendous contrasts: a different geometry, a different formulation of the laws of physics, and the transformation in knowledge from possible but unconfirmed to known although subject to some uncertainty in detail. We now know for certain that planets are common in the Universe, whereas until the early to mid 1990s we had only a reasonable suspicion that this should be so.

Familiar with unfamiliar worlds?

Is it just this deceptively subtle difference that made this revolution from one planetary system to an astronomical number go relatively unnoticed, with immediate acceptance that, yes, other worlds were common, and so what? Or did our culture change so much that we were already prepared for it? In 1859, Jules Verne (1828–1905) had already taken us to the Moon in "De la terre à la lune" in one of his *Voyages Extraordinaires* (see Figure 1.3), but that is just a hop away from Earth. Still, a fascinating and insightful story about the politics of power, 1966's *The Moon is a Harsh Mistress* by Robert Heinlein (1907–1988), depicts how different life would be up there. Many invented extrasolar worlds had already been explored in written fiction, and some in movies, when, also in 1966, we were introduced to many alien worlds (albeit decidedly Earth-like) in the voyages of *Star Trek*'s ship *Enterprise* in the USA and the comparable but short-lived endeavors of the spaceship *Orion* in Germany. (I saw the latter well before *Star Trek* came to the Netherlands, where I grew up, although I had a hard time with the German language.) We have seen the famous planet Arrakis—source of the mysterious and incomparably valuable "spice"—in Frank Herbert's *Dune* series; and the tropical planet Dagobah—where the Jedi master Yoda had taken refuge—and

Figure 1.3 Illustration from Jules Verne's *From the Earth to the Moon: Direct in Ninety-Seven Hours and Twenty Minutes; And a Trip Round It* from 1874. The travelers, having reached the Moon in a capsule (shown in the foreground) that was blasted toward the Moon by a giant cannon, are surveying the surface and breathing imagined air.

the frozen Hoth—the home base of the rebel army—along with many other worlds of the *Star Wars* movie franchise, to name but a few. Yet most of these Earth-like planets also seemed to have breathable atmospheres, even if it is not obvious why they should have. For example, the mineral-rich lava world Mustafar from *Star Wars* has no discernable reason to have oxygen in its

atmosphere. The same is true for the black-hole-orbiting planet Miller in *Interstellar*, where massive tidal waves sweep around the planet that destroy anything in their path, and that threaten to also do that to the travelers visiting from Earth (in what appears to be plain daylight coming from a black hole).

Then again, why should we extrapolate from a single shot that the entire planet would be like that? Is Mustafar's shown location as characteristic as the active volcanoes like Etna or Kilauea are of the entire Earth? Is Planet Miller covered all over with water, and even if it were, could there not be algae building up an atmosphere breathable to humans? After all, much of Earth's oxygen is generated by plant life in the shallow parts of the oceans. The realization that we generalize a local condition to be a ubiquitous one may just bring to our conscious attention that we need to be mindful that visualizing an exoplanet—as is done by research agencies and the media, and an impression is offered by me in Figure 1.4—could mean collapsing something as rich in diversity as Earth into one or a few vistas that may or may not be characteristic of much of the planet. After all, snapshots of New York's Times Square, Marrakesh's Jemaa el-Fnaa, or Tokyo's Ginza, although each iconic, do not capture what Earth looks like to most of its inhabitants, although one may argue that such images show one feature that makes Earth unique: human activity. Having said that, one might point out that any such landscape is atypical of Earth because most of our planet is covered by ocean.

There are, of course, far less Earth-like "planets". For example, I remember being fascinated when reading *Dragon's Egg*, written by Robert Forward (1932–2002) in 1980, about the fast-developing culture of a race living under the terrifying gravity of the surface of a neutron star being visited—at a moderately safe distance in an artificially modified gravity—by humans from Earth. But then, that, technically, is not a planet. And it isn't a star either, at least not any more.

Figure 1.4 Artist's impression of an icy exoplanet near a cool dwarf star. The star's low surface temperature would bathe everything in a deep red light. If close enough to the star, the daylight into a sufficiently dense atmosphere might keep (some) surface water liquid, but that proximity to its star would at the same time mean that the exoplanet's rotation would likely be synchronized to its orbit by tidal forces. Consequently, one side would always be in daylight and the other always under a night sky, leaving the planet in what is called an eyeball climate state (as imagined in the inset). Somewhere on the dayside, at the edge of the open water, a red star might shine down low on the horizon on a partially frozen ocean, as imagined in the main image. Many of the cool dwarf stars are magnetically active. The proximity to such a star will frequently cause auroras to appear in the exoplanet's sky, possibly shining blue-purple in a nitrogen-dominated atmosphere even at non-polar latitudes on the planet (See Plate 2 for a color version).

Insert credit: NASA/JPL-Caltech.

Nonetheless, in our literature and cinematography planets outside our own were already abundant decades ago. We were treated to planets with multiple "Suns"—such as Luke Skywalker's home planet Tatooine—before astrophysicists found them. We were engaged with the post-stellar life forms of *Dragon's Egg* before we realized that a planetary system can as a matter

of fact exist after the catastrophic demise of its star. And yet, imagining such worlds is different from knowing that such worlds indeed exist, and far from learning how they form, evolve, and in the end may disappear. Those are the stories that chapters in this book will go into: not only imagining, but reviewing the evidence, while in the process revealing that nature has still more surprises in store for us "than are dreamt of in your philosophy" (borrowing from Shakespeare's *Hamlet*).

Faraway suns

This is not only a story of planets; it is, by necessity, also a story of their stars. In ancient Greek times, going back to Aristotle and Ptolemy, the stars were, well, simply unknown and unknowable, but considered to be somehow affixed to some large sphere that encompassed the known world, including the Moon, Sun, and planets (see Figure 1.1). A difference between the Greek and early Renaissance views might be that either the sphere spins about in a daily rhythm or that the Earth spins about within a motionless sphere. But in the majority of views, whether the Earth spun or the stellar sphere spun, and whether the planets orbited Earth or all orbited the Sun, the stars were supposed to be distinct from the Sun: they were distant, differed somehow fundamentally from the Sun, and were not endowed with planetary systems. Then there was Giordano Bruno (1548–1600). He advocated for a view of an infinite universe full of planetary systems like our own, orbiting their own stars (their "Suns"), each system far from the others, going on forever without end and thus populating a universe without center. This is not quite right, as we know that stars are clustered into galaxies, but otherwise it is close to what we now hold to be correct. This late medieval concept is certainly different from that of Ptolemy who did not like the concept of a large, empty universe with widely separated bodies, writing in his *Planetary Hypotheses* that it "is not conceivable that there be in Nature a vacuum, or any meaningless and useless thing."

In Bruno's days, one practical problem hampered proving his case. If stars had vastly different distances to Earth then this should be revealed through something known as parallax, owing to the fact that the Earth orbits the Sun. Imagine, for example, driving in a car and looking at the landscape: nearby things obviously move by quickly, but they also shift in projection seen against buildings in the background and those in turn against yet more distant mountains. The same principle applies when the Earth moves around the Sun: Nearby stars should appear to move somewhat relative to more distant ones, so that after a full orbit, nearby stars would seem to have moved on a small ellipse compared to the most distant ones. As this was not observed in Bruno's days, the infinitely repeating structure of his concept could not be confirmed.

In fact, the apparent absence of parallax was taken to mean that all stars were at the same, although unknown, distance. In reality, however, we now know that it simply meant that the measurement tools were not accurate enough because the stars were vastly further away than seemed conceivable. And they were found to be more and more numerous with the increasing quality and power of telescopes; it would indeed be hard to imagine the numbers of worlds needed in Bruno's vision . . . although now we are essentially accepting the vastness of the Universe and the unimaginable, although estimable, number of stars within what is visible to us. But it would take until long after 1838, because not until then was the very first measurement published of the distance to a star. This was achieved by Friedrich Bessel (1784–1846), director of the Königsberg observatory in Prussia, Germany, who found "the distance of the star 61 Cygni from the sun 657700 mean distances of the earth from the sun" (correct to about 10 percent), which by its very magnitude was a convincing argument in establishing that the Sun is in fact comparable to billions of stars surrounding it, because the Sun seen from that distance would only have the brightness of a star. Note that Bessel's was the first published, not the first measured: Thomas Henderson (1798–1844), working at the Cape of Good Hope observatory, had made the

measurements needed to determine the distance to α Centauri in 1832 and 1833, but did not work them into a distance estimate until after he saw Bessel's work on 61 Cygni published not only in Bessel's native German in *Astronomische Nachrichten* (which still exists as a scientific journal) but also in English by the Royal Astronomical Society.

It would be a point of discussion well into the twentieth century just what stars were about, and whether there were other planetary systems. The year before the start of the seventeenth century, Bruno's beliefs led to his conviction for heresy by the Catholic Inquisition at the end of a trial that lasted seven years: he was burned at the stake on February 16, 1600, found guilty of multiple charges, among them his belief in the existence of many worlds and their everlasting nature.

It appears that, by the end of the seventeenth century, opinions had shifted. One observation on that is by Isaac Newton (1643–1747): "The Universe consists of three sorts of great bodies, Fixed Stars, Planets, & Comets … The fixt Stars are very great round bodies shining strongly with their own heat & scattered at very great distances from one another throughout the whole heavens." Another was written fifteen years later by Christiaan Huygens (1625–1695), in 1698 (in a translation back into Dutch from his original Latin writing). He wrote that whereas he realized that Kepler accepted that stars are distributed throughout a volume rather than affixed to a sphere, Kepler could not accept that they each had a system of planets around them. But Huygens disagrees and asks "Want wat weerhoud ons nu te gelooven, dat een yder van die Starren, of Zonnen, zoo wel als onze Zon, rondom haar Dwaalstarren heeft, die wederom met hare Manen verzeld zijn?"—or, in modern English: "After all, what prevents us from believing, that each of the stars, or suns, as well as our Sun, has around it planets, that themselves are accompanied by their satellites?" So, by then even exoplanets orbiting stars were acceptable. But the true nature of stars was not understood until well over two centuries later, and 297 years would have to go by before the existence of exoplanets was demonstrated:

astronomy, and the technologies with which it works, take their time.

We shall see that we now consider the formation of stars and of their planetary systems inseparable, with the formation of planets statistically estimated to accompany just about every formed star. We are realizing that planets can migrate through a planetary system, so they can orbit now where they did not form. Their interactions can cause other planets to be destroyed, ejected from their original system, or tipped and pulled into oddly eccentric orbits out of the original plane of rotation of the star and planets. We know they can form in or around double stars, triplet stars, ... multi-star systems. And they can be rocky and perhaps Earth-like or vast and heavy, some so heavy that they are nearly but not quite stars themselves.

Advancing astrophysics

All that knowledge about exoplanetary systems was gathered in a span of two decades, vastly faster than the transition from geocentric to heliocentric, which took centuries. Vastly faster also than the remote and in-situ exploration of the Solar System. And faster even than the realization of the enormity and age of the Universe itself, which took a century. Just like other natural sciences, modern astrophysics is moving fast, with findings rapidly communicated and hypotheses tested, and subsequently—depending on the outcome—adopted, modified, or rejected. The way modern astrophysics is done is fundamentally different from how it was done in the past. This change has come fairly gradually, primarily because of a shift in societal perception of the value of science, combined with a rapid increase in the efficiency and speed of communication.

Throughout the Middle Ages, progress in science was slow. There were advances in architecture and the art of warfare, but these were slow, spread out over centuries. Other sciences largely stalled. Chemistry (or alchemy) remained rooted in the belief

in four terrestrial elements—water, earth, fire, and air—and a fifth, quintessence or aether, that was supposed to permeate the Universe outside of Earth's domain. Medicine relied on ancient Greek writings more than actual investigation and study. Ancient texts were copied and it was bad form to question their contents. Therefore, scientific study overall was mostly limited to the interpretation of arguments by established and generally long-dead authorities.

Astronomy, for a long time essentially synonymous with astrology, gradually changed its role as it became evident that this was a very useful science for navigation. For a long time, however, it remained a descriptive science. The nature of stars was rarely questioned, at least in public. But there was a shift that occurred around the time of the work of Nicolaus Copernicus: the demise of the eastern Roman Empire with the fall of what was then its capital city, Constantinople (earlier named Byzantium, and nowadays Istanbul). The influx of refugees from that realm into, among other countries, Italy, triggered the beginning of the Renaissance.

A new curiosity emerged in religion-dominated Europe. It became tolerated, then allowed, and ultimately fashionable to question and to study, not only from books, but from direct observations. Science became more mathematical, seeking mathematical elegance and no longer enforcing the perception of geometrical perfection seen in circles. It was in this era that Kepler, looking for mathematical foundations in the world around him, noted that describing planetary orbits as ellipses was much more graceful mathematically than attempting to put "perfect" circle upon circle upon circle (using the "epicycle" idea going back to Ptolemy in ancient Greece). In that same effort, adopting Copernicus's notion that Earth also orbited the Sun, but on its own ellipse rather than Ptolemy's circles, made for an effective quantitative description. This was, initially, defended in writings as a pragmatic choice, but over time the picture shifted to demonstrating that it was, in fact,

true that the Earth was moving as one of multiple planets around the Sun.

Mathematical elegance has a lot to do with the way the Solar System works. This became clear with the miniature equivalent of the Solar System that Galileo discovered as he traced the orbits of the four largest moons of Jupiter which, early in the seventeenth century, he was the first to see with the recently invented telescope. And it became clearer still after Isaac Newton's quantification of gravity and motion, as he expressed his famous "laws" in equations.

At the same time, the other natural sciences also shifted from ancient knowledge and accumulations of description to systematic experimentation and integrating interpretation. But such a transformation took time.

In the seventeenth and eighteenth centuries, few individuals had the time or money to pursue science. Some learned about science in university, taught by experts, but most did so by reading books and by communicating with others, often across borders through letters. And, of course, they would publish books. That was much easier than in earlier centuries following the invention of the printing technique that used pre-made shapes to print individual characters set into lines and assembled into entire pages: this "movable type," which became available in Europe around the end of the fifteenth century, made it possible to mass-produce books where before labor-intensive manual copying or printing with laboriously cut wood blocks had taken a long, long time and had thus led to costly products that were simply not affordable to most.

As the desire for communication between scientists increased, societies sprang up where similarly inclined individuals could learn from others and could show off their own achievements. Letters to individual colleagues were now replaced by letters to the societies, soon to be published in printed form for circulation beyond those who could be present at the meetings at which these letters were read out. These publications started in the latter

half of the seventeenth century: among the first are the British *Philosophical Transactions* of the Royal Society of London, and the French *Journal des Sçavans* published from 1665, the Italian *Saggi di naturalezi esperienzi* in 1667, and the German (published in Latin) *Acta eruditorum* in 1682.

With time, these publications became longer and more specialized, started referencing other works, and transitioned into what we now know as the "refereed paper." Whereas originally it was essentially up to the secretary or president of a scientific organization to decide on whether to accept a letter for publication, or to limit publication to certain parts only, this now changed. The process of refereeing moved that decision to experts in the field that a particular manuscript dealt with: the journal's editor sought opinions from established scientists on whether the submitted manuscript lived up to certain standards. There is no censorship involved in this process: unusual and contradictory papers can and will be published. But any publication must adhere to the rules: the methods must be clearly explained; the author(s) must demonstrate awareness of existing literature on a subject; and they must articulate cogently why they consider their findings to be valid, particularly if they go against the prevailing understanding. These refereed papers now form the backbone of science, going back well over a century, archived for future reference, and in principle accessible to anyone.

The growth of the scientific community, the increasing availability of specialized teaching at universities, and the growing population of full-time professional scientists led to ever more specialization. Journals divided and subdivided into more and more specialized fields. The increased volume of publications led to the need for review papers in which dozens or hundreds of other papers were compared and integrated into bigger pictures that could be more easily digested by researchers. Journals developed "index volumes" in which one could search by keywords under general topics what had previously been published. Early in my career I spent many an hour in the library

with these bulky index volumes, running back and forth to the bound volumes of the actual publications, first in the labyrinthine library of the centuries-old observatory of Utrecht—one of the older ones in Europe—and later in the modern vastness of the physics library in the newly built campus in the grasslands into which the city was expanding; these days, computers allow vastly more efficient and up-to-date searches from the comforts of, well, wherever one happens to be with access to a networked computer.

As travel became faster and more commonplace, scientific meetings were organized: after national gatherings the international regional meetings became the norm, and eventually globe-spanning intercontinental meetings were the vehicle of choice to share and discuss findings before they were published in printed journals. The increasing speed did not end there, though. Now, researchers often make their papers publicly available on what are known as preprint servers. They do this typically after acceptance for publication by the editors following a refereeing cycle, but ahead of publication in the printed journals (although "printed" now generally means that an electronic copy is accessible that is professionally edited and laid out). In fact, more and more we see that publications are submitted to the preprint servers as the very first version is submitted for refereeing at a journal, so that they are still subject to what can be substantial changes before acceptance, or rejection, by a journal's editor, or on occasion retraction if a mistake is found.

All of this cranks up the speed at which a science can advance: findings are rapidly shared and responded to, so that the next step is often taken by a colleague somewhere in the world even before the written findings are formally archived into the repositories of the professional journals.

Tools of science

There is more to the efficiency of the progress of science than the size of the population and the efficiency of interactions within the population that can be reached. If we thought of it only in those terms, we would be drawing an analogy to chemical reactions in which the density of one of the two reactants involved in any reaction is multiplied by that of the others to estimate the rate at which the reaction will proceed, further controlled by the velocities of the molecules involved, which is what temperature describes. To speed up the advance of science there are at least three other crucial ingredients in addition to population size and efficiency of communication: technological tools, logic tools, and mathematical tools. These tools are like the catalysts in the reaction: they may speed up reactions, or fundamentally enable them, and in doing so they may also shift reaction networks into different pathways or products. Of these three catalysts, technology controls how well things can be observed and measured: better optics in telescopes and microscopes, access to space for X-ray to infrared observations without atmospheric effects, radio interferometers to mimic telescopes with "lenses" of multiple miles across, design and production tools, …; technologies developed for society's benefit are often adopted into science applications and vice versa.

The second catalyst, the tool set of logic, is also crucial. Let us look back at Asimov's assessment of the "relativity of wrong" for a moment, and take the example of stars and planets as viewed by Bruno, Newton, and Huygens. Huygens is very precise in his formulation that was quoted above, stating that there is no argument against stars being like the Sun, nor for them having their own planets. At the time, that would be the only position defensible, because there was no technology available to test that concept. Newton and Bruno were less careful in their positions, as they postulated the similarity of stars and Sun, but not having the means to prove or disprove this hypothesis with evidence that would be acceptable by today's standards. Bruno, in particular,

Table 1.1 Selected historical events that are part of the study of the Solar System, exoplanet systems, and their stars.

1522	First circumnavigation of the Earth, proving it to be a sphere.
1543	Nicolaus Copernicus publishes his heliocentric world view.
1600	Giordano Bruno burned for his ideas about exoplanetary systems.
1609–19	Johannes Kepler publishes his laws of planetary motion.
1610	Galileo Galilei discovers the four largest moons of Jupiter.
1655	Christiaan Huygens spots Saturn's moon Titan.
1687	Isaac Newton publishes his law of gravity.
1761, 1769	Venus transits of the Sun used to measure the Solar System.
1801	Ceres found as the first asteroid by Giuseppe Piazzi.
1838	First successful determination of the distance to a star.
1896	Radioactivity discovered by Henri Becquerel, and Marie and Pierre Curie.
1905	Albert Einstein publishes $E = mc^2$, general relativity, and more.
1908	George Ellery Hale measures magnetic fields in sunspots.
~1910	Hertzsprung–Russell diagram reveals patterns in stellar properties.
1912	Henrietta Leavitt uses Cepheid stars to measure vast distances.
1919	Total eclipse used to confirm Einstein's theory of general relativity.
1920	Arthur Eddington proposes nuclear fusion as power source for stars.
1925	Cecilia Payne finds that hydrogen and helium dominate in stars.
1929	Edwin Hubble demonstrates most "nebulae" are distant galaxies.
1957	First artificial satellites launched into Earth orbit.
1966	Russian Luna 9 is the first spacecraft to land on the Moon.
1969	First astronauts on the Moon: Neil Armstrong and Buzz Aldrin.
1970	First successful landing on Venus: Venera 7.
1972	Last astronauts on the Moon: Gene Cernan and Harrison Schmitt.

1976	Evidence found for a supernova near the forming Solar System.
1976	First landers on Mars: Viking 1 and 2.
1983	First planetary disk observed: β Pictoris.
1992	First exoplanets found, orbiting a deceased star, the pulsar Lich.
1995	First exoplanet found orbiting a Sun-like star: 51 Pegasi b.
1995	First brown dwarf found: Teide 1.
1999	First multi-planet exoplanetary system discovered: υ Andromedae.
2004, 2005	Dwarf planets Eris, Haumea, and Makemake found beyond Neptune.
2005	Huygens probe lands on Saturn's moon Titan.
2006	CoRoT satellite launched to look for exoplanet transits.
2009	Kepler satellite launched to enable a mass hunt for exoplanets.
2011	Free-floating nomad exoplanets found through gravitational microlensing.
2014	First Earth-sized exoplanet in the habitable zone found: Kepler-186f.

may have been more right about stars and exoplanets than his contemporaries, but for the wrong reason. In fact, one could argue he was just as wrong as the Inquisition that condemned him: both sides argued that the glory of a creator was greater for their particular world view, so they relied on arguments of imagination rather than on physical arguments of measurement and comparison.

The latter means of arguing is nowadays the acceptable tool: measurement, mathematical tools, analogies or contrasts with other situations or conditions, and tests against the universal laws of nature are acceptable defenses. Mere opinions are not, no matter how cleverly and eloquently they are argued. This shift in the tools of logic toward measurement and mathematics had

already started when Kepler used ellipses rather than nested
stacks of circles to describe planetary orbits. But it shifted
fundamentally and permanently when Isaac Newton (1642–
1726) and Gottfried Wilhelm von Leibniz (1646–1716) developed
infinitesimal calculus, which deals with changes in quantities
as a function of some variable that can be time, place, energy,
or really any quantity one cares to study. Differential calculus
and its opposite, integral calculus, aided Newton, for example,
to really utilize his second law to describe planetary orbits,
although he wrote his *Philosophiae Naturalis Principia Mathematica* (first
published in 1687) mostly using geometrical arguments in order
to reach a broader audience, and did not formally publish his
concepts of calculus for a further six years. The law states that
the acceleration of a body is equal to the force applied to it per
unit of its mass.^{C} But with infinitesimal calculus, this allows one
to compute how much a velocity changes when subjected to
a force, and thus to compute the path of an object subject to a
force that may depend on place or time. Or, as Newton combined
it with his first and third laws, it showed him how the balance
of inertia—the tendency to continue with a fixed velocity on a
straight line—and gravity—a force pulling toward the center of
a body—combined to result in orbits that are conic sections—
hyperbolic, parabolic, or elliptical—depending on the initial
momentum.

Newton's laws are still very much in use, although now
sometimes expressed in terms of Einstein's relativity in extreme
conditions or when exceptional accuracy is required. Other
laws were added over time, some expressing conservation and

---

^{C} Mass and weight: throughout this book, I use "mass" to indicate how much
matter is contained in a body. "Weight" is often used to mean the same thing,
but it does not: weight is the force that gravity exerts on a mass, so that the same
mass will have different weights on different planets and moons, depending on
their gravity, which is determined by their mass and size. I use the adjectives
"heavy" and "light" throughout referring to mass, not the weight as sensed
when subjected to gravity.

continuity of, say, energy or matter, and others applying to electromagnetism, chemistry, quantum theory, thermodynamics, and more. These mathematical formulations of relationships that must apply enable the determination of one quantity from another, and thereby comparisons between different quantities and different conditions. For this book, an example is that the orbital period of a planet provides a measure for the mass of its central star.

The use of these laws in probing and understanding the Universe is what differentiates astronomers and astrophysicists from astrologers. If we are precise about the terms, then astronomers are those who measure positions and characteristics of heavenly bodies; astrophysicists apply laws of nature to deduce the origin, structure, and evolution of heavenly bodies and their interactions; and astrologers use flawed reasoning to divine erroneous information about human affairs from the positions of heavenly bodies.

Anonymous renown

After ancient astrology led to astronomy, and astronomy transitioned to astrophysics from the Renaissance onward, the science of the Universe became a highly specialized field, requiring decades of education and training, generally focusing on a progressively narrower set of topics and phenomena. As science specializes and advances, more and more is demanded of the technologies behind the experimental, observational, and computational efforts. There is an increasing trend for studies to be co-authored by teams rather than individuals. By the year 2,000, for example, over 80 percent of publications in science and engineering were by groups of people, with an average number of co-authors also increasing per publication. This is a consequence of the ready contact with colleagues, the internationalization of work experience over the career of a scientist, and not in

the least because the equipment needed is increasingly complex and costly.

Personally, I think these trends are why we can no longer readily mention outstanding scientists of the present age. Almost everyone knows of Einstein, Galileo, and Newton, who all did their major work more than a century ago. We remember Christopher Columbus as the (accidental) discoverer of the Americas—or maybe it was Leif Erikson five centuries earlier. But do we know the discoverers of the new worlds in the Galaxy: Aleksander Wolszczan who chanced on the very first exoplanet, which happened to be orbiting a deceased star; or Michel Mayor and Didier Queloz who found the first exoplanet orbiting a living star? How many of us know William Borucki, David Charbonneau, Greg Henry, Geoff Marcy, or Alessandro Morbidelli, to name but a few of those who worked on finding and understanding extrasolar planets? These are just some of the many who were at the start of the exoplanet revolution and who changed our view of the dynamics of planetary systems. There are still visionaries, giants, and stars in the sciences (including that of the astrophysics of the exoplanet), but they work in teams, and discoveries follow each other so quickly that we can no longer name the individual scientists who made it all possible. That is unfortunate in a sense, because it also reflects that science is not really at the center of cultural interests: we can name movie actors, sports stars, musicians, and authors, but can we name living scientists? At the same time, it is in a sense fortunate; it is a consequence of so many being involved that science can advance so fast.

The changes in science are now so rapid that research moves much faster than the teaching of it. Consequently, many beginning scientists are educated in their specialty field by direct, hands-on work more than from textbooks, and many of the experts in a field came into it from another specialty. This phenomenon is particularly pronounced in the young field of exoplanet studies. The growth of that research

community is far faster than the inflow of new generations. Many researchers chose to move into this new, exciting field, while some may have been forced, or lured, by shifting funding priorities.

Competitors in a team effort

Funding priorities are the dominant mechanism by which present-day research is directed. Increasingly, any researcher who wants funds to support her or his salary while doing research must propose to some funding organization—if not directly for salary, then certainly for the means to support students, purchase equipment, travel, or for the publication of results. Proposals may be made to general funds, but more and more there are specifically formulated "opportunities" that narrow down the topics to which may be proposed. Competition is fierce, with often only one in five to one in ten proposals being accepted for funding. This may not seem very efficient, but it does lead to two things: the accepted proposals reflect the most valuable research efforts as judged by the peers who review the proposals, while the proposal formulation—including those that remain unfunded—is one way for researchers to educate themselves as they peruse the literature for support for their ideas and as they argue with their co-investigators to convince them to join in the effort.

Then there are also the large projects: many millions to billions of dollars worth may be set aside for large observatories on the ground or in space, and these offer substantial continuity for a small group of researchers as the project is built and executed. The price for that is, of course, years of devotion to the developmental phase of the project, and much-needed lending of support to the broader community that will use the observatory or satellite once it becomes operational.

Formulating proposals to compete for large or small projects keeps the community members on their toes, always striving

for innovation—but it also makes their colleagues their direct competitors. Scientists are increasingly entrepreneurs, marketing as well as reshaping their personal expertise in order to, quite literally, sell their ideas to the funding organizations.

Thus it is that shortly after a discovery, or after the development of an enabling technology, science rapidly adjusts and develops a new specialty field. The first confirmed planets were detected using the consequences of their repeating pull on their central stars as they orbited them, using the so-called Doppler effect, which can be by way of a cycling shift in color or in the timing of any regular pulsation signal. Other methods quickly followed. Most rely on some change in the light or the position of the central star. However, they may also involve direct imaging of the tiny dots that orbit their star with telescopes tuned to heat radiation rather than to visible light. The discoveries follow each other at a rapid pace: in exoplanetary astrophysics, the breakthrough of the detection of the first exoplanets in 1992–1995 led to five refereed publications with "exoplanet" in the introductory summaries by 2000, seventy-one in 2005, 300 in 2010, and 439 in 2015. Along with papers came scientific meetings, graduate students, and increasing mention in the media: a Google "news" search for "exoplanet" shows no entries for 2000, four pages of results in 2005, thirty-seven pages by 2010, and—with the "news value" perhaps wearing off a bit already, or maybe a statistical variation—twenty-five pages by 2015.

Such a new specialty field is not necessarily a permanent new subdivision of another field. In fact, for exoplanets it is fundamentally something different. It is rather a merger of existing fields that were drifting in the same direction anyway. On the one hand, exoplanet science is informed by, and conversely informs, the planetary sciences based on the Solar System. On the other hand, exoplanet science is a continuation of stellar studies toward ever smaller bodies. Thus exoplanet astrophysics is both a hyper-specialization and a field of interdisciplinary research. As the astrophysicists and planetary scientists re-educate themselves

about exoplanets, not only does their expertise shift, but also their collaborative organization transforms. Astrophysics is an active science in which frontiers between disciplines are blurry, evolving, and never stagnant for very long. But what else is new: every profession sees shifts and reorganizations as technologies and societal interest change.

Narrowing in on the Goldilocks zone

Possible life in other planetary systems is covered in the next chapter. Our current best hope of finding life is to first look for what we might recognize as such. That means it should be visible in some way, so at least occur at the surface, because if, say, it were hiding under miles of ice on a world like Jupiter's moon Europa we would, in all likelihood, not be able to detect it on far-away exoplanets using the best of imaginable telescopes. These telescopes would not see that life directly: exoplanets are simply too far away. Although science agencies and news media frequently show pictures of exoplanets and even exoplanet landscapes (and even advertise visiting them, see Figure 1.5), we should realize that none of these are real: the best that telescopes can do, and that only in a few cases, is to see "dots" orbiting a star. None of these dots are resolved into an image such as we can make of objects in the Solar System. In the vast majority of cases, we do not even know what the average color of the planet is. Unfortunately, the media do not always make that clear: when I followed Google's news items as I was writing this, almost all features included at least one image, but over half did not indicate whether the images were artist's impressions or images of the real world.

Because we cannot image most exoplanets, and certainly not resolve any of their features, the search for life necessarily focuses on signatures that would stand out even if we had only a single point of light to look at. Even if we were to narrow the search for life to signatures on exoplanetary surfaces and in exoplanetary atmospheres, it will be very difficult to choose which exoplanets

Figure 1.5 Four posters from NASA's "Exoplanet Travel Bureau". Despite these depictions, we have no idea of the surface conditions of these planets, and all are far beyond our reach with present-day technologies with the laws of nature as we know them. Each of these exoplanets is discussed in later chapters: 51 Pegasi b, the first exoplanet discovered around a Sun-like star; Kepler-16b, the first known planet to orbit two stars; Kepler186f, the first known rocky Earth-sized planet in the habitable zone; and TRAPPIST-1e, an exoplanet in a packed miniature system.

to point the required large, specialized telescopes at. That is why astronomers put a lot of emphasis on determining which exoplanets might have liquid water on their surface, simply because liquid water is needed by all life forms on the one planet, Earth, that has life that we know about. Of course, the possibility of liquid water does not mean that there actually is liquid water. But if the temperature is somewhere between the freezing and boiling points of water under the planet's atmosphere then there is a chance of water existing in liquid form. This depends on the thickness and composition of the atmosphere, both of which are at best poorly known. So, it really is an educated guess, and nothing more until we learn a whole lot more about the exoplanet under scrutiny. Astronomers refer to the range of distances from the star at which liquid water might exist on an exoplanet surface as "the habitability zone" or "the habitable zone", although sometimes the name "Goldilocks zone" is used. Some argue that this is confusing because in all likelihood exoplanets in the habitability zone are not habitable to most life forms that currently exist on Earth, and quite possibly to none. They argue that we should instead use the term "temperate zone" to clarify that it refers to only the temperature expected from the stellar illumination. But "habitable zone" has been adopted in the professional literature and in the media, so I shall do the same here. Just bear in mind: a "habitable" planet is not likely to be high on anyone's shortlist for a second home!

Before embarking on an exploration that is at the same time one of far-away worlds and of our own solar system, it is helpful to look at the history of our home in space and the local cosmos. This condensed timeline shown in Table 1.2 has two "clocks": one is in real time (in units of millions of years), while the other is a scaled time that fits the life of the Solar System into a human lifetime. This is done in an attempt to make things more comprehensible because astrophysical numbers tend to be incredibly large and span huge ranges: in this book, you will find timescales that range from fractions of years to multiple billions of years. The human

Table 1.2 Very approximate timeline of events in the history of the solar system. The first column shows "scaled time" in which one scaled human year contains 100 million years in real time. The second column shows real time. "Scaled time" thus passes at about three real years per scaled second.

Scaled time (divided by) 100 million)	Real time (millions of years)	
Jan 1, 1st year	0	Start of contraction of primordial cloud
Jan 7, 1st year	1.8	Nearby supernova in Sun's birth cluster
Jan 11, 1st year	3	50% probability that gas and dust clear from between forming planets
First few weeks	3–10	Jupiter mostly complete, Saturn growing
Jan 21, 1st year	3–10	Reorganization of giant planet orbits, clearing small bodies beyond Earth, and regrouping asteroids; Jupiter as near as 1.5 Earth orbits, then moves out with Saturn
In the first month	6	Solar System mostly clear of gas and dust
Early-mid May, 1st year	30–50	Earth's Moon possibly forms in collision
May–December, 1st year	30–100	Terrestrial planets mostly completed
Within 1st to 3rd year	100–300	Solar System leaves birth cluster
Around 6th birthday	600	Apparent Late Heavy Bombardment and rearrangement of the giant planets
Roughly 7th birthday	700	Possible first life on Earth
21st birthday	2,100	First significant rise in oxygen
24th year	2,400	Earth freezes over: Snowball Earth
25th year	2,500	Cells with internal organelles appear
37th year	3,700	Multicellular life likely formed
38th year	3,800	Earth freezes over twice: Snowball Earth; oxygen levels reach about present-day value; end of Venus's habitability?

Scaled time (divided by) 100 million)	Real time (millions of years)	
40th year	4,000	Diversification of life in Cambrian explosion; first true vertebrates; first exploration of land by animals
41st birthday	4,100	First vascular plants on land
41st year	4,100	First animals permanent on land
June, 41st year	4,150	First animals with arms and legs
Feb., 44th year	4,410	First flowering plants
46th year	4,600	_____ Present day _____
55th–65th year	5,500– 6,500	Anticipated loss of Earth's oceans; end of plate tectonics; end of photosynthesis on Earth
Mid 80s	8,500	Milky Way "collides" with Andromeda Galaxy
110–125 years	110,000– 125,000	Mercury and Venus engulfed in Sun, Earth possibly too; Sun goes through two phases as bright giant star
125th year	125,000	Sun ejects outermost layers, and shrinks into a white dwarf, forever fading

mind is not used to such enormous differences, so, in parallel to real time, I am also introducing a scaled time that expresses time relative to a human life span. In this scaled time, 100 million real years are mapped into one year of a human life, which makes scaled time pass at about three real years per scaled second. In that way, the existence of the Sun as a mature, rather stable, star will span the equivalent of about 100 years in a human life.

Notes

1 Claudius Ptolemaeus, who lived from about 100 CE to about 180 CE in Roman-occupied Egypt, developed his world system in a work with a title that translates into English as *Planetary Hypotheses*. About the Sun, Ptolemy wrote that he could not decide with certainty where it fit between these bodies, but that at least it was further than the Moon;

he wrote that: "there are three possibilities: either all five planetary spheres lie above the sphere of the Sun just as they all lie above the sphere of the Moon; or they all lie below the sphere of the Sun; or some lie above, and some below the sphere of the Sun, and we cannot decide this matter with certainty."

2 In a critique of what he calls "The great Copernican cliché," Dennis Danielson points out that when Copernicus proposed the Sun to be at the center of the Solar System he did not thereby demote the Earth but rather elevated its position: if at the center of the Universe, Earth would be in the lowliest of places, far from the heavens. Danielson writes about the pre-Copernican view that Earth was not viewed as the center of the Universe, but rather as being *at* the center. For example: "for Aristotle, the tendency of heavy things to fall down resulted not from the location of a certain mass but rather from the influence of the location itself, in this case the central location—and I mean not the center of the earth as such but the center, *period*. It is that central *place* itself, not a massive body, that draws heavy things to itself. As Aristotle says in Book 4 of the *Physics*, place itself "exerts a certain influence." And it is merely the fact that earth is composed of the heaviest element (earth being heavier than the other three: water, air, and fire, in that order) that explains why the body on which we live is motionless *in* the center of the universe." He notes that rather than a demotion away from the center of the Universe, the Copernican proposal that the Sun is at the center of the Solar System elevates Earth away from what was viewed as a "cosmic sump," where he paraphrases Galileo Galilei who wrote in 1610 that Earth no longer is to be viewed as "the sump where the universe's filth and ephemera collect."

2

One Step Short of Life

Is there life anywhere else on one of the hundreds of billions of planet-like bodies that are now estimated to exist in the Galaxy? Rocket scientist Wernher von Braun (1912–1977) once observed that "Our Sun is one of 100 billion stars in our Galaxy. Our Galaxy is one of billions of galaxies populating the universe. It would be the height of presumption to think that we are the only living things in that enormous immensity." And yet, despite all the things that we have learned about exoplanets, we remain thoroughly in the dark about whether there is life on other planets. There are many thoughts and investigations on this, including an entire new field of science that focuses on it: astrobiology started its ongoing process of rapid growth only after the mid 1990s, following the discovery of the first exoplanets. Not having found evidence of life does not, of course, mean it does not exist. We could have missed finding life, be it inside the solar system or outside it. And, of course, conversely, it may be that if there are advanced extraterrestrial life forms they could have missed finding us, or at least have been unable, or have not desired, to make us aware of their existence.

In a planetary system far, far away . . .

For any body within the solar system it is in principle possible to at least probe the atmosphere. For many such bodies it is possible, albeit expensive and time consuming, to visit their surface. There, instruments could look for the chemical signatures that are part of life's metabolism, or for fossil traces, or even for the life

One of Ten Billion Earths. Karel Schrijver, Oxford University Press (2018). © Karel Schrijver.
DOI: 10.1093/oso/9780198799894.001.0001

forms themselves. In fact, many spacecraft have already looked around on the frequently visited planet Mars, one has landed on Saturn's moon Titan with its dense atmosphere, and the ice-covered ocean-world of Jupiter's moon Europa is being seriously considered for a future visit. Finding life in the Universe beyond our Solar System, however, is complicated by a colossal challenge presented to us in the combination of the size of the Galaxy and the laws of physics: just about everything in the Universe is too far away for any space probe to get there in anything like a human lifetime.

It is hard to appreciate the vastness of the Galaxy, or even the distance to merely the nearest star, Proxima Centauri, located 4.24 light years away. Proxima Centauri is almost 270,000 times further from the Earth than the Sun is. Imagine we could shrink the Universe so that the distance from New York to Paris was about the thickness of a lentil. If you held that lentil in Paris, then the distance to Proxima Centauri in that scaled universe would be about the same as that to New York in the real world.

In terms of spaceflight, Proxima Centauri is very, very far away indeed: it is about 510,000 times further from Earth than Mars is even at its closest approach to Earth through their respective orbits. Getting to Mars using present-day technologies might take about six months, although that has not yet been achieved. But suppose we could do that, and suppose we could do that using the shortest route available. Then to get to Mars, we should be traveling at about 18,000 kilometers per hour, or 11,000 miles per hour. If traveling at the same speed as such an interplanetary mission within the Solar System, a one-way trip to the nearest known exoplanet, Proxima Centauri b, would take 255,000 years. In his vision for "Making Humans a Multi-Planetary Species" SpaceX CEO Elon Musk envisions making the trip to Mars in a month. With sufficient investment that might be feasible, although designs and cost estimates have yet to be made. At the speed that this shortened Mars transfer would

require, the trip duration to Proxima Centauri b could be brought down to just over 40,000 years, which is still a very, very long voyage.

In a 1984 essay, Isaac Asimov listed a few options from the science fiction literature to deal with the unbearable tedium of interstellar travel: "There are various strategies for dealing with this space-time problem—every one of them pretested in science fiction. Arthur Clarke's story '2001: A Space Odyssey' placed astronauts into frozen hibernation to wait out the long journey. Poul Anderson's book, 'Tau Zero,' was one of many that had astronauts take advantage of the relativistic slowing of time at high speeds [although that would need a propulsion system that has yet to be invented]. Back in 1941, Robert Heinlein described in 'Universe' a large ship—actually a small world—whose passengers were ready to spend millenia, and countless generations, to reach their starry destination." Whereas such options, if they could be realized, might make the trip bearable to the interstellar pioneers, those left behind on Earth might only have a vague cultural recollection of these travelers in evanescing legends.

Only if we could speed up our current technological capabilities of moving through space by a factor of 1,000 could a trip to the nearest star be completed within a human lifetime. Those left behind on Earth would have to wait another 4.25 years, by the way, for the radio signal from this hypothetical probe to reach Earth so they could learn about the findings of the expedition.

Advancing the achievable speed by any known propulsion system by a factor of 1,000 within the next few decades is beyond reasonable expectation. Moreover, this would only get us to the nearest of the multitude of exoplanetary systems that we should expect to have to visit in order to encounter life. In other words, with anything like the technologies that we can reasonably imagine to work given the laws of physics as we know them, there is no chance of us visiting any of the exoworlds, even if relying on purely robotic probes, within our lifetimes.

"Don't you ever wonder where everybody is?"

Should we be able to reach speeds in excess of that of light, then interstellar travel would become feasible. The laws of physics as we know them do not allow for that option, however. We may, of course, be wrong about the laws of physics, specifically the one that nothing can travel faster than light. But if that were indeed the case, then there is something called the "Fermi paradox" to consider. This paradox is named after nuclear physicist and Nobel laureate Enrico Fermi (1901–1954). In 1985, Eric Jones reconstructed how, some time around 1950, this paradox came to be formulated in a discussion between Fermi, Emil Konopinski, Edward Teller, and Herbert York, all four of whom had worked on the Manhattan Project that designed the nuclear bomb during the Second World War.

The surviving participants in this discussion remembered things differently (which is why scientists insist on documenting their experiments as well as they do, because after a while the details become muddled if not plain wrong in our minds), but among the three a consistent story emerged. They were talking about a cartoon showing aliens and a flying saucer. At some point during their conversation Fermi asked something like "Don't you ever wonder where everybody is?", referring to the aliens that might be expected to be abundant throughout the Galaxy and beyond, but that somehow had not shown up at Earth. Because if it were indeed possible to travel faster than light, it would be reasonable to expect that at least one civilization has chanced on how to do it. Such a civilization (or many of them, as the case might be) would certainly be inquisitive about the rest of the Universe, if not in fact interested in utilizing what they might consider valuable around the Universe. So, Fermi and his colleagues reasoned that if faster-than-light travel were possible, then one or more civilizations should by now have visited Earth. Why had we not yet been visited? The Fermi paradox has three ways out: (1) the aliens are not interested in contacting us or they

do not want us to know about them; or (2) no one can travel far if going faster than light or far enough if going slower than light; or (3) there are no advanced alien civilizations capable of (or interested in) interstellar travel either in person or by probe.

There is one rather unsettling side note to be made here that goes back to the principle of mediocrity which could imply that the last option should, in effect, be truncated to read that perhaps there are no advanced alien civilizations in the Galaxy, or at best that they are rare and therefore distant from one another. This rather pessimistic argument, formulated by Daniel Whitmore, in essence rests on applying the principle of mediocrity twice. The first time, it is focused on the overall development of advanced life, and goes like this: Our technologically advanced civilization appears to be the first to have developed on Earth after 4.5 billion years of existence of the planet. We can most readily imagine technologies to be developed on land, much more so than in the oceans. Earth has seen the rise of our civilization within a few hundred million years of animals existing on land and within dozens of millions of years from the earliest primate-like species. Overall, Earth will likely support habitability for a period of very roughly six billion years, lasting for another 1.5 billion years or so into the future. By principle of mediocrity we should assume that we are in a state that should be typical of all possible states, which leads to the inference that (a) whereas it apparently takes a long time for life to evolve to our current advanced state, (b) the development from the first land animals or early primates to that present state is relatively fast. If the latter inference were incorrect and it would typically take a much longer time, then statistically speaking we should have found ourselves more toward the end of the period of habitability. The second time that Whitmore applies the principle of mediocrity in his arguments, he is looking at our technological status. We are within a century or two of the first use of advanced technologies. Because the principle of mediocrity would suggest that we are most likely somewhere in the mid range of being a technologically advanced society, we

should conclude that this state will not last for very much longer than a few more centuries, if that.

By combining these two invocations of the principle of mediocrity, Whitmore concludes that technologically advanced societies rapidly destroy themselves, and moreover that they destroy their habitat so thoroughly that re-establishing the advanced technological state through biological and technological evolution is likely to take billions of years. After all, if only moderate disasters struck, not setting back the biosphere by a very considerable amount, then evolution looks to be capable of quite rapidly leading once again to an advanced civilization. Whitmore therefore suggests that the principle of mediocrity implies "that the typical technological species becomes extinct soon after attaining a modern technology and that this event results in the extinction of the planet's global biosphere." His is a statistical argument based on a single case, and therefore I hope that we can assume that there is considerable uncertainty about the conclusion so that we need not be too concerned about this gloomy outlook of looming catastrophe. On the positive side, in contrast, the relatively short time it took to develop life on Earth in the first place by the same principle of mediocrity suggests that life in general should be quite common on exoplanets throughout the Galaxy.

Back to the Fermi paradox: regardless of which of the three ways out of the paradox is the actual solution, one clear conclusion can be drawn for the benefit of terrestrial astrobiologists: if there is life outside the Solar System, it has to be found by the traditional methods of astronomy, namely remote sensing with telescopes. That brings us to the second challenge in the search for life, which reaches into philosophical issues: What is life? There are many attempts at answering that question through formulating definitions. Many of these rely on properties that we associate with life as we know it on Earth. For example, when I googled "What is life" a dictionary definition came up: life is "the condition that distinguishes animals and plants from inorganic matter, including the capacity for growth, reproduction, functional

activity, and continual change preceding death." The immediate problem with this definition is that it contains multiple implicit and explicit references to itself and to life as we know it on Earth: Should we assume that "animals and plants" are the only forms life can have (allowing for, for example, archaea also in this although they are not formally either animal or plant)? Is the phrase "inorganic matter" supposed to mean that life must be based on molecules with a lot of carbon in them, which is the definition of what makes molecules organic to a terrestrial biochemist? That seems unnecessarily restrictive. Or does it just mean that life contains some chemicals, which one could refer to as "organic," and excludes others? What about "death"? Is "death" a necessary characteristic of all life forms? Would we call something "alive" if it could only live up to the above definition with the help of other life forms, such as is the case for viruses? And would thinking about this last question concerning viruses not lead one to wonder whether "life" is about a matrixed network of organisms that may be viewed as quasi-independent but that are in fact quintessentially interdependent, being able to live individually only because they are part of a collective?

Looking for life

Within the Solar System, we can in principle send probes out with clever instruments, or even make them return soil samples back to Earth, to assess whether there is anything going on like "growth, reproduction, functional activity, and continual change." The probes sent out thus far and the meteorites found on Earth that are fragments from other bodies in the Solar System have not revealed any of such signatures that unambiguously constitute "life," whether presently existing outside Earth or having existed in the past. Maybe we shall find "life" in places yet to be explored, such as under the thick ice shells of Titan. Reaching such places is not easy, and it will easily take a decade of interplanetary travel after a decade of planning and designing

such missions. What do we do in the meantime? We observe from a distance both the collection of solar-system bodies and the multitude of exoplanets.

What would you look for to find "life" in the Universe? Life could take on a great diversity of forms. Our cultural thinking was initially quite limited to forms much like what we knew on Earth. In science fiction movies this was at first also the case: actors needed to portray the extraterrestrials and those were necessarily of human form. Advances in animatronics and computer graphics have lately enabled movie makers to go off in imaginative directions, not only in terms of body size and appendages but even in the basic kind of life that is being portrayed. So our views have gone beyond the largely humanoid forms in *E.T.*, *Star Wars*, and *Avatar*, although the reptilian forms encountered in *Men in Black*, *Independence Day*, or the *Alien* sequence do not stretch our imagination very much. The squid-like beings in *Arrival* are a step away from the usual, but are still reminiscent of terrestrial life. Really out-of-the-ordinary forms include the vast jellyfish-like beings traveling in space in *Star Trek*'s 'Encounter at Farpoint' episode and *Dr. Who*'s star whale, or beings made out of gas clouds or liquids, out of pure energy, or comprising collectives of tiny creatures (although the latter are reminiscent of colonies of ants and hives of bees). And then there are the ones we never get to visualize: the Martians from *The Martian Chronicles* by Ray Bradbury (1920–2012) who project a telepathic image but never show themselves to earthlings, or the deep-sea threats in *The Kraken Wakes* by John Wyndham (1903–1969) who might have migrated from one of the giant planets to resettle in the high-pressure environment of Earth's deepest ocean trenches where no human probe can venture. Even in real life, on Earth, we could find an array of diverse life forms that are quite different from what we are generally familiar with: fibril underground fungi, ice worms, corals, jellyfish, the amazing diversity of uncanny plankton species, etc. Whereas we should allow for the possibility that life elsewhere might have considerable similarities to life on

Earth, this is unlikely to be the case. Yet our imagination is limited by what we are accustomed to: we simply do not know enough about how life might work to stimulate our creative power as to what might be out there. Nonetheless, imagination as a first step is key in science: it may be proven wrong, but it drives research. Carl Sagan (1934–1996) said it like this: "Imagination will often carry us to worlds that never were, but without it we go nowhere."

The above example of a definition of life is but one as it appeared in one particular dictionary. There are many other definitions of life, in a multitude of distinct forms, often involving the foundations of the above example: metabolism, procreation, adaptability, and impermanence. Other definitions involve a subset of these. Some combine them with other characteristics. Many of these attempts at definitions are, ultimately, not so much definitions per se but rather inventories of phenomena associated with living organisms on Earth. Unfortunately, our thinking about a definition of life is at present unavoidably biased by the fact that we only know life as it is shaped on present-day Earth. Formulating a general definition for other worlds, or even for other times on planet Earth itself, that is based on a single example in space and time is hazardous, even if that one example comes in a great diversity of forms, as life on Earth does. Until we have a broader vision of what forms life has taken on other worlds, if indeed it has done so, we may be in a more advantageous position if we do not define life as yet: in the absence of an agreed-upon definition, our thinking is less constrained by the attributes that would be encoded as essential to life in such a definition. Moreover, if we consider how life may have started on Earth, we may need to allow for a continuum of states rather than a bifurcation: are life and lifelessness clearly separable states? Whereas it is tempting to think in terms of unambiguously binary states for life and lifeless, we should realize that physics has shown us multiple examples that such a simple separation may not be how nature works: matter and energy are exchangeable, time and space affect each other, and light behaves both as a particle and as a wave at the

same time. This possible smooth and fuzzy transition between lifeless and living has been called the continuity thesis.

However, no one should claim a discovery of extraterrestrial life based only on the attitude that "I'll know it when I see it": in order to claim success in finding extraterrestrial life we do need to agree on what signatures a claim of discovery may be based on. The growth of astrobiology in parallel with the discovery of exoplanets has made the need for a definition all the more pressing. In reality, the search for life and for an appropriate definition of it are going on simultaneously. On the one hand, scientists are discovering new chemical pathways in natural settings and are thinking of new biochemical systems imagined to function in the great variety of exoplanet settings and in the settings of Solar System bodies such as the planet Mars and the giant-planet moons Enceladus, Europa, and Titan. On the other hand, definitions of life are thought about by scientists in disciplines as distinct as philosophy, biology, and computer sciences.

Regardless of what we imagine individual extraterrestrial life forms to be like, we may well not see any of that kind of detail on even the nearest exoplanet. So that is not what astronomers look for. Imagine yourself as an astrobiologist trying to decide whether something seen from a great, great distance is a signature of life or instead a product of lifeless natural goings on. How could this scientist possibly prove that something like "growth, reproduction, functional activity, and continual change"—as required in the above definition—were actually occurring on an exoplanet? After all, even with the best of telescopes, we cannot hope to reveal details down to the size of even the biggest life forms we know of on Earth, at least not in the next few decades. Consequently, even if there were a clear, agreed-upon definition of "life," it would likely be of little help to present-day astronomers and astrobiologists who have, at best, access to signals from exoplanets as a whole.

Rather than being philosophical, researchers aiming to find "life" beyond Earth need to be pragmatic. After all, their first goal

is not to identify all life-bearing exoplanets, but rather to find the very first exoplanet that shows any life to exist outside Earth. So they set themselves a different goal, one that can be acted upon with the technologies of the late-twentieth and early-twenty-first centuries: find signatures of atmospheric chemistry or technological activity that are not reasonably expected to occur in nature but that instead are attributable to either one particular living species or to the entire network of life forms on that exoplanet, but in a way that could not occur in the absence of life. And that is a very complex challenge indeed.

One search for extraterrestrial civilizations that started decades ago focused on the detection of radio signals. For us on Earth, radio signals provide an efficient means of long-distance communication, readily crossing oceans and continents, and they are even used to reach Earth-orbiting satellites and robotic probes to other planets. The primary reason why space agencies communicate over long distances by radio is the same reason that a cell phone can work for a long time on a small battery: the Universe is incredibly dark at radio frequencies, so a low-power transmitter can be easily picked up and a very-long-distance connection can function. Radio signals are just one of the many forms of light that range from the low-energy radio signals through microwave and infrared to the visible and beyond to ultraviolet, X-rays, and gamma rays. All things in the universe glow most brightly at a wavelength—or "color" in the broadest sense of the word—that is set by their temperature, like the yellow of the Sun's surface at a few thousand degrees to the X-rays from the Sun's million-degree outer atmosphere; also, the higher the temperature, the brighter the glow typically becomes, at least for opaque bodies. Radio waves have so little energy that their associated temperature is very close to absolute zero, which very few things in the universe are at, and those that are that cool are extremely faint. That is why the Universe is so dark at radio wavelengths. There are exceptions for uncommon physical processes, but precisely because these are uncommon, radio waves from far

away can be readily searched for against the mostly pitch-black heavens.

When looking at things in the Universe at colors not matching their temperature, they are much fainter than at their peak color. The Sun, for example, is at least a hundred billion times fainter when seen at radio wavelengths than in the yellow light in the visible range; the longer the radio wavelength is, the fainter the Sun shines. You can imagine what the world looks like through radio eyes: if you chose the wavelengths at which cell phones are used and looked from afar, you would see multitudes of tiny dots where these phones are, surrounded by brighter cell towers, under an almost dark sky dotted with the faintest hint of stars around an odd-looking faint Sun. Radio waves travel through what we perceive as solid and opaque in visible light: the fact that we can use radio waves inside buildings and vehicles is the clue that most walls and barriers we see with our eyes are pretty much transparent at radio wavelengths. However, we would not perceive things very crisply: radio waves have wavelengths much longer than visible light, and as we went from radio 'colors' at millimeter waves to kilometer waves our vision would rapidly deteriorate.

Crispness of images is not what the hunters for extraterrestrial technologies are after, however. Rather, it is the darkness of the Universe that they use: as natural processes are generally very "quiet" at radio frequencies, it should be easier to pick up signals from alien life, both because the signal is easier to pick up against any natural emitters and also because we assume that such alien intelligence would have noticed the same and would therefore be likely to opt to use radio signals for their communication needs.

The search for extraterrestrial life started in earnest in 1960 when Frank Drake (1930–) pointed the 26-meter Green Bank radio dish at two nearby stars, tau Ceti and epsilon Eridani, and found—decades later—that both have planetary systems. Drake wrote about this in 2011: "No signals were detected. Now you might think, 'wasn't it rather foolish to search only two stars—it's

very improbable that you would succeed'. Actually, back then, for all we knew, every star in the sky had a planet that was transmitting radio signals. It could have been that the observers would have succeeded the very first day! They did not succeed, and the experiment did show that radio-transmitting civilizations are not ubiquitous in our galaxy." Other scientists started using additional radio telescopes around the world to listen for signals that should not occur in lifeless nature. Modern instruments are vastly more capable than those used at the beginning of the search. In a review paper, Drake illustrates this: "the sensitivity of our telescopes has increased by about 1000 times. Furthermore, searchers have gone from monitoring one [radio] channel at any moment to hundreds of millions of channels. Putting it all together, the capabilities of our telescopes and systems of today are about [100 trillion] times more powerful than the best systems of [1960]". Still, despite looking at thousands of stars, so far there has been no success.

The Drake and Rare Earth equations

Frank Drake also did something else: in 1961 he looked into how many alien civilizations might be expected in the Galaxy by writing down what we now know as the Drake equation. It is a deceptively simple equation in which seven numbers are combined to come up with the number of such civilizations. It uses a concept generally used in estimating how often something would be expected. Say you want to estimate how many female red-haired babies might be born in a year in a country. How would you go about that if you did not have the means to query every birth in some detail? You could take the number of citizens, divide it by the average life span (which then gives you the number of new births if you assume that the number of people is not changing too much in a year, because then the number of births statistically equals the number of deaths), multiply by one half (because close to one half of all children are female), and multiply it by the

fraction of people in the population who have red hair (which you could estimate by sampling the people you encountered on a given day). Some of the numbers in such a statistical estimate will be well known, others are estimated, while still others may be guesses or assumptions. Along with determining the numbers going into the estimate, you would also have to look into how well each of these numbers is known, which in many cases you can estimate, but in some cases you cannot.

Frank Drake came up with this formulation: the number of civilizations in the Galaxy using radio signals should be roughly equal to: (1) the number of stars suitable for sustaining a stable environment for a very long time, multiplied by (2) the fraction of those stars with planetary systems, times (3) the number of planets in a planetary system suitable for life, times (4) the fraction of such planets on which life develops, times (5) the fraction of planets on which life develops intelligence, times (6) the fraction of such civilizations that develop radio technologies, times (7) the fraction of a planet's lifetime during which such civilizations persist in using radio signals. Whereas this should give you an estimate of the number of civilizations we might be able to find by listening to their radio signals, we simply have no clue as to what the values of most of these numbers would be. In short: nice equation, but not—presently—particularly helpful in deriving a well-founded number of civilizations within the Galaxy. But that was not the original intent of the equation: it was written down initially to start discussions on the topic of finding extraterrestrial civilizations, which it did quite successfully.

We now know that some of the numbers in the equation are helping to make the outcome large enough that there is hope: there are lots of stable and long-lived stars, which we now know to have lots of planets, including lots of Earth-sized planets in Earth-like orbits, and we now know that these planetary systems offer fairly stable settings to their planets for billions of years. But we have no clue as to the frequency at which life develops technological use of radio and how long such civilizations might

survive. We do know that, so far, no convincing evidence has been found of aliens using radio waves.

The average planetary system is estimated to be some three to four billion years older than our own. That should have given extraterrestrial life around the Galaxy plenty of time to advance to a level at which technologies have gone well beyond our own. So what does it mean that although just about every star has a planetary system, efforts like SETI (the Search for Extraterrestrial Intelligence) have found no evidence of "intelligent life"? Or, to be more precise, SETI has found no evidence of radio signals used by intelligent life that are strong enough to be detected. Again, Fermi's question pops up: "Don't you ever wonder where everybody is?". As an editor at the journal *Nature* observed: "Regardless of how exhaustively the Galaxy is searched, the null result of radio silence doesn't rule out the existence of alien civilizations. It means only that those civilizations might not be using radio to communicate," or more precisely that they would not, for some unknown reason, beam out a lot of radio power inadvertently (or intentionally) in the direction of Earth.

Should a civilization use radio frequencies at some point in time, it could do so for communication purposes only briefly. For Earth itself, strong radio (including TV) signals have only been used for about a century. Increasingly, communication is shifting away from powerful broadcast antennas to much weaker radio signals between cell phones and cell towers, while most television and radio station signals now find their way to consumers hidden inside the glass fibers and copper wires of cable TV and the Internet, or can be much weaker following the switch to digital and satellite-based. Radar signals, also part of the radio frequency range, have also become much weaker with better detectors and are more spread out over the frequency ranges. So any exoplanet civilization looking at Earth using radio signals would have detected our planet only during about a century out of 4.6 billion years of the Earth's past and likely not

much of its future. For an average present-day planetary age of some 7.4 billion years, being "radio bright" for only a century means only one part in 74 million. So, if every planetary system evolved life like here on Earth, and if we disregard the different ages of planetary systems for the moment, we would have to look at 74 million exoplanetary systems to have a chance of detecting one using radio wavelengths (assuming our telescopes are sufficiently sensitive) even if every one of them evolved as we did.

Not having detected extraterrestrial radio signals could mean that use of radio signals is limited, perhaps because of evolving technologies or perhaps—as discussed above—because advanced civilizations simply do not last long. It could also mean that intelligent life is a lot less likely to develop than we would like to think. In their book titled *Rare Earth*, Peter Ward and Donald Brownlee formulate their perspective in their subtitle: *Why Complex Life is Uncommon in the Universe*. In essence, they argue that the Drake equation should show more multipliers: in their view, it should also include (a) the fraction of stars that lies in what they call the Galactic Habitable Zone (which means safely away from where the star density is so large that collisions or supernovas would be too likely), (b) the fraction of habitable planets with large moons (which stabilize the spin axes of planets), (c) the fraction of exosystems with Jupiter-mass neighbors quite far from their stars because these help shape the planetary system, and (d) the fraction of planets with a very low rate of mass extinction events. I would argue that these additional factors should be viewed as implicitly included in Drake's equation in the terms that quantify the number of planets in a planetary system suitable for life and in the fraction of planets on which life develops intelligence, because the terms discussed by Ward and Brownlee quantify some of the conditions for that to occur. In that case, their factors explicitly illustrate that the challenge of quantifying Drake's multipliers is quite a bit more complex than its simple phrasing might lead one to think.

The communication challenge

If we detected a world with extraterrestrial intelligence, by the way, then communicating with such an exoplanet would be exceedingly tedious, even if they were actually "listening" for a signal from Earth: to encompass 74 million stars, we would need to look out to a distance of close to 2,000 light years. The average distance to a star in that volume, most of which lie within the Galactic plane, would be such that a signal would require three millennia to travel back and forth (and then you would have to figure out what the signal meant, illustrated in Figure 2.1). That is not a pace conducive to communication, so the best we can hope for realistically is to limit ourselves to attempts to establish that there is a technologically advanced civilization anywhere (in other words, to establish whether any rocky exoplanet at all is showing radio signatures that are not attributable to natural causes). Establishing whether there is life at all may be easier,

Figure 2.1 Communication with alien civilizations will be a challenge far bigger than between peoples (or just generations) on Earth: not only would both sides need to learn to interpret the messages, but the distance between inhabited worlds with advanced societies is likely to be so large that any reply to a message sent, if any comes at all, will be received after many decades, if not centuries, in which technologies, if not entire cultures, may have changed multiple times.

particularly once future-generation telescopes enable us to study the chemical makeup of planetary atmospheres that may reveal the existence of life. Some such studies are already underway with the most powerful telescopes, including NASA's Hubble Space Telescope, but even with them it is a challenge to obtain enough high-quality signal; Figure 2.2 shows a transit of Venus from which it becomes clear just how small a signal is searched

Figure 2.2 The planet Venus transiting the Sun on June 5, 2012, the last such transit visible from Earth until 2117 CE. This image was taken by the Japanese Hinode satellite, using its Solar Optical Telescope. The Sun's bubbling surface is seen underneath some wisps of its outer atmosphere. For other stars, the slight dimming of their light by exoplanets that transit them is frequently how exoplanets are found. Details of the chemical makeup of Venus's atmosphere are imprinted in the light that travels through its thin envelope, which reveals itself clearly in the bright ring off the Sun's edge. Analogously, the modification of the starlight traversing exoplanet atmospheres can in principle be used to analyze these atmospheres in the hunt for biosignatures, although that requires exceptional telescopes.

Source: JAXA/NASA/Lockheed Martin.

for, particularly for Earth-like planets: astronomers are currently analyzing planets whose brightness is between one part in 1,000 and one part in 100,000 of the star's brightness.

Not having found any alien life forms that use radio signals, we need to look for other signals. Doing this brings us back to a perfect analogy that the first exoplanet hunters were facing. They needed to answer two questions: What signature should we look for? And where should we look?

Is disequilibrium a signature of life?

The signatures of life, regardless of whether it is technologically advanced or no more than a single-celled microbe, are referred to as biosignatures. In their "Astrobiology Primer v2.0" Shawn Domagal-Goldman, Katherine Wright, and colleagues have a definition: "A biosignature is…any characteristic element, molecule, substance, or feature that can be used as evidence of past or present life and is distinct from an abiogenic [not formed by life] background." Astrophysicists can only detect what they can measure with their instruments so that the question of what biosignatures to look for is reduced to finding those that affect the light that comes from exoplanets. We already mentioned the light at radio wavelengths emitted by advanced alien technological civilizations. Another such biosignature to look for is the imprint of certain chemicals in the light from the central star that is reflected by the exoplanet or in the light that passed through that atmosphere as the exoplanet crosses in front of its star. What to look for? Atmospheric chemists call it signatures of "large-scale environmental disequilibrium," which means we should look for something that would not be compatible with a natural lifeless state of an atmosphere. The latter is a real problem in this quest: there are many reasons why something might be off a state of chemical equilibrium. This can happen whenever chemicals are transported from one environment to another, or modified by some external influence, faster than the chemical

reactions can follow: volcanic eruptions, fast winds, crossing freezing points, irradiation by UV light from lightning storms in stellar atmospheres, etc.

One signature that comes up naturally is to look for substantial concentrations of oxygen in an atmosphere. Earth's atmosphere contains some 21 percent oxygen which is maintained by plant life. Without plants, the oxygen levels would rapidly drop to very low, perhaps even undetectable, levels. What is even more telling about life on Earth is that there is also some methane in the atmosphere. Donegal-Goldman and colleagues formulate it this way: "The mixture of oxygen and methane is far from 'atmospheric equilibrium,' and on short timescales the two species [here oxygen and methane, counter each other's existence as they] react to produce carbon dioxide. The short lifetimes of these gases mean that they have to be constantly replenished at a fast rate in order for them to be maintained at detectable concentrations. In other words, their presence implies large sources of these gases. The most sustainable source of these gases at the required rates is life. However, disequilibrium is also caused by non-biological processes, such as photolysis [i.e. reactions driven by changes in starlight] and physical mixing [such as by volcanic eruption, winds, or ocean flows]. The 'disequilibrium' in this case is a change with respect to the state of the atmosphere without biological inputs." In the last two sentences, a key problem facing astrobiologists is introduced: in order to flag a chemical signature in the light from an exoplanet as being made unambiguously by life you need to be able to rule out all the possible natural processes on an exoplanet as sources.

For example, oxygen could exist in an atmosphere like that which once existed on ancient Venus: being closer to the Sun than Earth, a relatively higher dose of UV light from the Sun would more efficiently break up water molecules, allowing the freed hydrogen, as the lightest of gases, to evaporate off the planet into outer space, which could leave enough oxygen, the hydrogen atoms' former partners in water molecules, in the atmosphere

to be detectable with remote telescopes. Methane is known to form in hydrothermal environments—perhaps offering an explanation for the 1.4 percent methane in the atmosphere of Saturn's moon Titan. These examples illustrate how the studies of exoplanet climatological chemistry can also offer increasing insights into the natural processes that create and destroy chemicals, subject to the light and particle radiation from their central stars and to the surface and atmospheric environments, guided by what we learn from the nearby examples that can be studied in some detail—the Solar-System planets and any of their moons with atmospheres—but there is still a long way to go. That is how science progresses, though: observe, try to interpret what is observed, understand it in the broad scientific context, rule out the impossible, and study the probable versus the improbable,...carefully and methodically advancing understanding of what we are learning.

Where to start?

The second question facing the hunters for extraterrestrial life is: where to look? As long as we are looking for the very first confirmation that life exists outside the Solar System, we should optimize our chances in recognizing life if we were to see it. We do, at least, know how to recognize life on Earth, so it is natural to look for something vaguely similar to that to start with. The chemistry of life uses the elements that are readily available on Earth. As these are the elements most readily available throughout the Universe, looking for, for example, worlds with carbon on them is hardly necessary. Chemically speaking, all life will require an energy source. On Earth that is either the Sun or geothermal energy. Any exoplanet that orbits a star will at least have starlight, although it may be faint, while internal planetary energy—as we shall see later—is likely to exist just because of the origin and formation processes of planetary systems. What makes Earth unique, at least within the Solar System, is what

enables all life on Earth: the availability of liquid water in the layers immediately above and below the solid surface. Finding evidence of liquid water in combination with a solid surface is therefore the first step in present-day searches for life. And that starts with finding the planets that are likely to be able to keep liquid water around for a long time, say at least a billion years. This guides much of the exoplanet research being done around the world.

In view of that, it is not surprising that the selection of exoplanets to which much time on powerful and costly telescopes is to be given is based on the concept of what has become known as "the habitability zone": for starters, pick those exoplanets that have the largest likelihood of having liquid water on their surfaces. That criterion narrows down the number of targets to those that offer the highest likelihood of having life on them in a form that we might recognize and that we would have a chance of detecting using available remote-sensing tools.

Among other interesting lists, Stephen Kane and colleagues compiled an inventory of exoplanets that are expected to be in the liquid-water habitable zone even by a conservative analysis, of which the properties are fairly well known, that are at most twice the size of Earth, and that seem to be in settings where any other (known) exoplanets in their planetary system are not likely to upset the orbits too much for long periods of time. They selected twenty such exoplanets from the long list found by the NASA Kepler satellite that monitors the brightness of many tens of thousands of stars to find planets that modulate their stars' light by crossing in front of them during their orbits as seen from Earth. That makes these planets particularly interesting because future studies could then look at how the starlight is modified by passage through the exoplanet's atmosphere, which could reveal biosignatures.

Although we have not yet detected life anywhere but on Earth, we have made great strides toward being able to detect it. With the development of technologies to detect exoplanets and the

resulting realization that planets are common we are now, in a very real sense, only one step short of the detection of extraterrestrial life, if indeed it exists. We have also learned that the habitability of any planet changes over time as the central star changes, as planets move through the planetary system, and as the planets see their atmospheres change with evolving interior dynamics and—in the case of Earth—life. For over a century biologists have thought about the fundamental processes of natural selection in how life evolved. As we learn more and more about exoplanet systems, astrobiologists are expanding that concept to natural selection that is not limited to Earth only but as something that plays out for planetary systems as a whole.

As I am writing this, the odds seem to be that for every ten stars in the sky we should expect to have one Earth-sized planet orbiting at a distance from its Sun-like star such that liquid water could exist on its surface. With over 100 billion stars in the Milky Way, that means there are ten billion planets in just the one galaxy that we live in that could be vaguely similar to Earth, and yet in all likelihood be completely different: in our Galaxy, our Earth is one of very roughly ten billion kindred, yet distinct, worlds. We do not know what these worlds look like. But we can distill what we do know into visions of imagined worlds, and these imagined worlds can give us new hypotheses to test, lead to bold ideas to pursue, and suggest designs for advanced telescopes. Every step we take in the real world of observations and in the virtual world of the computer takes us toward a better understanding of stars, their planets, and the life forms that they may support.

3

Exploring the Solar System

When astronauts Neil Armstrong (1930–2012) and Buzz Aldrin (1930–) spent a few hours as the first humans on the Moon's surface, Mike Collins (1930–) was alone in the Apollo 11 command module. Orbiting the Moon in absolute solitude he recalls looking back at Earth and that "the overriding sensation I got looking at the Earth was, my god that little thing is so fragile out there," (see Figure 3.1) despite the fact that the Earth covers an area in the Moon's sky that is almost 14 times larger than that of the Moon seen from Earth. Humans have yet to travel beyond the Moon, although they have traversed the Solar System vicariously far and wide with a variety of space probes. Through spacecraft and telescopes we have seen a diversity of worlds within the Solar System—in other planetary systems none can be imaged any better than as a dot, and that only rarely! Even Jupiter, by far the largest of all bodies in the Solar System, is minute when compared to the whole: the space populated by the planets and their retinues is first and foremost emptiness, dotted only here and there with something substantial.

The Solar System comprises a whole bunch of different objects that occur in regions with names that reflect attributes of the objects they contain; these may be the names of their discoverers, or may have roots in ancient history or extinct languages. We shall get to all of these components in time, but just so I can describe them to embark upon this journey, I need to put some on the table right here. Nearest to the Sun, reaching out as far as thirty times the distance from the Sun to the Earth, orbit the planets. Closest to the Sun are the four so-called terrestrial

One of Ten Billion Earths. Karel Schrijver, Oxford University Press (2018). © Karel Schrijver.
DOI: 10.1093/oso/9780198799894.001.0001

Figure 3.1 Earthrise over the region of Smyth's Sea (Mare Smythii) on the near side of the Moon. This picture was taken in July 1969 by astronaut Michael Collins, as the Apollo 11 command module was orbiting the Moon just prior to the descent and landing by Neil Armstrong and Edwin "Buzz" Aldrin. The Mare is the relatively dark plain formed by ancient basaltic lava flows, subjected to subsequent impacts by meteorites causing craters upon the solidified surface (See Plate 3 for a color version).

planets, carrying that collective moniker because they are very roughly like Earth—Terra in Latin—in both size and internal composition: Mercury, Venus, Earth, and Mars. Then there is a band of relatively small clumps that are referred to as asteroids (from the Greek word for star-like) in what is called, helpfully, the asteroid belt. These components are depicted in Figure 3.2.

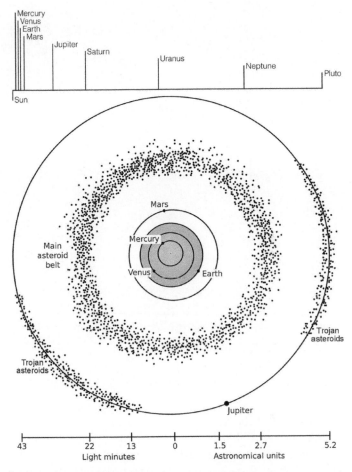

Figure 3.2 Top: Relative sizes of the average distances to the Sun marked on the horizontal bar. Bottom: Main elements of the asteroid populations compared to planetary orbits in the inner part of the Solar System; note that the dots representing the planets are shown much too large to match the scale of the orbits. Even the small dot in the center that represents the Sun should be shrunk somewhat to be to scale. The larger shaded area that fits Earth's orbit indicates the approximate size to which the Sun will expand during its end-of-life giant phase. The bar that goes with the bottom figure shows distances. Going from '0' to the right it shows distances expressed in Sun–Earth distances, otherwise known as "astronomical units." Going from '0' to the left it shows these same distances expressed in minutes of travel time for light.

Further out than the asteroid belt orbit the four giant planets: Jupiter, Saturn, Uranus, and Neptune. Beyond them are relatively small bodies (compared to planets) that include dwarf planets (such as Pluto); these are all chunks of rock and ice, all cold so far from the Sun, and relatively small compared to planets. Most of these are members of what are known as the Kuiper Belt and the still more distant Oort Cloud. All of these are collectively referred to as minor planets, except the comets which are called, well, comets, although the visible difference with the minor planets is mostly based on their appearance should they ever fall closer to the Sun: the solid cores of comets, when warmed sufficiently by sunlight, emit gases that form their famous tails. We shall get to these last populations later on, but for now I concentrate on what is closer to the Sun than, say, fifty times the Sun–Earth distance. That is as far out as Pluto travels in its orbit among its sibling Kuiper Belt objects.

Where did it all end up?

The Solar System is commonly represented in either a diagram, or a table, or in words by showing its contents in order of the distance of the planets from the Sun. Such a representation may include some of the various other objects that are part of the whole: natural satellites that are generally referred to as moons, asteroids, dwarf planets, comets, Kuiper Belt and Oort Cloud objects, down to planetary rings (that exist around all the giant planets) and dust. Because we are looking at the Solar System as one of many planetary systems, let us look at it from a different perspective here through the answer to these questions: where has the material ended up out of which the Solar System was originally made, and how is it partitioned over the Sun, the various types of planets, and the other objects that orbit around them? The answers initially guided how to look for other planetary systems. Now that many such systems are known, the characteristics of the Solar System form a foundation against

which new insights are tested to advance our understanding of stars and their planets.

Part of the material in the pre-Solar-System cloud is no longer here, having been blown out of the Solar System a long time ago, or having been thrown out in chance encounters of relatively small objects with a partially or fully formed larger planet. Astronomers are still learning what fraction of the original mass of the cloud out of which the Solar System formed has been ejected. Therefore, for now, let us limit the question to where the mass is that is still here, somewhere in our Solar System. The mass that resides in the many small bodies very far from the Sun is actually not known very well, nor do we know how many there are. They are expected to contain relatively little of the total mass of the Solar System, though, with possibly less than 10 percent of an Earth mass in all the Kuiper Belt objects combined. I do not discuss them further here.

Nowadays, the masses of the various bodies in the Solar System can be readily looked up with Google or Wikipedia. But the numbers collected there were painstakingly put together over the course of several centuries. In order to measure the mass of a body in space, you have to use the law of gravity. That was not known until 1687 when Isaac Newton put together the arguments that led him to formulating the law: gravity causes a mutual attraction between two bodies that is determined by the product of the masses of the two bodies, divided by the square value of their separation, with a multiplier in front that, in modern terms, is called the gravitational constant. Although the gravitational force is pretty weak between any two objects from everyday life on Earth, one can in principle measure that force between objects in a laboratory if one were clever enough to devise a very sensitive apparatus to do so. That measurement was not accomplished, however, until Henry Cavendish (1731–1810) built such a device: a pair of heavy lead balls was used to pull sideways on a measuring bar with two smaller balls mounted on its ends, while Cavendish measured how much the bar was rotated by the mutual attraction

between the large and small balls and then determined the force needed to torque the bar as far as he measured it to move. But Cavendish did not perform this experiment until 1797, over a century after Newton had first published the law of gravity. This left Newton with a big problem when he applied the law of gravity to the orbiting motions of the planets: he did not know the distance to the Sun, he did not know the gravitational constant, and he did not know the mass of the Earth. Nonetheless, he made a lot of progress by using an estimated Sun–Earth distance, by using ratios of masses rather than their individual actual values so that he could do without the gravitational constant, and by combining this with other findings.

How far to the Sun?

Newton had derived three other fundamental physical laws that would enable him (and others since then) to move ahead in estimating masses of bodies in the Solar System, albeit subject to an uncertainty in another important piece of information: the distances between the Sun and the planets. Establishing these was a tough problem. One of the greatest leaps toward knowing these distances was made when Johannes Kepler (1571–1630) put together his three laws of planetary motion, first published in the first two decades of the 1600s. This was around the time when Galileo Galilei spotted the four largest moons of Jupiter with the newly invented telescope and described their orbits around that planet as a small version of the much wider orbits of the planets around the Sun. Kepler used the measured movements of Mercury, Venus, and Mars against the fixed background stars. He based this largely on measurements made by Tycho Brahe (1546–1601) that were of unprecedented accuracy in his days. Kepler came upon the idea that planets orbit the Sun not in circles—as people assumed until then—but in ellipses. For Mercury and Venus, both closer to the Sun than Earth, he used measurements taken over years of the furthest angular distance

of these planets from the Sun as seen in the sky. He then combined that with the shape of ellipses to determine the orbits, and thus established how many times larger Venus's orbit was than Mercury's, both relative to the size of Earth's orbit. For Mars he had to use another procedure, necessary because Mars moves outside Earth's orbit so there is no largest angular separation. In fact, Mars moves around the entire sky as seen from Earth, but because both Earth and Mars are moving, the directions in which Mars is visible from Earth did also, in the end, allow Kepler to deduce Mars's elliptical orbit. Eventually, at the beginning of the seventeenth century, the relative geometry of the orbits of the terrestrial planets was known.

Note that the "relative geometry" did not help Newton enough: Kepler had established the sizes of the orbits of several planets relative to the orbit of the Earth: all distances were expressed in units of the Sun–Earth separation (known later as the "astronomical unit"), but until about 15 years before Newton published his *Philosophiæ Naturalis Principia Mathematica* no one knew how to express these as an actual length, no matter whether in feet or in kilometers. For that to be possible, at least one distance would have to be actually measured. The first to claim a rather successful determination was Giovanni Domenico Cassini (1625–1712). He used what we call parallax: if two well-separated observers look at a relatively close object from two different positions, then that object will appear in a somewhat different position compared to the distant background. In this case Mars, although far away, was observed against the background of the very much more distant stars, separating the two observers by a long distance on Earth: Cassini sent a young colleague, Jean Richer, off to Cayenne, in South-American French Guyana, while he stayed in Paris, and they both observed Mars against the starry skies. The goals of this expedition were strictly scientific and it appears that it was the first, or maybe only the second, major expedition like that to ever be undertaken.

The measurements of Mars's position against the distant stars were performed in 1672 at a time of Mars's closest

approach to Earth in their respective orbits. This provided the best configuration for the measurement to have the largest sensitivity and thus smallest possible error. Even at this minimal interplanetary distance it would be hard enough. Of course, they had to observe Mars at the same time, which was a problem because clocks were not particularly precise in those days. Moreover, they had to know the exact distance between Cayenne and Paris, but the ability to measure such distances also was in need of improvement. For both these things they needed a truly good way of determining time. To achieve this, they selected a natural clock that would be error proof: they used the movement of the moons of Jupiter to ensure that, when they met again in Paris and looked through their respective series of measurements as recorded in their logbooks, they would be able to select a moment of simultaneous observation Richer completed the voyage, it having taken a year and a half, despite the fact that he was forced to return early because of illness; Richer's assistant actually died on the expedition, undertaken in a time when travels around the world were still very hazardous. In the end, Cassini's result was off by just under 8 percent when compared to the modern-day value. In getting to the result, however, he had to make corrections for the bending of light through Earth's atmosphere and in retrospect it seems he got lucky in choosing a nearly correct value for that.

The same parallax method was used by Hipparchus (190–120 BCE), an astronomer from ancient Greece (born in what today is Turkey), to measure the very much smaller distance to the Moon to the same relative accuracy. Hipparchus did that some 1,800 years earlier than Cassini worked out the distance to Mars, but for a target that was 142 times closer and thus much less demanding on the required precision of the measurements. Knowing that distance helped Newton, because the Moon orbits the Earth, and only together with the Earth does it orbit the Sun. This was important because Newton's second problem was that, because at the time no one knew how to measure the gravitational

constant, he could not express the Sun's mass in actual physical units like kilograms or pounds. He figured out, however, by also using the information on the Moon's orbit around the Earth, that he could use his laws to come up with a mass of the Sun expressed in Earth masses. The value he came up with relied, of course, on how well the scale of the Solar System was known. Around the time Cassini made his measurement, others did so too, among them Flamsteed, Halley, Lahire, Picard, and Newton himself. Their results differed by up to more than a factor of two. The first edition of Newton's book in which he used his own observations had the mass of the Sun expressed in Earth masses quite a bit off. For the subsequent second and third editions he realized that other measurements of the distance to the Sun were more accurate, so he revised his numbers, and came up with a solar mass, expressed in Earth masses, that was much closer to what we have as the present value.

Other important opportunities to establish the basic unit of distance in the Solar System more accurately came in 1761 and again in 1769: Venus would cross in front of the Sun, in what is known as a transit. With this method the light-bending problem of Cassini's method can be avoided, as noted by Edmond Halley (1656–1742, after whom Halley's Comet is named) who came up with the method. Unfortunately for him Venus transits are rare because of the different tilts in the orbits of Earth and Venus so that alignments between Earth, Venus, and Sun rarely take place: they occur in pairs with eight years between them, but these pairs occur separated by over 100 years. So, it was not until nineteen years after Halley's death that the first opportunity to apply his method arose. Multiple expeditions were sent out to observe the Venus transit from different locations on Earth, including, for example, that of James Cook (1728–1779) to Tahiti. This measurement still relied on triangulation, but used an intermediate object passing in front of the Sun, eliminating many of the possible sources of error. From these transits resulted a determination of

the Sun–Earth distance that proved correct to within 3 percent compared to the present-day value.

Not until Henry Cavendish calibrated the strength of gravitational forces in his laboratory, twenty-eight years after the second of these two Venus transits, could the Sun's mass be expressed in absolute kilograms rather than relative to Earth's mass. And so could ultimately most other masses in the Solar System. Modern methods of measuring distances and masses include precise measurements of the orbits of man-made satellites through the gravitational fields of Sun, planets, and moons, and even by bouncing radar signals off planet surfaces.

These methods based on orbits of relatively small bodies orbiting a much larger one work well for the Sun and for planets with moons. But what about moons themselves? In particular, what about our own Moon? Various methods have been used to weigh the Moon, from a comparison of the differences in ocean tides caused by the Sun and by the Moon, to orbital measurements based on the realization that the Moon does not orbit the center of the Earth but that both orbit a common center of mass that is offset from the center of the Earth. The latter uses the fact that this offset is sufficiently large that there is a measurable wobble in the movement of the Moon as seen against the stars. In recent decades, man-made satellite flybys have refined the lunar mass estimated earlier. For other moons and asteroids the mass determination remains a combination of satellite flybys and detailed computations of the orbits around the Sun that take into account not only that the Sun or the nearby planet exerts gravity, but that all bodies in the Solar System pull on all others. Over time, these measurements and computations became so accurate that they led to the discovery of the large planet Neptune: that body pulled slightly on the known outer planets so that orbital calculations ignoring its presence were somewhat off. Ingenious calculations then directed observers on where to look, and distant Neptune was eventually found in 1846.

A manifest of mass

Once astronomers had a way to determine masses of bodies in the Solar System based on their gravitational interaction with the rest of the Solar System, they could establish where how much matter went during the formative phases. Some of their findings stand out so much that they may be part of the collection of forensic clues important to understanding the planetary systems overall. Here is the summary: (1) the Sun outweighs everything else in our Solar System combined by about 740 to one; (2) 99 percent of the mass in the Solar System outside the Sun resides in the four giant planets, which together weigh over 200 times more than the four terrestrial planets (Mercury, Venus, Earth, and Mars) combined; (3) each of the planets weighs at least 4,000 times the combined mass of all of its natural satellites, or moons, except for the Earth which weighs "only" eighty-one times as much as its solitary Moon; and (4) Mercury and Mars are really lightweight planets with masses that are only two to four times larger than the largest moons of Jupiter and Saturn (see Figure 3.3 for a comparison of the sizes of selected bodies in the Solar System). Now for more details:

The bulk, by far, of the mass of the Solar System resides within the Sun: 1,998,550,000,000,000,000,000,000,000,000 kg (where all the zeros are really just rounded off: we do not know the values of those digits, nor do we need to). Here we immediately run into the problem with astronomical numbers: they are often so big that they convey no intuitive meaning. As we shall be talking about planets throughout this book, it may be more helpful to express the masses throughout in relative terms (which is exactly what Isaac Newton did in 1687, not having Henry Cavendish's measurements available): although the mass of the Earth itself is a large number, most planets and exoplanets weigh from a few percent of the Earth's mass to a few hundred Earth masses. Taking the Earth mass as the relative unit, the Sun weighs in at about 333,000 times as much. That

Figure 3.3 Comparison of sizes to scale: the Sun (1,391,400 km in diameter; observed with the NASA Solar Dynamics Observatory), the gas giant Jupiter (at a diameter of 139,820 km it is ten times smaller than the Sun; here seen from underneath the southern polar region by the NASA Juno spacecraft), Earth (12,740 km across, 109 times smaller than the Sun) behind its Moon (3,475 km; with the NOAA DSCOVR satellite, 400 times smaller than the Sun), and dwarf planet Ceres (945 km; from the NASA Dawn spacecraft, 1,470 times smaller than the Sun). Comet nuclei are too small to show on this scale: Halley, for example, is sixty times smaller than Ceres (See Plate 5 for a color version).

is still a large number, but at least much closer to what we can relate to.

Excluding the Sun, the rest of the Solar System—as best we know at present—comprises approximately 450 Earth masses, or only 0.135 percent of the total. Most of that resides in the giant planets: Jupiter comes in at 318 Earth masses, Saturn at 95, Neptune at 17, and Uranus at 14.5. That leaves some five Earth masses for everything else (but note that there is a considerable uncertainty about the mass of what lies furthest away: the objects in the very distant Oort Cloud). Next, in decreasing order of mass, come the remaining planets: Earth (guess what, at 1 Earth masses), Venus at 0.82, Mars at 0.11, and Mercury at 0.055 Earth masses. Together, these four so-called terrestrial planets are good for a total of 1.98 Earth masses, or merely 0.44 percent of the mass of the Solar System outside of the Sun!

After that—in terms of mass—comes a long list of satellites and dwarf planets. In order of decreasing mass, the first nine are: Ganymede (a moon of Jupiter), Titan (a moon of Saturn), Callisto and Io (orbiting Jupiter), the Moon (at Earth), Europa (at Jupiter), Triton (at Neptune), and Eris and Pluto (the two most massive dwarf planets, weighing 0.0028 and 0.0022 Earth masses, respectively). Anything lighter weighs less than one thousandth of an Earth mass.

The Solar System has 178 known, officially confirmed "moons" or natural satellites that orbit planets (see Figure 3.4 for a comparison of the larger ones) and at least another 300 or so that orbit the minor planets, including the dwarf planets and asteroids. All of these planetary satellites weigh less than the least massive planet, Mercury, although the difference is not large for the heaviest among them: Ganymede (the heaviest moon in the Solar System) at Jupiter and Titan (only a little lighter than Ganymede) at Saturn weigh only just under half as much as Mercury. The combined masses of the satellites of each of the planets[1]—now in order of their distance from the Sun—amounts to at most 0.025 percent of their corresponding

Figure 3.4 Select moons in the Solar System, shown in their actual relative sizes, compared to Earth shown in the background. Pictured are; top right: Earth's Moon; upper left: Jupiter's Galilean moons: Io, Europa, Ganymede, and Callisto; bottom left: Pluto's Charon; bottom center: the five major moons of Uranus: Miranda, Ariel, Umbriel, Oberon, and Titania; lower right: the seven major moons of Saturn: Iapetus, Enceladus, Titan, Rhea, Mimas, Dione, and Tethys; and below Earth on the right: Neptune's largest moon, Triton. For comparison: the planet Mercury is smaller than both Ganymede and Titan, but slightly larger than Callisto (See Plate 7 for a color version).

Source: Modified after a figure by NASA.

planet's mass, with the exception of Earth for which it is
1.2 percent.

In the asteroid belt between the orbits of Mars and Jupiter we
find a large population of small Solar System bodies. For a long
time, the objects in the asteroid belt have been called asteroids in
order to contrast them with the larger planets, of which—until
a few years ago—we counted nine in the Solar System. Pluto had
been classified as a planet since its discovery in 1930. But then,
early in the twenty-first century, things got confused: in 2005 Eris
(2,325 km across) was discovered as a new "planet" orbiting the
Sun. It turned out to be rather comparable in size and mass to
Pluto (2,380 km across) and in an orbit similarly far from the Sun
as Pluto. Around the same time other such bodies, only somewhat
smaller, were found far out in the Solar System, including known
fairly large ones called Haumea (found in 2004) and Makemake
(discovered in 2005). These are members of what is called the
Kuiper Belt, a second asteroid belt that lies beyond the orbit of
Neptune, extending from about 30 to 50 Sun–Earth distances.
After Pluto, the next Kuiper Belt object (KBO) was not found until
1992. We now know that there is a whole flock of them of all sorts
of sizes out there, with over a thousand objects seen and up to a
hundred times more expected to exist. The KBO size distribution
turned out to overlap with sizes of bodies in the asteroid belt
between Mars and Jupiter. Two of the five known dwarf planets
are KBOs: Haumea and Makemake. They were discovered after
astronomers had already observed several thousand exoplanetary
systems; that shows how hard it is to find "small," dark things
even relatively nearby. Another dwarf planet, Eris, in an orbit
from 38 to 98 Sun–Earth distances, moves through the Kuiper
Belt in part of its orbit, but is now classified as a member of the
"scattered disk" which reaches twice as far as the Kuiper Belt and
has objects with orbits inclined well outside the plane defined by
the orbits of the main planets.

There was a short period around 2006 in which the International Astronomical Union (IAU)—the keeper of names and

definitions in matters astronomical—worked to formulate new definitions that were in line with what we then knew of the Solar System. The result was that for a body to be called a planet it needed to have three properties: (1) to directly orbit the Sun and not do so indirectly by traveling along with a larger body, (2) to be heavy enough that its own gravity would have pulled it essentially into a spherical shape, and (3) to be heavier than other objects in its orbit so that, in effect, its gravity would have "cleared the neighborhood" around its orbit either by capturing such nearby objects in collisions or by catapulting them away in something like a slingshot orbit when close encounters occurred. Pluto did not meet the third requirement, and was demoted to "dwarf planet". By the way, the reclassification of Pluto came seven months after the NASA New Horizons mission was launched in January of 2006 to make a close flyby of what then was the ninth planet, but was designated a dwarf planet when the spacecraft completed its nine-year mission to its primary target (Figure 3.6).

At the time of this writing there are four other bodies in the category of dwarf planets that are formally recognized by the IAU: Ceres (within the asteroid belt, orbiting at 2.5 to three times the Sun–Earth distance; see Figure 3.5), and Eris, Haumea, and Makemake which are all orbiting beyond the furthest planet (Neptune, at thirty Sun–Earth distances) never coming in closer than thirty-five Sun–Earth distances in their highly non-circular orbits around the Sun. Expectations are that there are hundreds like them, or maybe many thousands more, far, far from the Sun.

A dwarf planet orbits the Sun, so is not a natural satellite, or moon. It is large enough to collapse into a spherical form under its own gravity. But it has not cleared the vicinity of its orbit of other bodies. The latter is clearly true of the asteroids: they orbit in a wide band, weaving between each other as they swarm around the Sun on somewhat distinct elliptical orbits and with slightly varying velocities (see Figure 3.2). Only the largest, Ceres, is nearly spherical because of its own gravity, which makes it a dwarf planet as well as an asteroid.

Figure 3.5 Part of the dwarf planet Ceres, photographed by the Dawn spacecraft. The largest crater in this image is the Occator Crater, 92 kilometers (57 miles) wide and four kilometers (2.5 miles) deep. The bright patches within it may be salt deposits formed after briny water seeped onto the surface, froze, and then sublimated to leave behind the shiny salt. Ceres is the largest object in the asteroid belt and the only dwarf planet there. It was the first asteroid discovered, found by Giuseppe Piazzi (1746–1826) in 1801. Ceres has a girth about thirteen times smaller than Earth (compared in Figure 3.3), spins around once every nine hours, and orbits the Sun every 4.6 Earth years about 2.8 times further out than Earth.

The estimated masses of all of the asteroids combined add up to 0.0005 Earth masses, more than half of which resides in the largest four: Ceres, Vesta, Pallas, and Hygiea. The number of asteroids is unknown, but there are probably over a million that are larger

Figure 3.6 Dwarf planet Pluto pictured by the New Horizons spacecraft just after closest approach on July 14, 2015. Looking back in the direction of the Sun, the image shows the ice plains of Sputnik Planitia on the right and mountains reaching 3.5 kilometers (11,000 feet) on the left. The setting Sun lights up various layers in Pluto's atmosphere. The typical temperature on the surface is below −200 degrees centigrade (or −400 degrees Fahrenheit). The atmosphere of primarily nitrogen has a pressure of no more than one ten thousandth of Earth's atmosphere. The field of view is about 380 kilometers (230 miles) across.

Source: NASA/JHUAPL/SwRI.

than 1 kilometer (0.6 miles) across, and some 30 million that are larger than 100 meters (about 300 feet) across.

With this sketch of where the mass of the Solar System is— primarily within the Sun, a tiny fraction in the largest of the planets, and a sprinkling in a gazillion small Solar System bodies— another question arises: How are the chemicals that make up that mass distributed? Knowing the answer to that question is as important a clue to the history of the Solar System as knowing how much of the total is where: the processes of differentiation of material are sensitive to the history of conditions within the Solar System. Among a multitude of possible physical processes, some win out to shape the past and future of the Solar System.

But the speeds at which things develop do not always come in the same order of dominance: which one wins depends on local conditions at the time. Such conditions include densities, temperatures, movements, gravity, magnetism, radiation, and transparency. These conditions change with time, so other processes may take on the dominant roles, then others again, and again. To unravel the history, we are back at forensics: gather facts, and sift out those that are important.

The composition of major and minor planets

Once the mass of an object is known and its diameter measured, the average density can be derived by dividing total mass by total volume. That tells us something about the mix of chemicals that can be expected to dominate hidden under the surface or underneath clouds. For comparison, here are some characteristic densities on Earth's surface: water comes in by definition at one gram per cubic centimeter, rocks somewhere between 2.3 and 3.1, while pure iron and nickel weigh in at 7.9 and 8.9 times the density of water. Any gas has a wide range of possible densities because, in contrast to liquids and solids, gases are highly compressible: the more pressure is applied to them, the denser they become. For example, what we know on Earth as the lightest of gases, hydrogen and helium, that weigh many thousands of times less than water for a given volume, occur in the very center of the Sun in a mixture that is compressed to 150 times the density of water. Solids and liquids are somewhat compressible also, but that is noticeable only under very high pressures, such as in the deep interiors of planets, and then by a factor of no more than two.

The four terrestrial planets have the highest average densities of all the planets at 3.9 to 5.5 times the density of water, with Earth having the highest value. The oceans do help to lower that average, but if they were thought of as equally spread out over a perfectly smooth sphere the size of the Earth, they would have a thickness of not quite one part in 2,400 of the radius of the sphere.

That is clearly not enough to make a substantial difference: the average density of Earth is determined by what lies below the surface! That means that although the outer layers are formed of various types of rock with densities somewhat below Earth's average, deeper down there must be substantially denser stuff because otherwise the average value for the planet as a whole could not come out as high as it does. This reasoning turns out to be correct. For Earth, geologists have shown that the planet's core is mostly made up of somewhat compressed iron and nickel. Planetary scientists have determined that the cores of the other terrestrial planets are also chemically dominated by iron.

After these four densest bodies in the Solar System come Earth's Moon, the moons of other planets, and the asteroids. They all have densities between 1.5 and 3.5 times the density of water. So they likely contain at least some rocky materials whereas those with densities less than about twice that of water are likely to contain a lot of water (either in liquid form or frozen into solid ice) along with their rock and metal content to bring the average density down to the measured value.

But when we look at the giant planets, we find average densities of 0.7 to 1.6 times that of water (with Saturn coming out as the Solar System body with the lowest average density). Now, their measured volumes include their thick, opaque gaseous atmospheres into which we simply cannot see. Their low densities must thus mean they have thick atmospheres. Studying these reveals that they are mostly composed of hydrogen and helium, but mixed with substances that make them opaque. Deep below these opaque envelopes lie realms of liquids and solids, not visible from the outside.

Hydrogen and helium not only dominate in the outer atmospheres of the giant planets, but are also the dominant elements within the Sun. In fact, these two combined account for 98.1 percent of the Sun's mass. Hydrogen and helium, along with many other substances that we know as gases, are known in astronomy as volatiles. This means they require very low temperatures or

very high pressures in order to freeze out or to form liquids. Chemicals that are, in contrast, solid up to high temperatures, and that thus require high heat to evaporate, are called refractory. By the way, the Sun is so massive that the 1.9 percent of its mass that is not hydrogen or helium still puts over 6,000 times the Earth's mass in heavier elements within the Sun.

The planets closest to the Sun, in the region that is warmed up most, are composed mostly of refractory elements. In contrast, those planets furthest from the Sun, in the cooler reaches of the Solar System, are composed to a large extent of volatiles. It seems to make sense to have volatiles dominate in the part of the Solar System that is furthest from the Sun, and thus coolest, but it does lead to a question. If all planets formed out of the same gas mixture that made the Sun, then where did all the volatiles end up that should have been around in the terrestrial planets? To unambiguously answer that question we had to learn how other planetary systems and their stars form.

Expeditions to the neighbors

Close encounters, actual visits, or samples of Solar System material brought to laboratories on Earth tell us many more details about the planets, of course, than we can learn by looking at them from afar. That requires travel to the planets, or at least an approach to within a reasonably small distance. Reaching space became a reality in 1957 when the Soviet Union put a small satellite in orbit around the Earth on October 4 of that year. It was nothing more than a metal sphere with a few antennae. Nonetheless, this Sputnik 1 (Russian for "satellite 1") was a turning point in history because it proved that the technology of rocket propulsion into space could work. It was a mere four years after that when Yuri Gagarin (1934–1968) became the first human in space. Seven years after that, NASA's Apollo 8 orbited the Moon with three astronauts. In the early years, it proved a difficult enterprise to get to the Moon: of the first fifty spacecraft launched

to the Moon, be it to orbit or to land, only fifteen were successful. Even nowadays, launches, engine burns to adjust course, close flybys, and controlled soft landings still fail with some regularity as probes now aim to reach more distant targets than the Moon. But many space probes have successfully flown across the Solar System, from close to the Sun to beyond the planetary domain to probe interstellar gases. By the way, spacecraft do not, in fact, "fly": there is no air in space, so nothing to beat wings against to provide lift like birds do. And because there is no air, in any sense of the word, there is no sound either. Spacecraft are mostly "thrown" through space: a rocket motor burns for a short while to accelerate or decelerate the craft, but apart from those "burns" the craft simply moves silently like a rock, just following a ballistic trajectory determined by the engine burns. When it gets close to a planet, the craft "falls" into its gravitational pull, sometimes using a burn to put itself into a bound orbit like a satellite going around Earth, or it may be hurled through the planet's pull toward some other goal. These trajectories are computed by highly specialized orbital engineers who find ways to go far while minimizing the weight of the fuel that has to be brought along.

In 1969, excitement enveloped the globe when the Apollo 11 Eagle lander put the first two people onto the lunar surface: Neil Armstrong and Edwin ("Buzz") Aldrin. Many of us are familiar with Armstrong's words upon stepping onto the Moon, about which he wrote: "It wasn't until after landing that I made up my mind what to say: That's one small step for man, one giant leap for mankind." Armstrong afterwards reminisced: "The sky is black, you know. It's a very dark sky. But it still seemed more like daylight than darkness as we looked out the window. It's a peculiar thing, but the surface looked very warm and inviting. It was the sort of situation in which you felt like going out there in nothing but a swimming suit to get a little sun" (see Figure 3.7).

I had an opportunity to shake hands with Aldrin once in the early 2000s: I was waiting in the lobby of NASA Headquarters to participate in a planning meeting when he walked up to the guard

Figure 3.7 Apollo 17 crew at Shorty Crater in the Taurus–Littrow highlands and valley area. The image of a sunlit lunar surface under a pitch-black sky was taken by Gene Cernan, who described the lunar surface as a place of "magnificent desolation." It shows fellow astronaut Harrison Schmitt standing with the Lunar Roving Vehicle at the edge of the crater. At this location, Schmitt stumbled on color in the otherwise colorless lunar surface material. Analysis of a sample of the orange moon dust that was taken to Earth showed it to be tiny beads of colored glass likely formed in an ancient volcanic fire fountain. The picture was taken in 1972 during the last manned mission to the Moon, by the last astronaut to leave it (See Plate 9 for a color version).

and said who he was, that he had worked at NASA for a long time, but that he had forgotten his badge that day. The guard, much younger than both of us, looked at him and said she did not know who Aldrin was so she would have to call someone. I walked up to Aldrin and told him that I certainly remembered him and

that his work was one of the reasons I became an astrophysicist: I remember sitting spell-bound at our tiny black-and-white TV in the middle of the night—it was very late in the Netherlands because lunar activities were scheduled for daytime throughout the US—as they traveled to the Moon and at last landed there. Aldrin did not mind waiting for someone to vouch for him at NASA HQ that day; I wondered how he felt at that moment about being simultaneously world famous and yet unrecognized.

Aldrin has characterized the Moon as "magnificent desolation": there is no life, little color,[2] no haze to help estimate distances, no brightness to the black sky and therefore no light at all in the shadows. He noted the odor of lunar dust. Not out on the surface, of course, as there is no atmosphere and he was in a space suit. But the dust that they carried into the lander stuck to their boots and suits did smell: "pungent, like gunpowder or spent cap-pistol caps. We carted a fair amount of lunar dust back inside the vehicle with us, either on our suits and boots or on the conveyer system we used to get boxes and equipment back inside. We did notice the odor right away." Perhaps some day someone will comment on the odor of Martian soil, but so far the Moon is the only Solar-System body other than Earth itself that has been walked on by humans.

Synchronizing the Moon

As we are on the topic of the Moon and its astronauts, it is a good moment to digress to a peculiar property of the Moon that is now being recognized as important to many of the exoplanets that have been found: the Moon always has its same side turned toward the Earth. This means that the Moon rotates exactly once for each time it goes around the Earth. You can visualize this by imagining that you are walking around someone else: in order to keep facing that person, you need to keep turning as you go around so that this other person stays in front you but everything else around you reveals that you yourself are, in fact, turning as you walk around.

For heavenly bodies, this is called spin–orbit synchronization. For the Moon, it is a consequence of tidal forces from the Earth. Gravity decreases as a function of distance, so the Earth's pull on the near side of the Moon is stronger than that on the far side. The orbital centrifugal force that is the result of the orbital motion around the Earth, in contrast, increases with distance, so it is larger on the far side of the Moon than on the near side. The combined tug on the Moon by these two forces is weakest right in between the points nearest to and furthest from the Earth. Together, these forces work on the Moon, slightly deforming it with bulges pointed toward and away from the Earth: if the Moon were spinning at another rate, then it would spin under these bulges that form a wave on the spinning Moon that remains fixed along the Earth–Moon line. Consequently, if not synchronized, the Moon's surface would stretch and contract slightly every time it rotated relative to Earth's direction. The same forces work on the Earth, and these cause bulges there, too: these are known as the lunar tides that cause the ocean levels to rise and fall twice a day as the Earth spins underneath the Moon, rising once when the Moon is overhead and rising once when the Moon is underfoot.

This tidal flexing takes energy away from the rotation. A large body like the Earth has a lot of rotational energy in comparison to the work that the Moon can do, so it has kept spinning, although it is, in fact, slowing down ever so gradually. But the Moon's rotational energy is much smaller and the forces exerted by the heavier Earth are larger, so its spin rate changed until the rise and fall of the tides vanished and the bulges were fixed in the Moon's shape. In other words, the Moon changed its rate of rotation until it always had the same side facing the Earth, and thus also has the Earth just hanging in its sky.

Although the Moon always has the same side facing the Earth it continues to orbit the Earth, and therefore it does still spin. If you were somewhere on the Moon where the Apollo astronauts were, you would see the Earth in an almost fixed location even as the Sun would appear to go around: once per twenty-seven days

the Sun would rise, set after about two weeks, and then for two weeks it would be dark. On Earth, we see the consequence of this in the changing faces of the Moon: we see the same hemisphere, but lit differently by the Sun during the lunar "day." This means that there is a "dark side of the Moon," but it is the far side only part of the time: once every twenty-seven days, on days of a New Moon on Earth, the dark side of the Moon is in point of fact the hemisphere facing the Earth, and at that time, on the Moon, you would see a "full Earth." Such "tidal synchronization" results whenever a relatively small body is close to a much larger one: it is important for other moons near their planets in the Solar System, and is important for habitability of exoplanets close to their stars, as we shall see in Chapters 9 and 10.

Gene Cernan (1934–2017), was the last astronaut to step off the Moon to board the Lunar Module of the Apollo 17 mission about five hours before it launched back into space on December 14, 1972. He noted the consequence of the locked spin and orbit of the Moon as he left: "But when I climbed up the ladder for that last step, and I looked down, and there was my final footsteps on the surface, and I knew I wasn't coming back this way again, somebody would—and somebody will—but I knew I was not going to come back this way. I looked over my shoulder because the Earth was on top of the mountains in the southwestern sky. Never moved for the whole three days we were there."

Voyagers, explorers, rovers, and probes

It is truly remarkable how many bodies in the Solar System have been visited by an impressive sequence of interplanetary probes, some performing flybys en route to somewhere else, others orbiting these bodies or landing on them. Despite all these missions, samples of surface material have been taken to be brought to Earth from only two bodies: the Moon and an asteroid.

The Solar System bodies visited[3] by orbiting or landing (or deliberately crashing onto the surface) include the Moon, all of

Figure 3.8 Images to scale of nine asteroids: 4 Vesta (525 kilometers, or 350 miles across), 21 Lutetia (about 100 km diameter), 253 Mathilde (50 km), 243 Ida (60 km), 433 Eros (about 34 km along its longest dimension), 951 Gaspra (18 km), 2867 Šteins (not quite 7 km along its longest axis), 5535 Annefrank (comparable to 2867 Šteins in size), and 25143 Itokawa (only 0.5 km in diameter, sample returned by the Hayabusa space mission). For the naming of asteroids, see note 4 following this chapter.

the planets, two dwarf planets, thirteen asteroids[4] (see Figure 3.8 for a few snapshots[5]), eight comets (see Figure 3.9 for a comparison of several to scale, and Figure 3.10 for a close-up of 67P), and over a dozen moons of Jupiter and Saturn (pictures of Saturn's Pan and Tethys are shown in Figures 3.11 and 3.15).

Spacecraft have landed on[6] two planets, two moons, two asteroids, and two comets. The first landing occurred on the Moon with an unmanned Russian Luna 9 spacecraft in 1966, followed by eighteen more landings. After initially reaching for the Moon, spacecraft have landed successfully on Solar System bodies far beyond the Moon and have functioned for some time to send back information.

Figure 3.9 Comparison of the solid cores of comets (approximately to scale; using the sharpest images available). Clockwise from the top left: Tempel 1 (just under 8 kilometers or 5 miles across at its largest) from NASA's Deep Impact mission, Borelli from NASA's Deep Space 1 probe, Wild 2 from NASA's Stardust, Hartley 2 from Deep Impact, and 67P/Churyumov–Gerasimenko imaged by the ESA/Rosetta/NAVCAM.

Some of the landers on the Moon and on Mars have been equipped with rovers able to provide wide-ranging information during their surface explorations. The unmanned Luna 16, 20, and 24 missions returned a total of 326 grams. Twelve astronauts walked on the Moon, half of them drove around on the surface,

Figure 3.10 This picture of the comet 67P/Churyumov–Gerasimenko was taken by an instrument on ESA's Rosetta spacecraft. The spacecraft was launched on March 2, 2004 and navigated around the inner Solar System for a decade before catching up with the comet on August 6, 2014. The comet (67P for short) is in a 6.45-year orbit around the Sun, has a size of about 4.3 by 4.1 kilometers (or 2.7 by 2.5 miles), and is comprised of two lobes on either side of a connecting neck. Illumination by the Sun causes volatile material to evaporate (sublimate) carrying less volatile dust with it; scattered sunlight shows these materials as beam-like brightenings around the solid core of the comet.[7]

and in total they brought back over one thousand times more material than the unmanned probes, collecting altogether 382 kilograms of rock samples to study. The Hayabusa spacecraft returned a sample of asteroid material to Earth. Samples of dust in comet tails and floating around in interplanetary space are captured as they descend into Earth's atmosphere by specially

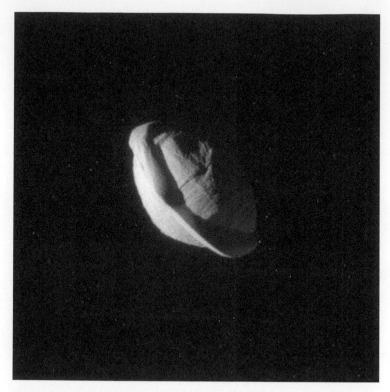

Figure 3.11 Saturn's moon Pan photographed by the Cassini spacecraft in 2017. Pan is about 35 by 23 kilometers (22 by 14 miles) in size, orbiting in a gap in Saturn's outermost large, bright ring, known as the A ring. The odd shape may be due to material that was piled up in the enveloping ridge after being scooped up from the rings surrounding Pan's orbit. Cassini was launched on October 15, 1995, entered orbit in the Saturnian system on July 1, 2004, and was sent crashing into Saturn's atmosphere on September 15, 2017.

equipped stratospheric airplanes. A sample of comet dust has been brought back from deep space by the Stardust spacecraft that flew past comet Wild 2. And even samplings of the interplanetary dust and of the solar wind have been returned, by the Genesis spacecraft.

Although space probes have traveled to, or flown by, many bodies in the Solar System, all their sampling still provides only

limited snapshots about the conditions throughout the Solar System: the Earth remains the most extensively studied body in the Solar System. On Earth, geologists can go anywhere on continents, they can visit all the diversity of islands, and they can traverse the ice shelves in the far north and south. The deep seas and oceans may not be directly accessible to humans, but instruments can probe the undersea mountain ranges and trenches.

Sounding Earth

Our direct access to the interior of the Earth is limited, however, to the near-surface layers. We cannot travel far into the Earth: the deepest mine reaches down no more than four kilometers out of the 6,371 of the planet's radius, or just 0.06 percent of the distance to the center of the planet. Some of the interior sometimes comes to us in the form of lava (which is molten rock at the surface, but is generally called magma when below the surface) during volcanic eruptions, but even that appears to originate from the bottom of the rocky crust at no more than five to 50 kilometers in depth (0.08 percent to 0.8 percent of the way to the center) in oceanic or continental crust, respectively. Compare that to the shell of the chicken's egg or to an apple's skin, which have a typical thickness of about 0.5 percent of the whole: in relative terms, these thin outer layers are as thick as a continent is compared to the whole of the Earth. In other words, all that we can directly access, with much digging or by getting samples from erupting volcanoes, gives us information only on the skin of the Earth. How can we know about the rest of that big sphere?

Earthquakes send waves rattling and vibrating through deep layers of the interior of the Earth. These waves bend back up as they are refracted in their propagation in the hotter deeper layers. Exactly how they propagate reveals the internal structure of our home planet. It is the science of seismology that enables us to probe down to the core.

What causes earthquakes was subject to much speculation for many centuries. By the beginning of the eighteenth century, there were hypotheses about fires and chemical explosions occurring deep in the Earth. By the middle of that century, the concepts shifted correctly toward the interpretation of earthquakes as being the result of sudden movement of layers of rocks, but why that would occur remained mysterious. Early in the twentieth century the concept of moving continents was proposed, but it would take until the 1960s before the final convincing evidence supporting it was encountered—in the form of magnetic striping mirrored on either side of mid-oceanic ridges from where newly formed ocean floor moves outward—and the theory of plate tectonics was formulated.

The realization that entire continents were slowly moving about pointed to the existence of fluid-like motions deep inside the Earth that cause the continents floating on top to be dragged about like rafts on a river current.

This is where seismology came in. The theory behind it was essentially developed over the course of a century, between about the 1850s and the 1950s. Although complex mathematics is needed to apply seismology, the principle is easy to understand: if you strike an object, the sound you hear depends on the structure and shape of the object and on the materials that constitute it. This is what enables us to tell apart gongs from drums, trumpets from flutes, and violins from cellos. Each of these three pairs carries primarily one particular type of wave: surface waves, pressure waves, and transverse waves, respectively. The surface waves are often the dominant contributor in damaging buildings by the shaking of the ground during earthquakes. Where deep Earth research is concerned, pressure and transverse wave types are the most important.

Pressure waves propagate because a material is compressible. Compressing such a material induces a response that is, effectively, an increased counter-pressure that builds up with compression. This counter-pressure then pushes against neighboring

material through which the material tends to relax back from a preceding compression. This relaxation often overshoots the equilibrium, so that another relaxation follows, again overshooting so that another compression follows again. This oscillatory cycle causes the wave to propagate as a compression followed by a decompression, which is generally repeated a number of times to create a propagating wave. The propagation of such waves is dependent on the temperature, pressure, and density of the medium through which it propagates. With the large number of seismometers deployed around the Earth, seismologists use the recorded shaking to map the structure of the Earth using these pressure waves, or P waves as seismologists call them.

Transverse waves, in contrast, are the result of bending a material, such as in a tight string on a piano, guitar, or violin. These waves can propagate in general through any solid that flexes somewhat if moved quickly sideways. Such waves, called shear waves or S waves for short, cannot propagate through a gas or a liquid because these can only carry P waves, but they propagate efficiently through solid or mostly solid materials. This difference in propagation behavior enabled seismologists to determine, at the beginning of the twentieth century, that the Earth has a deep liquid layer within its core.

The types of waves, their strengths, and their arrival times, are all used to deduce the internal makeup of the Earth. For the largest earthquakes, such waves may go around the Earth multiple times and thus meet up with themselves. This causes global 'ringing' as in a bell. When a bell is first struck, there is a sharpness to that sound caused by so-called overtones, but when these dampen away, a purer, warmer sound dominates. This ringing continues for quite some time and only gradually fades away. Seismometers measure a similar behavior of waves inside the Earth, but far slower than would be audible to us as sound, associated with very weak tremors that only sensitive instruments can pick up. The modern seismometers can detect this ringing in the Earth after the strongest earthquakes for days up to a few weeks.

The result of seismic measurements combined with laboratory and computer-based experiments shows the Earth to comprise three primary regions. Innermost is the core which extends over 53 percent, or just over half, of the diameter of the Earth. That core, mostly solid in its inner third and mostly liquid in the remainder, is composed primarily of iron and some nickel. The densities in the core range from ten to thirteen times the density of water; the pressure of the overlying layers is so large that even the iron–nickel mixture yields somewhat and is compressed to a higher density.

Above the core lies an almost equally thick mantle, taking some 45 percent of the depth of the Earth. The mantle is predominantly silicate rock. This type of rock comprises a large collection of minerals in which silicon and oxygen are essential components. The simplest combination of silicon and oxygen contains only these two elements to form what—in one crystalline form—we know as quartz. Combined with other elements it creates a very diverse group of minerals that makes up most of what in everyday life we refer to as rocks and sand.

The mantle is solid, but under conditions of high pressure and temperature the solids are deformable. It behaves not unlike, say, iron: we know it as a very strong solid at room temperature, but it can be hammered or rolled or drawn into desired shapes from metal sheets to cable strands. Similarly, the mantle material can move subject to the enormous stresses within the Earth, but very, very slowly, typically only a few centimeters per year, literally inching along.

The outermost layers of the Earth, the ones we live on and can at least partially probe, are called the crust. Depending on location, the crust is somewhere between 5 to 70 kilometers (3 to 45 miles) thick. It is thinnest below the oceans and thick where it forms the continents. The crust, too, is mostly composed of silicate rock. It is densest under the oceans, comprising a lot of dense rock types, such as basalt. The crust is less dense in the continents where a preponderance of lower-density rock types,

such as granite, enable the continents to essentially float on top of the mantle.

The separation of the dense iron–nickel core and the lighter silicates into core and mantle must have taken place a long time ago, relatively early in the life of the planet. The details of when and how that came about remain under study, but the principle is well understood: if two (or more) liquids are mixed and left to themselves, they will sort themselves out by chemical composition such that the heavier chemical sinks to the bottom and the lighter floats to the top. In order to enable chemicals to separate (or, using the technical term, to differentiate) these bodies must at some point have been liquid (and they may still be that in part, as is the case inside the Earth). The interiors of these bodies would not differentiate strictly by element, because different elements will chemically bind to others with particular preferences, and then would have to move together with their chosen partners. For example, molten iron and nickel do not mix well in a chemical sense with molten silicates, resulting in the eventual separation into the iron-dominated core and silicate-dominated mantle and crust above that.

The process of differentiation that took place in the Earth raises many questions, two of which stand out clearly where the formation and history of the Solar System are concerned: (1) do we have evidence that such differentiation happened elsewhere in the Solar System?, and (2) what caused Earth and other bodies that differentiated to be so hot that they melted? The second question will be addressed in Chapters 5 and 6. The affirmative answer to the first question was encountered lying around on Earth, and subsequently confirmed by interplanetary probes.

The stories of impacts

Evidence for the differentiation of extraterrestrial bodies comes in the form of meteorites, the solid remainders of objects fallen from space that impact the Earth. Meteors are the visual signature

of something entering the Earth's atmosphere (or in principle any other atmosphere) at great speed: as a body enters the atmosphere, the friction with the atmosphere causes the outer layers to heat up and evaporate. The gases are so hot that they glow and cause a streak of light, most easily visible in a clear night sky, that is referred to as a meteor. If the body is large and dense enough, it stands a chance of surviving the crossing of the atmosphere and thus of impacting Earth on land or, more commonly because there is more of it, in an ocean. That body, or any fragment thereof, is called a meteorite.

Meteorites can be grouped by what dominates them chemically or by what differentiates them by their internal appearance. The appearance is often characterized by whether they are mostly comprised of packed grains or are more uniform. The technical term for the grains is chondrules. These vary in size from over 1 cm (or half an inch) to just dust-like at microns across, with an average size of less than a millimeter (1/25 of an inch). Their name derives from the ancient Greek word for grains: chondros. Meteorites comprised predominantly of chondrules are called chondrites. Approximately 86 percent, so by far the majority of meteorites falling to Earth, contain such grains. They are comprised largely of rocky silicate materials. Astronomers argue that the chondrules are ancient material that solidified out of the gas from which the Solar System formed. In due course these stuck together, mixed in with other materials, and formed ever larger bodies in the Solar System.

At the other end of the spectrum are those meteorites that have little or no chondrules. Most have a chemical composition comparable to those with chondrules. These rocky meteorites are classified as achondrites, a name made by adding the Greek prefix for 'not' or 'without' to chondrites. Some 8 percent of all meteorites falling to Earth are achondrites. The fact that sizable meteorites exist without chondrules tells us something very interesting: these meteorites must originate from bodies that at some point in their lives were hot enough to be melted to a

substantial degree, which would allow the chondrules to dissolve and blend with their neighboring materials, and their chemicals to mix.

Approximately 5 percent of all meteorites from which material has been recovered are composed of a mixture of rock and iron–nickel or are composed largely of iron–nickel. Now, this not only requires that they were at some point melted to dissolve the chondrules, but that they were once part of a body sufficiently large to experience what Earth has: the differentiation of chemicals by density subject to gravity, so that iron sank down to form a core surrounded by the floating mix of rock. These iron or stony-iron meteorites thus provide very important forensic information: they must have been part of large, hot bodies. At some point in time, a cataclysmic collision with another sizable body must have caused fragments to fly around the Solar System. These were either already solidified when this occurred or they solidified after they were ejected into space. In any case, when these fragments fell to Earth, they maintained the signature of their origin in large bodies that were, at least at one time, hot enough to melt.

The percentages of the types of meteorite as they occur in the Solar System are rather uncertain, because in order to be classified, they have to be found and reported. And because it is easier to find clumps of meteoric iron–nickel lying around on Earth than it is to identify rocky materials as meteoric rather than terrestrial in origin the iron–nickel-rich ones are likely to be substantially over-reported and over-counted. It is the very existence of iron–nickel meteorites, more than their precise number, that provides a clue about the history of the Solar System. By the way, some of these iron–nickel meteorites proved very valuable to certain cultures who had no other access to iron: for example, eighth-century inhabitants of Greenland—Dorset people who lived there centuries before the Inuit arrived—already used to hammer fragments off iron meteorites they found stemming from an impact that occurred thousands of years earlier, to make iron tools for hunting and for trading with distant arctic tribes: the unique

chemical mixture of individual meteorites makes it possible to trace traded iron tools back to the same origin, revealing that some were traded up to 2,500 kilometers (1,500 miles) from the impact site.

Interestingly, some of the meteorites recovered have a chemical composition very much like the rocks brought back from the Moon by the Apollo astronauts. Others are like the surface of Mars. Although no rock samples have ever been brought to Earth from Mars, the landers and rovers have given geologists a lot of information on structure and composition over the past few decades. From this they concluded that, indeed, there are Martian rocks on Earth in the form of meteorites.

How did these rocks from Moon and Mars get here? The last bit of their trip here is not hard to put together: Earth blocked their path, and after glowing brightly as meteors on their descent to the surface, their remains were thereafter found and identified as meteorites. What about the beginning of their trip through space, though: how did they end up being launched beyond the reach of the gravities of the Moon and Mars in the first place?

The question of how meteorites from the Moon and Mars came to be hurled into space loops back onto itself: meteorites. Large meteorites impact a planet or moon at speeds that may be as high as multiple kilometers (or miles) per second. And they can be large. The largest meteorite found, the Hoba meteorite, comes in at 60 tons, but the original bodies can be much larger. They either fragment or explode as they traverse the Earth's atmosphere, or—if they survive atmospheric travel largely intact—they shatter into pieces as they impact the surface.

Large objects that impact Earth create correspondingly large impact craters. All bodies in the Solar System are covered by impact craters (compare Figures 3.1, 3.5, 3.8, and 3.15): all the planets, moons, and asteroids—if we can see through their atmospheres—show the characteristic round basins surrounded by an elevated ridge, generally overlapping with or embedded within others. The meteorites causing these impact craters have

a wide range of sizes, from the smallest that can make it to the surface to many kilometers (or miles) across. Upon impact, they cause an explosion that blasts material outward and upward, causing a central depression, a ring-shaped pile-up as the outer edge of the crater, and a surrounding area covered in rock debris shattered into pieces that range from dust up to large boulders. Very large impacts can be associated with a rebound of the crater bed leading to a central elevation. The largest impacts can cause some of the interior to melt or can even reach into magma reservoirs should these exist within the impacted body. This leads to a partial or complete flooding by lava of the interior of the crater, creating a relatively smooth plain (like the mare on the Moon, such as the example in Figure 3.1) that may have a different composition to the surrounding terrain.

With relatively small binoculars, we can already see a variety of impact craters on the Moon. At the day–night boundary of the Moon, we can see the light-and-shadow patterns of the high surrounding ridges. The large areas of slightly darker color are signatures of the largest impacts, in which the basins were flooded with lava leaving the terrain smoothed and with a different hue. Astronomers estimate that almost all of the surface of the Moon is some 3.9 billion years old, so that we see essentially the entire history of meteorite impacts on the Moon stacked into a complex pattern of superimposed craters of all sizes. The largest impact basin visible on the Moon is some 2,500 kilometers (or 1,550 miles) across. That is large, particularly for the Moon: these impact basins span up to some 70 percent of the Moon's width.

Earth also has its impact craters. A very pretty example is the Barringer Crater (or Meteor Crater) in the American state of Arizona: this 1.2 kilometer (half-mile) crater was created 50,000 years ago in a desert climate that has left its features very well preserved. The largest, and also oldest, impact crater of which features can be recognized on the Earth's surface is the Vredefort crater not far from Johannesburg, South Africa: it has a diameter of some 300 kilometers (almost 200 miles) and an estimated age of

2 billion years. Terrestrial weathering wears down impact craters fairly quickly by astronomical standards. The oldest craters are simply wiped off the Earth by geological processes: erosion of mountains, volcanism, and tectonic motions. In contrast to the long meteorite-collecting time of some 3.9 billion years for most of the Moon's surface, 80 percent of the Earth's surface is no more than 200 million years old—less than 5 percent of the age of the Moon's surface.

Large impacts wreak havoc with a Solar System body. On Earth, one well-studied example is the impact that created the Chicxulub crater off the Mexican Yucatan peninsula. This impact, with a crater only just under the size of the Vredefort crater, took place some 65 million years ago. The impact shattered the land around it. The associated dust clouds spread around the Earth, and tsunamis reached far coasts all around. The effects of these, and the long-lasting impacts on weather, are thought to have eliminated some three quarters of all species then existing on Earth, including all the large dinosaurs.

But that was a very long time ago, of course. Impacts still occur, and the consequences of more recent ones are much better known, and some are documented as they occur. There was, for example, the Chelyabinsk meteor over Russia on February 15, 2015. In this era of smartphones, its movement, initially at almost 20 kilometers per second, was recorded from multiple vantage points. Its brightness exceeded that of the Sun and it was visible over 100 kilometers (60 miles) away. This object did not actually impact the solid Earth, but exploded at a height of some 30 kilometers (20 miles), resulting in a blast wave that broke windows, blew in doors, caused damage to thousands of buildings, and wounded many hundreds of people by the indirect effects of the blast wave. The original body, estimated to have weighed about 10,000 metric tonnes, exploded into fragments that rained down over a wide area.

A century earlier, on June 30, 1908, another meteor similarly exploded in the atmosphere with tremendous force over central

Siberia, over the upper Podkamennaya Tunguska river, in what is known as the Tunguska event. Its explosion ("airburst") knocked down over 80 million trees over an area of more than 2,000 square kilometers (almost 800 square miles). The sound of its explosion was heard up to 1,000 kilometers from the event. It was not until years afterwards (following the First World War and the Russian Revolution) that scientific expeditions documented the impacts on the remote, inhospitable marshy forests of the primary site directly underneath the explosion. One in 1927, sponsored by the Academy of Sciences of the USSR, eventually came upon an area shaped much like a crater surrounded by chains of hills. Their conclusion was that it appears not to have been caused by an actual solid-body impact, but rather by the primary blast wave. All the trees in this area had been knocked over, lying neatly parallel to each other, pointing away from the center, stripped of branches and their bark, and charred by tremendous heat up to over 10 kilometers (six miles) from the center. An eyewitness of the event felt the heat of the blast so intensely he had to step under cover; the blast wave that reached him somewhat later threw him a few yards away and knocked him unconscious. A reindeer farm near the blast was found utterly destroyed, and of its 1,500 reindeer only some burned remains were ever found. Of the meteorite itself, only fragments were found; the object may have been largely composed of ices and dust, with an estimated weight of perhaps a million tonnes. The dust released in the explosion spread around the Earth and affected the transparency of the atmosphere for months. Astronomers in California, for example, noted the strongest effect more than a month later.

True meteorite impacts into solid Earth, such as the one that caused the Chicxulub crater, can be so large that they eject rocks into orbit. Astronomers estimate that Earth and the other terrestrial planets are struck, on average, by an asteroid of up to two kilometers (over a mile) in size once every 200,000 years. If such an asteroid were to hit the Moon, its impact would enable some

material from the Moon to go into orbit, with a slight chance of falling down to Earth.

The combination of studies of the Moon, from orbiting satellites to the Apollo rock samples and lunar meteorites, show that the Moon's average composition is different from that of Earth. But the Moon's chemical makeup looks a lot like that of the Earth's mantle, except that many of the compounds that easily evaporate when heated (the so-called volatile materials) are less abundant.

This and other evidence has led to the concept that the Moon formed when the Earth was struck, when still in its infancy, by an asteroid the size of Mars (which is 1/7 the volume of Earth, and roughly twice as wide as the Moon). That collision would have thrown material into orbits over a wide range of distances from Earth. Hot and molten debris probably orbited Earth for years to centuries after the catastrophic impact. No one was around to see what happened next, but modern-day computer simulations show that the wreckage far from Earth would have clumped within decades and cooled efficiently. Rubble closer in would have felt strong tidal forces from Earth that kept it from clumping. There would also have been many collisions between the bodies in the debris ring. High-speed collisions transform their energy of motion into a lot of heat which would have slowed the cooling of the debris. The inner material, remaining hot for a sustained period, saw the relatively volatile elements essentially evaporate from the debris. That material would thus have lost much of its potassium, sodium, and zinc, among others.

Then, over subsequent centuries, the volatile-poor rubble migrated under the influence of gravitational tides, either falling back to Earth (where it would be recycled and mixed into the interior by plate tectonics) or onto the still growing Moon, there to form a stable mantle of which we now see the surface impacted by many more generations of lesser meteorites over the billions of years since then.

On a side note, it is worth pointing out that the relatively heavy Moon in close proximity to Earth continues to drive the ocean tides—along with the more distant but heavier Sun—and to stabilize Earth's spin axis as it has over the past billions of years. However, it remains unclear how important these are to life on Earth as well as how frequently such a planet-plus-heavy-moon system exists in the Galaxy.

Whether this scenario by which the Earth's Moon formed is correct remains under study, but time and again it is found to be consistent with a variety of observations. Exactly when it took place is still being debated, but it must have been at least some 3.9 billion years ago, and probably more. The first evidence for this came from the lunar samples returned from six Apollo landings and three Lunokhod missions. The rocks were dated using radioactive elements that are present all over the Solar System, using a technique we come back to later. The landing sites from which rocks were collected were all on the Earth-facing side of the Moon because radio contact had to be maintained with Earth at all times. This still covers a large area, though: the locations of the Apollo landing sites are spread out over an area of nearly 1,800 kilometers (1,100 miles). Despite the diversity of sampling locations, almost all of the age measurements from individual rocks came in between about 3.8 to 4.1 billion years; only a few were older, including one dated to 4.5 billion years. Comparison with lunar meteorites on Earth confirmed that these, too, were generally about the same age as the majority of lunar samples. These lunar meteorites were likely to have originated from a variety of locations on the Moon, with probably around half of them having been blasted upward by impacts on the far side.

This general uniformity inferred for the ages of lunar rocks is very curious and suggests that something particular befell—at least—the Moon. Equally telling is the contrast between the age of the lunar surface and the age of the Solar System and the Sun overall: astrophysicists estimate the age of the Solar System to be about 4.6 billion years. Together, these have led to the hypothesis

that some 3.9 billion years ago, the Moon, and likely other bodies in the inner Solar System, were bombarded by intense showers of asteroids. As this would have taken place well after the beginning of the Solar System, this has been called the Late Heavy Bombardment. More recent evidence suggests that this bombardment may have extended over many dozens, possibly several hundreds, of millions of years, so that it may not have been quite as cataclysmic as first thought. An intense, relatively short-lived Late Heavy Bombardment is still among the possibilities, but even if a more sustained and less intense bombardment turns out to be what actually happened, the initial hypothesis did lead researchers to other interesting concepts, as noted in Chapter 6.

By the way, the very method used to date the lunar rocks, which is of course also used on rocks on Earth, relies on the existence of radioactive materials in the Earth, the Moon, and all other Solar-System bodies. That, too, is a clue to the history of the Solar System.

The tales of comets

Most bodies entering the upper atmosphere and seen as meteors are not asteroid fragments but are what should be referred to as comets. Comets are distinct from asteroid meteorites primarily in their composition: they do contain rocks down to the scale of dust, but are also very rich in frozen volatiles. In fact, very roughly and depending on the comet, the dust component and the ice fraction are comparable in total mass. The most common volatile frozen into comets is water. Others, in lesser quantities, include carbon monoxide, carbon dioxide, methanol, ammonia, hydrogen sulfide, methane, and ethane.

How do we know the composition of comets in general? Those comets that end up as meteorites have been substantially affected by their trip through the Earth's atmosphere, the impact on solid Earth or in oceans, and their stay on Earth prior to being found. So to learn about comets best we have to look at them while in

space. How do we obtain the information on their composition? Mostly from the analysis of sunlight scattered off the gas cloud around their solid core and off their extended gas tails. The bodies of comets fall toward the Sun from the outer reaches of the Solar System. When they approach the Sun, the sunlight warms their outer layers and ices of various compounds begin to evaporate. As the gas streams away, dust and ice fragments can be released also. Because a comet has only a very weak gravitational pull, much of these gases are not held onto the comet's solid body, or nucleus, in a contained atmosphere but can stream away if they don't refreeze onto the comet. At first, close to the solid core, the gas and dust cloud is still fairly dense, and sunlight scatters off it efficiently creating a foggy halo that is called the coma. Then sunlight and the solar wind push the gas and dust out of the coma into what we know best of comets: their tails. Tails can span a wide arc in the sky as seen from Earth, and then such comets are a great spectacle, such as comet McNaught C/2006 P1 shown in Figure 3.12.

Figure 3.12 The comet McNaught (C/2006 P1) seen over the Pacific Ocean in 2007, viewed at sunset from the Paranal Observatory in the Chilean Andes (See Plate 10 for a color version).

Source: ESO.

Astronomers unravel the light scattered off the coma and tail of comets, compare it with the sunlight that fell onto it, and then look in detail at fine structures in the color change from sunlight to comet light. This special science, called spectroscopy, enables us to determine the composition of the volatile component of the comet because each chemical leaves its own colorful fingerprint in the light. As the gas streams from the solid core, it may already be changing its chemical structure as it is exposed to the Sun's ultraviolet and X-ray light. Inferring what is inside the comet's solid core from what is actually measurable thus also includes understanding the chemical reactions that occur once gas leaves the core and before light is observed by a telescope. For a few comets, spacecraft flying close to the comet have analyzed the gas and dust directly with specially designed instruments.[8]

Pictures of the solid cores of comets that have been approached or orbited by spacecraft show a collection of highly irregular shapes, such as those seen in Figure 3.9. Comet "landscapes" are distinctly other-worldly: for example, the Rosetta spacecraft made many stunning and surprising images of 67P/Churyumov–Gerasimenko (Figure 3.10) showing a barren world of boulders, dust fields, and cliffs, with jets of gas and dust lighting up in the stark sunlight.

Although these objects have sizes as large as entire mountains (dozens of kilometers) they are fragile, strangely shaped worlds: their masses are not large enough to pull themselves together into spheres, while relatively mild perturbations can cause them to disintegrate. That can be the result of collisions with other objects, but far more probable is that they fall apart or are pulled apart.

One way in which they can fall apart is because they spin. As a result, sunlight warms the lit "dayside" part of the solid core, which then loses frozen compounds through evaporation; then things cool down on the "nightside," and that repeats over and again. Any loosened material may be thrown off the solid core by its spinning motion. Or, if a large amount of frozen

solids is evaporated, the core may become unbalanced, so that the forces associated with its own spinning can pull the remainder apart. This affects many comets, some experiencing it while being observed by powerful telescopes.[9] Alternatively, they can be pulled apart when passing a heavy other body in the Solar System. That is what befell comet Shoemaker–Levy 9 (see Figure 3.14) in 1994 as it passed too close to Jupiter. The gravitational pull of a planet decreases with distance, so that the near side of the comet's solid core is pulled toward the planet more strongly than the far side. If the difference between these pulls exceeds the strength by which the core hangs together, it will tear the core apart. This occurs on close approach to planets, but also on close approach to the Sun, such as happened to the well-observed comet C/2012 S1 ISON.

Coming too close to the Sun is a hazard to the very existence of a comet. In each pass close to the Sun in their generally elliptical orbit the solid core loses mass through evaporation. In the end, that causes comets to break up into pieces. These pieces move on slightly different orbits so that over time they increasingly move apart. Yet they retain their family ties: their fragments can become a sequence of comets, showing up one after the other on their successive approaches of the Sun. I got to know one such family with a particularly hazardous orbit in 2011 during routine daily inspection of solar observations made with a special set of telescopes on NASA's Solar Dynamics Observatory, a satellite designed to keep continuous watch on the Sun. Multiple times a year it would be my task for some days in a row to carefully inspect observations of the Sun's outermost, hot atmosphere, called the corona. I would look for and catalog special events within the corona, and would check the quality of the images and thereby the health of the satellite and its instruments. Observations by another solar-looking spacecraft with a much wider field of view had shown a particularly bright comet diving toward the Sun against the starry sky. Not expecting to be able to see such a small, faint object against the large, bright Sun, I nonetheless estimated

Figure 3.13 The comet C/2011 N3 seen very close to the Sun just minutes before it completely evaporated in the Sun's heat. The image was taken by a telescope on NASA's Solar Dynamics Observatory tuned to the Sun's extreme ultraviolet light that shows only the million-degree outer atmosphere of the Sun. The Sun's surface does not emit such light and therefore shows up as a black sphere underneath the atmosphere. The comet, then a meteor in the Sun's atmosphere, glowed just bright enough to be seen—the first such solar meteor ever to be observed. The arrow points to the comet, and its direction roughly indicates the direction of motion.

where the comet would approach the Sun and looked, at first just above the Sun's edge where the Sun's signal was weak. To my surprise, the comet showed up in the extreme ultraviolet light to which the telescopes were tuned (Figure 3.13). And not only off the Sun, but also in front of it as the comet descended further. Within about ten minutes, however, the signal disappeared: the heat of the Sun had evaporated not just the frozen volatiles but apparently the entire solid core of the comet. Comet C/2011 N3 (SOHO)—briefly a meteor in the Sun's atmosphere—was no more.

That comet was part of the Kreutz family, named after the German astronomer Heinrich Kreutz (1854–1907) who figured out that a series of comets coming very close to the Sun were all related and on very similar orbits. After this first spotting of

Figure 3.14 Fragments of comet Shoemaker–Levy 9 prior to impacting Jupiter. The image, a composite of photographs taken by the Hubble Space Telescope, shows that the comet's solid core had been pulled apart into twenty-one visible fragments by the time of this observation, on May 17, 1994. The fragments impacted Jupiter between July 16 and 22, 1994. The shadow of Jupiter's moon Io is seen as a small dark spot on the Jovian clouds.[10]

a comet within the solar atmosphere, a second bright Kreutz comet, C/2011 W3 (Lovejoy), was observed later that year also plunging into that hot atmosphere. This one, however, survived to past the time of its closest approach to the Sun, although only for some hours before it, too, disappeared. These two sun-grazers were the first ever seen entering the Sun's atmosphere. Their observation there told us both about the comets' properties

and also about the Sun's magnetized atmosphere. Astronomical instruments are used to observe comets no matter where these can be seen in the skies, though, collecting more and more information on their properties.

Water in the Solar System

All of these processes, from comas and tails to disintegrations or sampling, show that the solid cores of comets are rich in water. The core of comet Halley, for example, is estimated to be 80 percent water. More typical is about half of the total. This is in contrast to the present-day water content of the terrestrial planets: Mercury, Venus, and Mars are essentially devoid of water, although some remains frozen beneath Mars's surface and in its polar regions, and some ice has been spotted at Mercury's poles. Earth, the wettest of the terrestrial planets, is still essentially dry: its total water content is estimated to be no more than 0.023 percent by mass.

Far out in the Solar System there seems to be an abundance of water. The giant planets have thick atmospheres made up of hydrogen and helium that make it hard to know in detail what lies underneath. But there is abundant water on the moons in the outer Solar System. Most of this is frozen, but there are oceans of liquid water in several places. For example, Ganymede, Jupiter's largest moon, may have as much water ice as silicate rock around a fairly small metallic core. Layers of solid and liquid (salt) water may be as thick as 800 kilometers (500 miles), or 30 percent of the moon's size. Callisto, Jupiter's second-largest moon, has an old, solid surface full of impact craters. But deep in its interior, over 100 kilometers (60 miles) below the surface, there appears to be liquid water enveloping an ice–rock deep interior; overall, ice and liquid water (if indeed that is there as a liquid) could be as deep as 300 kilometers (or 200 miles). Europa, the fourth-largest moon of Jupiter, has a water–ice crust of which the smoothness suggests that a liquid water ocean lies underneath it, supported also by

Figure 3.15 Saturn's moon Tethys observed by the Cassini spacecraft. This face of the moon is dominated by the large Odysseus crater, measuring 450 kilometers (or 280 miles) across, large compared to Tethys's diameter of 1,071 kilometers (665 miles). The crater is the result of a large impact onto this body that has a density of 0.97 times that of water, and is thus plausibly composed largely of water ice. Tethys, which was discovered by Giovanni Cassini in 1684, orbits Saturn once every 45.3 hours and always presents the same face to the gas giant.

the observation of geyser-like vapor plumes. Overall, the layers of water as ice and liquid may be about 100 kilometers (some 60 miles) thick. Io, the third-largest but innermost moon of Jupiter, in contrast has very little water and is composed mostly of silicate rock around an iron core. It has a lot of volcanic activity: over 400 active volcanoes dot its surface (see Figure 12.2). As there is so much water on the other large moons of Jupiter, it appears that

Io somehow lost it, possibly by heating induced by Jupiter's glow (much brighter when the system was young) and gravitational tides.

Further out, moons of Saturn also contain a lot of water ice. To name a few: Titan, Saturn's largest moon, may be half water. Even Enceladus, Saturn's sixth-largest moon, contains much ice and perhaps a liquid ocean of two dozen kilometers in thickness. Mimas, just under 400 kilometers (about 250 miles) in diameter, is among the smallest objects in the Solar System for which gravity is strong enough to pull it into a sphere. Its density suggests it is largely water ice with only a small fraction of rocks.

The tipping point in where bodies in the Solar System contain a lot of water and where they have very little lies somewhere between Mars and Jupiter, in the asteroid belt. Asteroids exist in very different families where their composition is concerned. Those that orbit between three and 3.5 times the Sun–Earth distance, in the outer edge of the asteroid belt, are predominantly (over 80 percent) as black as coal. That makes it easy to remember that they are referred to as "C type," for all their richness in carbon. Closer than three Sun–Earth distances, many are still C type, but some 60 percent are grey, mostly composed of silicate rock, so they are "stony," and thus—again fortunately for us to remember—called "S type." Then there are also those that are largely metallic, which, guess what, are called "M type." There are also other types of divisions into classes and subclasses but we need not go into those here.

Many asteroids contain a lot of water, mostly soundly frozen, and often but not always deep in the body. For example, the largest asteroid, Ceres (a C-type asteroid; see Figure 3.5), was visited by the Dawn spacecraft. The craft had a special instrument on board that led astronomers to conclude that water is abundant, maybe representing as much as 30 percent of the total mass of the asteroid. The asteroid and dwarf planet Ceres possibly contains a few times as much as all water on Earth combined. Water resides also in the multitude of other, smaller asteroids.

The coal-black C-type asteroids often have surfaces with water-rich (hydrated) minerals; some in fact have water ice on their surfaces. The grey S-type ones often do not. As a rule, C-type asteroids have densities that are about half those of S-type asteroids, to be expected as they contain relatively more water compared to rock. The two populations are largely but not completely mixed: there is a significant tendency for the water-rich ones to orbit further from the Sun than for those with less water.

Spinning and orbiting

One very important factor in the formation and evolution of the Solar System that has not yet been mentioned is movement. Almost everything in the Solar System moves or spins in the same direction. Seen from above Earth's north pole, the Sun rotates in an anti-clockwise direction, and all the planets orbit the Sun in that same direction. In fact, they orbit the Sun in an almost flat plane: all planetary orbits align with the Sun's equator to within 7.2 degrees, with Earth's orbit the most inclined of all eight planets. The asteroids have an average orbital inclination within the range of the planets, but with a much wider distribution, some reaching an inclination of 30 degrees.

Most of the planets also spin in that same direction, and most of their moons orbit them likewise. Venus, however, is the contrarian of the bunch; it spins the other way around. Its rotation is referred to as retrograde, meaning opposite to the norm, or backwards. Venus's spin rate is also remarkably slow: it takes 243 Earth days to complete one rotation, which is longer than the 225 days that Venus takes to orbit the Sun: Venus's day is therefore longer than its year! Another outlier in terms of rotation is Uranus: it spins essentially lying on its side with its north–south rotation axis lying within eight degrees of the average plane of the planets.

Of the large moons, only Triton, the largest moon of Neptune, orbits in a retrograde direction. Many of the smaller ones, however, orbit against the rotation direction of their planet. Such

moons with retrograde orbital motions are members of the population of "irregular moons," which are often distant from their parent planet and have orbits that are retrograde, highly inclined, highly eccentric, or a combination of those. About half of all moons in the Solar System are irregular moons (giving exact numbers is impossible because a few dozen objects have yet to be classified officially as moons, and many more small moons may yet be found). Most of the irregular moons are either relatively far from their host planets, or they are small, or both.

Patterns in orbits within the Solar System

Now, we come to distances: all planets in the Solar System are far from all other planets and far from the Sun. For example, the Earth is over 11,500 Earth diameters away from the Sun. The closest approach of the nearest planet to Earth, Mars, puts that planet almost 4,300 Earth diameters away. Even the many millions of rocky bodies in the asteroid belt have not prevented multiple spacecraft from safely traversing it without running into anything. The Solar System, despite its zoo of objects, is mostly empty space.

People have long hunted for a pattern in how far the planets are from the Sun. The best-known such pattern is known as the Titus–Bode law, after the two first proponents. It works to within a few percent for the inner seven planets, but not nearly as well for Neptune. Its apparent success was why people started looking for a planet between Mars and Jupiter where the "law" predicted there should be a planet but none was known. This led to the discovery of the asteroids, with Ceres coming first, which is the largest asteroid and was eventually promoted to dwarf planet, not too far from the distance expected from the Titus–Bode law.

This pattern in orbital diameters should actually not be referred to as a "law" in the true physical sense, but rather be called a "rule." A physical "law" is a mathematical summary of a relationship between quantities that has been tried and tested

over and over and to which no exceptions have been found. However, there is no commonly accepted theory in the formation of planetary systems that explains the "Titus–Bode law," or in fact that predicts any particular pattern in planetary orbit sizes, and the Titus–Bode rule does not even hold throughout the one planetary system for which it was formulated. This is where observing exoplanet systems may eventually help us realize whether the Titus–Bode rule is a chance quasi-regularity for our own Solar System, or part of a pattern recognized more widely. So far, it seems we should assume the former, although exoplanets are, typically, more separated the further they are from their star, as is the case for the planets in the Solar System.

One more property of the planetary orbits within the Solar System should be noted: the orbits of the planets are nearly, but not quite, circular. As Kepler already noticed, they are best described as ovals known as ellipses. The shape of an ellipse is quantified through the eccentricity of the orbit. For a circle, which is a special case of an ellipse, the eccentricity is 0. An eccentricity larger than 0 means that the orbit is an ellipse for which there is a point of closest approach to the Sun (called perihelion for the Sun in particular, or periastron for any star) and, half an orbit later, a point of maximum distance (called aphelion or apastron). The value of the eccentricity is the difference between the largest distance from the Sun and that at closest approach, divided by the largest diameter of the orbit; it thus measures how stretched the orbit is.

For Earth, the orbital eccentricity is 0.017, which means the orbit is very nearly circular. The Earth is closest to the Sun in early January and furthest in early July. The Sun thus is somewhat closer to Earth in the middle of the winter on the northern hemisphere, but the low eccentricity means that only just over 3 percent more solar energy falls onto Earth than does when the planet is furthest from the Sun. Mars has a considerably larger eccentricity: 0.09. Because that means that almost 20 percent more solar energy warms Mars at its perihelion than at its aphe-

lion, the seasons on Mars's northern and southern hemispheres are noticeably different. The winner is Mercury, however, with an eccentricity of 0.21, leading to 45 percent more solar energy at closest approach than at furthest separation. The average eccentricity for the Solar System planets is 0.06.

The active Sun

At the center of the Solar System resides the Sun. It provides a remarkably stable and long-lived source of energy for all the planets, moons, and asteroids. But not all about the Sun is benign: apart from the visible light, the Sun emits X-rays, ultraviolet light, particle radiation, and a hot, magnetized, gusty solar wind. All of those phenomena are related to the Sun's magnetic field. That field is generated deep inside the Sun by what is called the solar dynamo. Strands of it buoy to the surface where they form sunspots where the field is strongest with attendant weaker "magnetic plage" in their wide surroundings. Wherever the field penetrates the solar surface, it heats the overlying atmosphere to between thousands and millions of degrees. These heated gases shine in ultraviolet and X-ray colors; that glow beams out at most one part in 10,000 of the energy that comes out as visible light, but it carries a dangerous punch for molecules in planetary atmospheres that can break their chemical bonds.

Where the magnetism in the Sun's outer atmosphere is relatively weak, the push of the hot material may cause a breach in the confining magnetic field from where the hot gas escapes to form the solar wind, which moves at speeds of typically 1.5 million km/h (1 million mph) but can reach three or four times that during eruptive coronal mass ejections. Although this wind cools somewhat before reaching the planets, it is still at hundreds of thousands of degrees when it passes by the Earth. It continues to flow outward, past the outermost planets until, somewhere over 100 Sun–Earth distances out, it collides with the interstellar gas.

The Sun's magnetic field is ever-changing, from an overall 11-year pulsing down to day-by-day changes as new field emerges onto the surface and existing field decays away. The emergence of new field and its interaction with pre-existing field often leads to instabilities. At such times, explosions occur that heat yet more gases to still higher temperatures and greater X-ray and UV brightness, and that eject gases into interplanetary space. The larger of these phenomena, called solar flares and coronal mass ejections, can pack energies equivalent to billions of hydrogen bombs going off at the same time. Generally, we are shielded from the detrimental consequences of these occurrences by the distance to the Sun, the Earth's magnetic envelope, and the Earth's atmosphere. Sometimes, however, these phenomena, collectively known as space weather, cause spectacular auroras, damage spacecraft, affect the electric power grid, and present a threat to astronauts in space. Rarely do they present a direct hazard to humans on Earth, but in the long run, measured over billions of years, all aspects of the Sun's magnetic activity conspire to erode planetary atmospheres: Venus lost its water, Mars is left with but little of its original atmosphere, and Earth, too, is gradually being stripped of our breathable environment—I return to all that at the end of this book. Most stars in the Universe have analogous signatures of magnetism; only the heaviest adult stars—some two percent of all stars—do not.

Characteristics of our Solar System

After this description of the Solar System we have a number of facts that we should keep in mind when looking at other planetary systems. With the principles of uniformity and mediocrity in mind, these facts outline in large brush strokes what astronomers were anticipating to be true also for other planetary systems:

1. The combined mass in all planets of our Solar System is small compared to the mass of the central star.

2. Almost every heavy body in the Solar System orbits or spins in the same direction. About half of the small and distant moons of the large planets orbit in the opposite direction.

3. Close to the Sun, planets are composed mostly of rock and metal. Far from our star, they are mostly composed of volatiles, much in gaseous form.

4. Close to the Sun, planets are much less massive than the more distant ones.

5. Water (liquid and frozen) is very rare near the Sun and very common beyond about three Sun–Earth distances.

6. Between Earth and Jupiter lie surprisingly light bodies, including the small planet Mars, the dwarf planet Ceres, and the multitude of small asteroids that all together contain but a small fraction of the mass of a planet.

7. Beyond the planets in the Solar System, we find multiple dwarf planets and many comets.

8. Dwarf planets and planets in the Solar System were essentially entirely liquid some time in their lives, allowing for gravitational separation of chemical compounds in their interiors, letting heavy materials sink toward the center and lighter ones float to the surface. For terrestrial planets this led to iron cores and rocky mantles, while for smaller moons this created rocky interiors and icy (or watery) shells.

9. Metal and rock deep inside several planets and moons in the Solar System remain liquid, despite billions of years of cooling down. Water remains liquid in some moons of the giant planets despite the cold of space so far from the Sun.

10. All bodies in the Solar System contain limited amounts of radioactive materials. The heat released by radioactive decay supports, at least for some time in the terrestrial planets, tectonic activity.

Many of these characteristics were found to be in need of reformulation, however, to be valid more generally, as we shall see

next. It is through the combination of information on our "home system" with new and often surprising facts about exosystems that we can better understand how planetary systems form, how they evolve, and in particular how our own came to be as it is.

Notes

1 Masses of the satellites for each of the planets: nothing for Mercury and Venus; 1.2% of an Earth mass for Earth's Moon; 0.024% of Mars's mass for Phobos and Deimos at Mars; 0.02% of the mass of Jupiter for the sixty-nine moons of that planet (at least up to the most recent discovery of two moons in 2017); 0.025% of Saturn's mass for its sixty-two confirmed moons; 0.02% of Neptune's mass for its fourteen known natural satellites; and about 0.01% of Uranus's mass for its twenty-seven known moons.

2 Astronauts' excitement about coming across orange material on the Moon, from the Apollo 17 Lunar Surface Journal (with hours:minutes:seconds of mission time): 145:26:25 Schmitt: "Wait a minute..."; 145:26:26 Cernan: "What?"; 145:26:27 Schmitt: "Where are the reflections [in the helmet visors]? I've been fooled once. There is orange soil!!"; 145:26:32 Cernan: "Well, don't move it until I see it."; 145:26:35 Schmitt: (Very excited) "It's all over!! Orange!!!"; 145:26:38 Cernan: "Don't move it until I see it."; 145:26:40 Schmitt: "I stirred it up with my feet."; 145:26:42 Cernan: (Excited, too) "Hey, it is!! I can see it from here!"; 145:26:44 Schmitt: "It's orange!"

3 The Solar System bodies visited by flying close by, orbiting, or landing (or deliberately crashing onto the surface) include the Moon, all of the planets (in chronological order in which the planets were first visited: Venus, Mars, Jupiter, Saturn, Mercury, Uranus, and Neptune), the dwarf planets Ceres and Pluto (one in the asteroid belt, the other far beyond it), the asteroids (in chronological order from their first and often only visitor) Eros, Itokawa, Vesta, Gaspra, Ida, Mathilde, Masursky, Braille, Annefrank, Šteins, Lutetia, APL, and Toutatis (see Figure 3.8 for a few snapshots), and eight comets (also in chronological order from first visit) 21P/Giacobini–Zinner, 1P/Halley, 26P/Grigg–Skjellerup, 19P/Borrelly, 81P/Wild 2, 9P/Tempel, C/2006 P1 McNaught, 103P/-Hartley 2, and 67P/Churyumov–Gerasimenko (see Figure 3.9 for a comparison to scale, and Figure 3.10 for a close-up of 67P). Other

spacecraft crash-landed or were designed to function only during the descent through an atmosphere; the Galileo Probe is one such experiment descending into the atmosphere of Jupiter. Some Solar-System bodies were visited in passing during a flyby en route to somewhere else. Such visits would be brief and not necessarily very close, but far more informative than inspection by telescope from distant Earth. A set of moons has also been explored by flybys by satellites that either orbited their planet or were underway to other targets. These include Jupiter's moons Io, Europa, Ganymede, and Callisto, and Saturn's moons Dione, Enceladus, Hyperion, Iapetus, Mimas, Pan (see Figure 3.11), Pandora, Rhea, Tethys (see Figure 3.15), and Titan.

4 Asteroid names: Asteroids with confirmed orbits are given a number in order of discovery. Formally, Ceres, the first to be discovered, would be called "(1) Ceres", and the fourth, Vesta, is formally "(4) Vesta" or often "4 Vesta".

5 Figure 3.8 shows images to scale of nine asteroids. Their types, sizes, discovery years, and the spacecraft visiting them are: 4 Vesta (a V-type main-belt asteroid—similar to S-type, but somewhat distinct, of which Vesta is the prototypical example, hence the name—, 525 kilometers (326 miles) across, second in mass only to asteroid and dwarf planet Ceres; found in 1807, visited by the Dawn spacecraft in 2011–12), 21 Lutetia (inner-belt M(etallic)-type, about 100 km diameter, first observed in 1852, flyby by the Rosetta probe in 2010), 253 Mathilde (main-belt C-type, 50 km diameter, first spotted in 1885, flyby by the NEAR Shoemaker spacecraft in 1997), 243 Ida (S-type, 60 km along its long axis, discovered in 1884, visited in 1993 by the Galileo spacecraft flying by), 433 Eros (near-Earth S-type, about 34 km along its longest dimension, found in 1898, visited by NEAR Shoemaker in 1998 (flyby) and 2000 (orbit)), 951 Gaspra (S-type near the inner edge of the belt, 18 km along its longest direction, first observed in 1916, flyby by Galileo in 1991), 2867 Šteins (main-belt asteroid, not quite 7 km along its longest axis, found in 1969, visited by the Rosetta probe in 2008), 5535 Annefrank (inner main-belt asteroid, comparable to 2867 Šteins in size, discovered in 1942, flyby by Stardust in 2002), and 25143 Itokawa (Mars-crosser asteroid, only 0.5 km in diameter, discovered in 1998, sample returned by the Hayabusa space mission).

6 The first landing occurred on the Moon with an unmanned Russian Luna 9 spacecraft in 1966, followed by eighteen more landings:

Surveyor 1, Luna 13, Surveyor 3, 5, 6, and 7, then manned landings by Apollo 11 and 12, Luna 16 and 17, manned Apollo 14 and 15, Luna 20, manned Apollo 16 and 17, Luna 21 and 24, and—after a 37-year gap—Chang'e 3 in 2013. After initially reaching for the Moon, spacecraft landed successfully on Solar-System bodies far beyond the Moon and functioned for some time to send back information from two planets, two asteroids, a moon of Saturn, and two comets. These destinations include (listed in order of first touchdown) Venus (Venera 7 through 14, Pioneer Venus Multiprobe, Vega 1 and 2), Mars (Viking 1 and 2, Mars Pathfinder, MER-A/B Rovers Spirit and Opportunity, Phoenix, and Curiosity rover from Mars Science Laboratory), the asteroid Eros (NEAR), Saturn's moon Titan (Huygens), the asteroid Itokawa (Hayabusa, which returned samples from the asteroid's surface to Earth in 2010), comet 67P/Churyumov–Gerasimenko (Rosetta's Philae, although the lander communicated for only just over a minute and that only months after touching down), and comet Tempel (crashed into by the impactor of the Deep Impact satellite).

7 Scientists analyzing image sequences of comet 67P taken by Rosetta's cameras have noticed evidence of landslides (collapsing cliffs) on the comet, exposing ancient material to sunlight and causing outbursts of matter leaving the solid core. In fact, gravity on 67P is very weak and the shape of the body means that gravity's up and down are sometimes in odd, unexpected directions. The mean density is lower than that of ice, so that the core's interior must be quite porous.

8 Comet material was analyzed by several spacecraft: the Giotto spacecraft for comet 1P/Halley and the Rosetta mission for 67P/Churyumov–Gerasimenko. The Stardust spacecraft returned samples of the dust from comet 81P/Wild 2. In the case of comet 9P/Tempel 1 it was possible to study some of the material deep inside the solid core and not only what was being released from the surface: an impacter flying with the primary Deep Impact spacecraft hit the comet's solid core, and the spray of gas and dust was subsequently analyzed by instruments on the nearby main spacecraft.

9 Comets 73P/Schwassmann–Wachmann 3 and 332P/Ikeya–Murakami were observed to fall apart as they were watched by the Hubble Space Telescope.

10 When comet Shoemaker–Levy 9 was first spotted, on March 24, 1993, by Carolyn and Eugene Shoemaker and David Levy, it was

already captured into an orbit around Jupiter, and subjected to tidal and centrifugal forces that pulled it apart, likely starting during a close encounter with Jupiter in July of 1992, after which the fragments spread out along and around the initial orbit, with subsequent breakups occurring up to nine months after that.

4

Exoplanet Systems and Their Stars

A desert-wrapped planet, an inhabited neutron star, a water-world orbiting a black hole, a planet with two suns.... These assorted worlds of Arrakis, Dragon's Egg, Miller, and Tatooine are fabricated by writers and moviemakers and exist only in fictitious universes. The worlds that scientists are finding in the real Universe are just as diverse, however: they have found planets around double stars and deceased stars, half a dozen planets packed inside an orbit where our Solar System has but one, planets that are just forming and planets that are billions of years older than Earth, planets that have permanent daysides with lava oceans and permanently frozen night sides, giant planets so close to their stars that they are falling apart in the extreme heat, glowing stellar remnants with evaporated material of broken-up planets floating in their surface layers,... and Earth-sized worlds that could, in principle, sustain water in liquid form on their surfaces. The imaginary worlds of literature and cinema may have led science, but—as Isaac Asimov expressed it in the title of an essay—"The Truth Isn't Stranger than Science Fiction—Just Slower."

One development that is definitely going to be slower than fiction for a long time to come is the imaging of exoplanets. If by "imaging" we mean seeing the dots of exoplanets orbiting the dot of their central star, then some telescopes have indeed "imaged" exoplanets. But I bet that is not what most of us think of as "imaging"; we would expect a picture with at least some details of the exoplanet seen as a sphere, perhaps with clouds, ideally with surface details. There is no such image and there will not

One of Ten Billion Earths. Karel Schrijver, Oxford University Press (2018). © Karel Schrijver.
DOI: 10.1093/oso/9780198799894.001.0001

be one for at least the coming decades…unless the extraterrestrial inhabitants of an exoplanet have already sent us one that is currently racing toward us at light speed somewhere through interstellar space (not likely!). If you see an image of an exoplanet landscape or an exoplanet depicted as a sphere with some detail on it, it is a work of imagination: artful it may be, but it is not based on actual imagery. Having said that, we should not downplay "imaging" that shows only dots: over time this reveals the planetary orbit, and if sufficiently separable from the light of the central star, the light from these "dots" can be fed into spectrographs. That is one way to learn about the atmospheres of these exoplanets, or—if the atmosphere is sufficiently transparent—their surfaces. This important and challenging work is just beginning and therefore only touched upon here and there in what follows. Here, we shall focus not on landscapes or cloud covers, but on exoplanet orbits and masses, and on their central stars, addressing these questions: where are the exoplanets around their stars, where are the heavyweights and where are the underdogs, what kinds of stars do they orbit, and what do these orbits look like?

The diversity of worlds

Within two decades after discovering 51 Pegasi b, the first planet found outside the Solar System around a star like the Sun, we learned that planets are common throughout the Galaxy and that they come in a great variety. Any of the known exoplanets has a story of its own to tell, while together they can help us understand our own Solar System. So which do we choose to look at in some detail to find our way to the fundamental characteristics in their overall properties? NASA asked over sixty exoplanet scientists to pick their favorites on the occasion the twentieth anniversary of the finding of 51 Pegasi b, and *Scientific American* magazine published its own list of favorites in 2017. From these, I selected eight, each of them a "first" or record holder.

I added six other unusual cases. Together, these examples illustrate the bewildering diversity of exoworlds that have been found. All are flagged as "confirmed" exoplanets in the Extrasolar Planets Encyclopaedia.[A] By the way, you will see that the names of the exoplanets show an apparently confusing but nonetheless standardized mix of roots; more on that can be found in the Notes at the end of this chapter.[1]

Exoplanets orbit much heavier objects. Some of these are true stars, whereas others are not or may not be true stars any more. Here is the short guide to "stardom": A star is a sphere of gas in which energy is generated by nuclear fusion of hydrogen or, for older stars, in which products of earlier hydrogen fusion are fused into still heavier elements. When all fusion fizzles out, a star may transform into a black hole, or into a neutron star or into a white dwarf star (often called white dwarf for short). Although the latter two may have names with "star" in them, they are not formally stars any longer when they reach these retirement phases. Another family of objects with a confusing name is that of the so-called brown dwarf stars, which are often just called brown dwarfs. These are star-like spheres of gas that are not heavy and dense enough to ever have nuclear fusion of the most common form of hydrogen. Brown dwarfs are, however, just sufficiently heavy that there is a phase in their lives during which the heavy form of hydrogen—deuterium—does reach nuclear ignition. This yields relatively little energy, however: brown dwarfs are tiny, faint, and cool compared to true stars. Anything less massive than a brown dwarf—weighing less than 0.1 percent of the Sun's mass, or less than about 4,000 Earth masses—never has any kind of nuclear fusion in its interior; such objects are regarded as exoplanets.

By way of introduction to exoworlds, let's meet these fourteen fascinating exoplanets and exoplanetary systems.

[A] See http://www.exoplanet.eu.

1. 51 Pegasi b, the first exoplanet discovered around a Sun-like star The planet, discovered in 1995, weighs in at about 150 Earth masses, or half the mass of the giant Jupiter. It orbits its star in four Earth days. In other words its "year," which is the time it takes a planet to go around its star, is amazingly brief. The planet is likely to always have the same side facing its sun because of tidal forces like those that locked the Moon to always have the same side facing the Earth. The dayside atmospheric temperature of the planet is estimated to lie around 1,000 degrees centigrade (1,800 degrees Fahrenheit), enough to melt brass as well as several types of rock; most unpleasantly hot! Its orbit puts it at only 1/20 of the distance from the Sun to the Earth. Finding an exoplanet that large orbiting so close to its host star was not at all expected, and clashed with existing ideas of giant planet formation at the time it was first reported; 51 Pegasi b was an eye-opener that caused the concepts of how planetary systems form and develop over time to change fundamentally.

2. CoRoT-7b, the first known rocky lava-ocean planet This planet, discovered in 2009, was the first planet found that was neither a giant planet nor a super-Earth: it came in with a size only 60 percent larger than Earth's. Although that may mean that CoRot-7b is a rocky planet, it is a very strange world, even compared to 51 Pegasi b: it is so close to its star that it has an orbital year that is only 20.5 Earth hours long. Its sun is a bit smaller and redder than our Sun, but being that near to its planet means that it appears several thousand times brighter in the planet's sky than our Sun in ours. Surface temperatures on the planet are likely some 2,000 to 2,500 degrees centigrade (or 3,500 to 4,500 degrees Fahrenheit). All rocks on Earth are melted by 1,200°C, so the surface of CoRot-7b is presumed to be a deep lava ocean, at least on the side that is expected to always face inward toward its host star. After all, so close to its star, the planet probably orbits and spins such that it has a day side with everlasting sunshine and a permanently dark night side.

3. Kepler-186f, the first known rocky Earth-sized planet in the habitable zone The planet, first detected in 2014, orbits a star with half the mass and half the size of our Sun, being considerably fainter and much more red in appearance. There are at least four other exoplanets closer in to its star. Kepler-186f is probably only about 15 percent larger than Earth or may in fact be Earth-sized. It appears to orbit its star at a distance where its surface temperature may be neither too high to evaporate all water nor too low to freeze it all. Whether it actually has any liquid water depends on how the planet formed and what its atmosphere is. By 2017 it was one of only some twenty small, Earth-sized planets catalogued with orbits within the habitable zones of their stars.

4. Kepler-11, the first known compact multi-planet system the Kepler-11 system (Figure 4.1), discovered in 2011, has at least six exoplanets (named Kepler-11b through -11g) orbiting a star very much like the Sun: within a few percent, the star Kepler-11 has essentially the same mass, size, and surface temperature (and therefore also color) as the Sun. Five of the six orbit their star closer than the closest planet, Mercury, does our Sun, with orbital years between 10 and 47 Earth days. Only Kepler-11g orbits a bit further out, with an orbital period of 118 Earth days. For the innermost five exoplanets, masses were determined between most likely two and nine Earth masses; too little is known about the outermost planet to say much about its mass. All of the five inner exoplanets found have average densities well below that of Earth, with two of them less dense than water. So, these exoplanets are likely gaseous to a large extent, perhaps with a lot of water or ice underneath their atmospheres, and unknown but small rock and metal components. This would make them more like smaller versions of Uranus than scaled-up versions of any of the terrestrial planets in the Solar System.

5. Kepler-444, the oldest known compact multi-planet system The star, first announced to have a planetary system in 2013,

Figure 4.1 Artist's impression of the Kepler-11 system which has at least six planets orbiting a star very much like the Sun. Five of the six exoplanets orbit their star closer than the closest Solar-System planet, Mercury, does our Sun. The innermost five planets have densities so low that they are likely gaseous to a large extent and perhaps with a large amount of water or ice. This would make them more like smaller versions of Uranus than scaled up versions of Venus, Earth, or Mars. The inset on the lower left compares the orbits of the known planets in Kepler-11 with the innermost part of the Solar System.

Source: NASA/Tim Pyle.

is estimated to be approximately 11.5 billion years old, which is almost two and a half times older than the Solar System, and over 80 percent of the age of the Universe. Scientists working with the Kepler satellite have found five exoplanets, all with sizes between those of Mercury and Venus. That planets formed so early in the life of the Universe is interesting because of how chemical elements came to be formed after the Big Bang: anything heavier than hydrogen and helium was fused over many millions to billions of years deep inside stars that, at their life's ends, ejected much of these heavier elements into interstellar

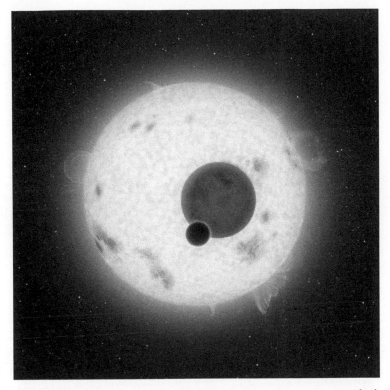

Figure 4.2 Artist's impression of Kepler-16, a planetary system in which at least one planet (Kepler-16b, rendered looking down at its dark night side with some light scattered through the top layers of its imagined atmosphere) orbits not one, but two stars with very different sizes and colors.

Source: NASA/JPL-Caltech/R. Hurt.

space to seed the clouds out of which subsequent generations of star–planet systems formed with what could condense into solid planetary embryos. The system's age also suggests that there may be vastly older civilizations around than ours, but not likely on these planets in particular: they are so close to their star that they cannot escape extreme heat.

6. Kepler-16b, the first known planet to orbit two stars

Because Kepler-16b, found in 2011, orbits a binary star (Figure 4.2)

it should officially be known as Kepler-16 (AB)b to indicate that it is a satellite of both stellar components A and B of the double star. Kepler-16b has a mass of about 100 Earths, so comparable to Saturn's mass, and a density that suggests it is about half gas and half rock and ice. It orbits its host stars in a year that is 229 Earth days long at 70 percent of the Sun–Earth distance from its stars. These stars are about two thirds and one fifth of the Sun's mass, respectively, and both are much fainter than the Sun, making for seriously cold weather on a gloomy Kepler-16 (AB)b, possibly only somewhat above −100 degrees centigrade or −150 degrees Fahrenheit.

7. HD 80606 b, a planet with a very eccentric orbit This is a large, heavy planet (discovery announced in 2001) weighing almost 1,300 times as much as Earth, or four times more than Jupiter. It has the distinction of having the most non-circular orbit of all exoplanets known by 2016: at closest approach it is only 3 percent as far from its star as the Earth from the Sun, while the largest distance in its elliptical orbit brings it to 88 percent of that distance, or just about 30 times as far away as at closest approach. The orbital motion near closest approach to its star would be swift, but not so swift as to avoid a rise and subsequent fall in the dayside temperature by over 500 degrees centigrade (or 1,000 Fahrenheit) in a matter of hours when the planet races through this phase in its orbit on its way to a much more unhurried pace further out.

8. V391 Pegasi b, the first known post-apocalyptic planet V391 Pegasi b, found in 2007, weighs over 1,000 Earth masses (over three Jupiter masses). It orbits an old, hot star with an orbital year of 1,170 Earth days. That star has perhaps half a solar mass and a quarter the solar diameter. Its surface is five times hotter than that of the Sun and consequently it glows bright blue with a luminosity that is fifteen times higher than that of the Sun. This star is perhaps more than twice as old as the Sun, and it has gone through a phase that awaits our Sun in some five billion years: when a star has processed all of its hydrogen fuel in the

core, it swells up to become a red giant star—that will make the Sun close to two hundred times larger than its current size. Some time around its largest phase, V391 Pegasi lost probably over a third of its mass in a period of extreme outflows that would have enveloped the planet (discussed in Chapter 5). After that, it shrunk again for some time into the stage that V391 Pegasi is in at present, with its nuclear furnace fueled by helium rather than hydrogen, and a surface heated up again to shine bright blue.

9. Proxima Centauri b, the nearest exoplanet known (by 2018) Proxima Centauri b was found in 2016 a mere 4.224 light years from Earth. Well, that is "mere" by astronomical standards; it is still 267,000 times further away than the Sun is from Earth, or 8,900 times further than the most distant planet orbiting the Sun, Neptune. Unfortunately, not very much is known about it because it does not cross in front of its star to reveal its size and mass: it is heavier than 1.3 Earth masses and probably less than three Earth masses, but how heavy it really is remains uncertain. Its star is small and cool, and consequently not very bright. This exoplanet is, however, much closer to its star than Earth is to the Sun, so that its daytime temperature might be pleasant—but we don't actually know because we do not know the properties of its atmosphere, or even whether it has one. The planet appears to be formally in the habitable zone, but its actual water content—if there is any—remains unknown.

10. HR 8799, the first directly imaged multi-planet system The star HR 8799 is very roughly some 30 million years old, weighs in at about 1.5 solar masses, and is some five times brighter than the Sun. It is surrounded by a ring system of dust particles that extends out to 1,000 Sun–Earth distances within which four giant exoplanets were directly spotted as unresolved dots. These four planets have similar masses ranging from 1,500 to 3,000 times that of Earth, orbiting at 15, 24, 38, and 68 Sun–Earth distances, with orbital years of 50, 100, 190, and 460 Earth years. The planets

could be imaged (first in 2008) because they glow markedly in the infrared, still radiating away the heat of their formation this early in their lives.

11. HD 13189 b orbits an evolved, cool giant star The giant star's mass is estimated to be between two and seven times the mass of the Sun, and that of HD 13189 b between 2,500 and 6,000 Earth masses. The companion, first reported in 2005, is consequently too lightweight to be a true star, but it is unclear if it should technically be called a brown dwarf or a true planet. The companion orbits the star somewhere between three to five stellar diameters, so really close in compared to Earth which orbits at just over 100 solar diameters. This system is among those with the heaviest stars with a confirmed substellar companion.

12. PSR B1257+12, the first detected, confirmed exoplanetary system The central object is a neutron star—so an ex-star without current nuclear fusion—that spins at an absolutely dizzying rate of 160 times per second. This "pulsar" is a heavy stellar remnant with a mass of about half a million Earth masses compressed into something with a diameter of about 20 kilometers (or 14 miles). There are three confirmed exoplanets, found in 1990, all within a distance from the pulsar of about the orbit of Mercury around the Sun. They have masses of 0.02, 4.3, and 3.9 Earth masses, and orbital years of 25, 67, and 98 Earth days, respectively. None of these exoplanets are thought to be survivors of the disastrous phases in which the neutron star formed: the neutron star probably formed by the coalescence of two stars dying violent deaths that would have likely destroyed any existing planets. The exoplanets we now see probably formed as second-generation planets after these catastrophic phases. How exactly that happened remains unsolved, for now.

13. K2-22b, a disintegrating rocky planet The exoplanet, discovered in 2015, orbits a low-mass, red (M-type) dwarf star with a size of 0.6 solar diameters. It has an orbital year of only 9.15

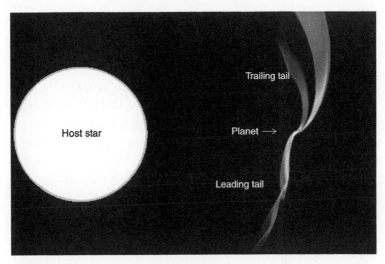

Figure 4.3 Computer simulation of the formation of leading and fol-lowing tails in the disintegrating planet K2-22b. K2-22b orbits its star so tightly and fast that it has an orbital period (planetary "year") of only 9.15 Earth hours. It orbits its star just over one stellar diameter above the surface of the star. That proximity causes the planet's surface, likely a few thousand degrees hot, to disintegrate into dust that escapes to form what is, in essence, a very large comet. This computer experiment shows how the matter that leaves the exoplanet's gravitational pull moves both ahead and behind the exoplanet. The perspective is from above one of the star's rotational poles, and such that the exoplanet moves clockwise around the star. The intensity shows the density of the matter.

Source: After Sanchis Ojede and colleagues (2015).

Earth hours (see Figure 4.3) at 0.9 percent of a Sun–Earth distance, or about just over one stellar diameter above the surface of the star. The proximity to the host star causes its surface, likely a few thousand degrees hot, to disintegrate into dust that blows off the exoplanet to form what is a very large comet orbiting very near to the star.

14. TRAPPIST-1, a packed miniature system The central star of this system—itself not discovered until 1999—is merely 1/8 the size of the Sun, only slightly larger than Jupiter at a width

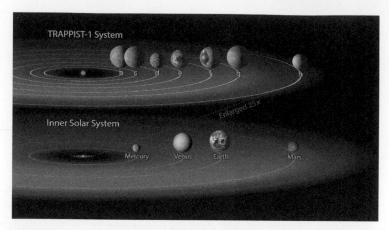

Figure 4.4 A visualization of the TRAPPIST-1 exosystem orbits compared to the Solar System. Note that the orbital scales for TRAPPIST-1 were magnified 25 times in order to show the system despite the extreme closeness of all the exoplanets in this system to their star. The exoplanets are shown much exaggerated in size for clarity, but they are all on the same scale compared to each other and the planets in our Solar System. Imaginary images are shown for the exoplanets; we have no idea what they actually look like. The green bands are estimated liquid-water habitability zones around TRAPPIST-1 and the Sun; where the band turns red it may be on the warm side, where blue it may be on the cool side (See Plate 4 for a color version).

Source: NASA/JPL.

of 0.12 solar diameters, and has a brightness that is almost 2,000 times less than that of the Sun. The star is orbited by seven known exoplanets (first published on in 2016), much like Earth in size and mass, but all very close to their star (see Figure 4.4). At least three of these seven planets are estimated to orbit within the liquid-water habitable zone.

Recognizing biases and the patterns in exosystems

These fourteen selected exoplanet systems offer but a glimpse of the diversity of what has been found. Looking through the

properties of these particular objects and systems, you would get the impression that the Galaxy is full of systems in which there are many planets very close to their stars, and that many such planets are heavy like Jupiter or heavier than that. Whereas these obviously exist, they do not appear to be the most common properties of planetary systems, though. What you see when your list includes many "firsts" is distorted by what is most easily detected: large, heavy planets close to their stars are simply far more easily detected than those further out. By carefully reviewing just how difficult it is to detect smaller planets or planets further from their host stars it becomes possible to know how many with what properties may be missed. And then one can infer something about the actual total population by combining what we know of the population that was observed with estimates of what would simply not be observable with a given method and instrument. For example, knowing the orbital geometry of the Solar System enables us to say that no hypothetical observer looking at our system from an exoplanet in a random direction would ever be able to detect more than three transiting planets, while thirty-nine in forty would never see any of our planets transit the Sun.

After accounting for how difficult or easy it is to spot exoplanets depending on their size and orbit, it seems that planets occur less frequently the further out they orbit, and that that change is—statistically speaking—quite smooth. For example, planets around a Sun-like star with orbital periods between 10 and 20 Earth days are as frequent as planets with orbital periods between 150 and 300 Earth days. Because giant planets are easier to find still further out we know that for them this trend extends to orbital periods up to at least 2,000 days. What this means is that, on average, the further we go from a star, the larger the separation between the planets. We can use the third law of planetary motions as found by Johannes Kepler to put that in numbers: if we look at any exoplanet, the next exoplanet further out from its star orbits statistically 1.6 times more distant. This is roughly true for our Solar System also, although the spacing in

that may be a bit larger than this average trend, at least for the outermost planets.

Not many exoplanets are found with orbits below about 10 days, but they do exist, mostly with sizes that are at most twice as large as Earth. These planets, very close to the tremendous heat of their stars, are sometimes referred to as "lava worlds". Planets really cannot exist much closer than these lava worlds are to their stars: if they were substantially closer the gravitational tidal force of their star would pull them apart.

Beyond orbital periods of a few years, or some thousand days, exoplanet finds are rare, which is at least in part because most methods used cannot uncover any such distant exoplanets. We know from our own Solar System that there are planets out to orbital periods of that of Neptune at some 165 Earth years, or some 60,000 Earth days, so there is much to learn yet about exoplanets far from their host stars.

Detecting exoplanets

How are exoplanets found? The first exoplanets were detected by using a technique that relies on the Doppler effect. We are all familiar with the Doppler effect in our daily lives: whenever a sound coming from an object moving toward us is compared to the sound from that object moving away after passing by us, we hear that the pitch has gone down. We hear that when trains go by, when police sirens pass us, and when an airplane goes by overhead. The basis of the effect is this. Sound is a series of compressions and rarefactions in the air that travel away from the source. When a source is moving toward us, the next compression is emitted a little closer to us than the preceding one, so has less distance to travel to reach us, and therefore reaches us a little earlier than it would have if the source had not moved toward us. The consequence is that when something making a sound moves toward us, the sound waves reach us in a somewhat compressed form, which we hear as a higher pitch. Similarly, when it moves

away from us, the waves are somewhat stretched, and the pitch is correspondingly lower.

Any type of wave is subject to this change in pitch or frequency when movement is involved, including light. When a light source moves toward us, the result is a higher pitch of the light, which we experience as a little more blue. When a light source moves away, we see this because the light is a little more red. So, when a light source alternates between moving away from us and moving toward us we see its light shift toward the red, then toward the blue. A light source in an orbit that alternates between moving away from us for half of the orbit and moving toward us for the next half of the orbit thus periodically shifts from red to blue to red to blue and so on.

How much the tone or color shifts depends on how fast the source is moving relative to the natural speed of propagation of sound or light. In Earth's atmosphere, sound moves at about 1,200 kilometers per hour, so a car going at, say, 90 kilometers per hour (55 mph) on the freeway will shift the pitch by somewhat more than a full note as it goes by. Light moves at close to 300,000 kilometers per second, so that same car would shift in color by only one part in almost six million. That is far too little to notice by eye, so this Doppler effect on colors is not seen in our everyday lives. But astronomers do use this effect on colors to measure velocities all around the Universe using sensitive instruments called spectrographs. The police do, too: the Doppler effect is exactly what is used by police to catch speeding vehicles using radar signals, which are among the many forms of light that are invisible to the human eye.

If we listened to the sound made by someone hammering or to the repeating banging sound of a pile driver (with which you may be familiar if you live on soft, wet soils where new construction is going on), similar changes would occur. When moving toward the hammering you would hear it at one rate, but when passing the source, it would appear that the hammering slowed down

because it would take a little longer for the sound from each strike to reach you. Similarly, with pulsed light the pulses would appear to be closer together in time when source and observer are moving toward each other and further apart in time when moving away from one another.

The latter effect of pulses slowing down and speeding up in repeated succession was, surprisingly, the way in which the first planets outside our own system were discovered, in a place where they were not expected to exist at all. In 1992, Aleksander Wolszczan and Dale Frail reported on what they found in observations made with the 305-meter (1,000-foot) dish antenna of the Arecibo telescope in Puerto Rico. That telescope, famous for its role in movies such as *GoldenEye* and *Contact*, is essentially a metal bowl built into a sinkhole with a large secondary reflector suspended over the middle of the primary dish that dangles on tight metal wires connected to three high masts placed just outside the bowl. The two scientists were using the telescope during a time that it was being repaired to take care of metal fatigue in some of its parts, and it could not be pointed at most targets in the sky. During the repairs, instead of having visiting scientists use the telescope, the resident astronomers were given access for over a month, provided that they came up with a good idea for how to use the largely immobilized telescope. And they did: they proposed hunting for pulsars around the Galaxy. Pulsars are rapidly spinning stellar remnants (called neutron stars) that emit a radio beam that sweeps by the Earth every time the star spins around. In their search, they found a very fast-spinning pulsar, which was only the fifth pulsar known at the time to spin around hundreds of times per second. But what was really exciting was that follow-up observations of that interesting target revealed evidence of planets orbiting it. When Wolszczan realized this, he said "that was the moment when my knees started to wobble a little bit... I began to suspect I was getting into something really, really exciting and quite unusual."

A most unlikely place

A pulsar was just about the least likely place in the Universe to find planets because they would be orbiting a very peculiar object with a pretty violent history. In fact, a neutron star is not regarded as a proper star. A star is defined as a body in whose center nuclear fusion is sustained for a very long time. This no longer occurs in a neutron star: here, the nuclear furnace has run out of fuel and as a result what was once a star has run out of a heat source to keep its internal pressure up. With insufficient internal pressure, gravity would pull the star in onto itself until another force took over, this time a quantum-mechanical force that keeps particles from simply collapsing onto each other entirely. This is a force that acts deep inside all atoms, generally hidden from our macroscopic world. But it is seen to act also on a large scale under some extreme conditions that we are not at all familiar with in our daily lives. One of those is inside neutron stars where gravitational and quantum-mechanical forces come to balance in a ridiculously compressed material: densities near the core of a neutron star are over 100,000,000,000,000 times higher than that of water—that number really has fourteen zeros! The star that condensed into this neutron star transformed from something over a million times larger in volume than Earth to being almost 100 million times smaller than the Earth. Often this happens in conjunction with a massive stellar explosion at the end of star's life, in what we call a supernova.

But this particular neutron star appears to have yet another history. Astronomers have pieced together the evidence that it was once a binary system with two stars not unlike our Sun orbiting each other. Each was not massive enough to end in a supernova explosion, but instead petered out after a phase as giant stars, thereafter gradually contracting into what is called a white dwarf: a stellar remnant with the mass of a star compressed into something the size of Earth. Then these two white dwarfs spiraled

in toward each other and merged, then compressing yet further into a neutron star of only some 20 kilometers (12 miles) across. And because of this merger and compression, the once leisurely rotational motion sped up to an astonishingly dizzying one with a full turn every few thousandths of a second. That formed a millisecond pulsar, spinning lighthouse-like beams around so fast that the human eye would not even be able to see that and would only show us a blurry sphere.

The Arecibo observations of this pulsar showed that the timing between the pulses increased and decreased in a somewhat irregular pattern. It turned out that the irregularity was caused by several planets pulling on the neutron star. What happens is this: Although we tend to simplify things by saying that the Moon orbits the Earth, or that the planets orbit the Sun, this is not quite true. It is a very good approximation when one body is very much heavier than the other, but there is a difference that leads to a measurable effect. In fact, the Moon and the Earth orbit each other, and the planets and the Sun orbit . . . well, what do they orbit? This goes back to Newton's laws. Not just that on gravity, but another: every action leads to an equal but opposite reaction. As a result, the Earth and the Moon, for example, orbit a point in space that is called their mutual "center of mass" or barycenter. For the Earth–Moon system, for example, that center lies inside the Earth, but some 70 percent of the Earth's radius away from Earth's center in the direction of the Moon. Seen from afar, the Earth wobbles around that barycenter once per lunar orbit and precise measurements can reveal that.

The pull of the exoplanets on their pulsar caused the strictly regular blips emitted by the lighthouse star to show up with a change in their spacing as they were received at Arecibo. At first, two planets were thus inferred, but additional measurements revealed a third. The two first found have masses a few times that of Earth, while the third one appears only twice as heavy as our Moon.

Even imaginative astronomers have a hard time envisioning how these planets would survive the end phases of the lives of stars that led to the formation of the pulsar. Not the violent death throes of just one, but in this particular case of two stars that then merged in what is also not a gentle process. Consequently, it is thought that these planets did not form when the stars were born, but rather formed from the material that was thrown off them as they died.

Astronomers have found other ways to also use the Doppler effect on ordinary stars without such regular, frequent radio pulses to find planetary systems. This is difficult for planets with a mass like Earth's: the pull of such planets on a star like the Sun is not very big in relative terms to its mass because the Earth weighs only three millionths as much as the Sun. But for heavier planets, the signal is proportionally stronger.

For the Solar System, it turns out that the barycenter about which the Sun orbits lies outside the Sun some of the time primarily because of Jupiter's strong pull. If we looked at the Sun–Jupiter pair in isolation, their joint barycenter lies about 7 percent of the solar radius outside the Sun. But not only Jupiter pulls on the Sun: all other planets do, too. Consequently, the Sun describes a complex spiral reflecting the tug of all planets. Combined, the planets pull the Sun around at a typical speed of up to about five meters (or yards) per second, which matches the pace of a moderate bicyclist. That may sound like a lot for something as heavy as the Sun, but it is nonetheless really difficult to measure: the associated Doppler effect changes the Sun's colors by only one part in 30 million. This method that relies on the Doppler effect is sometimes also called the "wobble technique" because that is what is actually looked for.

A most unlikely place, again

The heavier the dominant planet is, and the closer to the star it orbits, the larger the velocity swing is and the easier the detection.

Not surprisingly, therefore, the first detection of a planet around an ordinary star using the Doppler effect on stellar colors revealed a heavy planet in a tight orbit. Two Swiss astronomers, Michel Mayor and Didier Queloz, were monitoring 142 Sun-like stars with an instrument that could detect Doppler signals about four times larger than Jupiter's fingerprint on the Sun's movement; they measured the color spectra of these stars over and over, for 18 months. Then, in 1995, they were able to report that they saw a repeating Doppler change in the color spectrum of the star 51 Pegasi (in the constellation Pegasus, named after the flying horse in Greek mythology). This color spectrum shifted by an equivalent velocity of 70 meters per second, almost twenty times more than Jupiter's pull on the Sun. It turned out that this planet, which is estimated to be at least 150 Earth masses, is in what was then a most unexpected orbit: nineteen times closer to its star than Earth is to the Sun. Being so close to a star means that the planet is subjected to a strong gravitational pull. In order to avoid falling into the star requires that that pull is countered by an equal but opposite centrifugal force associated with a fast orbital motion around the star, and the consequence of orbiting close to a star therefore is a short orbital period: 51 Pegasi b is so close to its star that its orbital year is merely 4.2 Earth days long.

Mayor and Queloz were excited by this, but they reigned in their enthusiasm to make absolutely sure. After all, careers can be made by a true discovery but they can also be derailed by the claim of an erroneous one that needs retracting. Michel Mayor had realized that very well and decided to wait to announce the discovery of the first planet orbiting another star until more observations could be made. In an interview with *Scientific American* he recalls: "We fed our data into the computer and saw we had something going around [the star] 51 Peg with a period of 4.2 days, meaning it had to be in a very close orbit. This was a surprise, because at the time the idea was that giant planets had to be more than five [times the distance from Earth to the Sun] from their star. That was in fall of [19]94." But despite their excitement,

they waited for another year before announcing their find: "It was so unusual, so unexpected, that we decided to wait for the next season of visibility for 51 Peg. We wanted to be sure the amplitude [or strength] of the variability was the same, the phase was correct [so that the timings lined up properly], and the period was the same [proving that it really was a planet in a stable orbit]. We didn't have telescope time again until July 1995. We saw that all the parameters matched up—that was the time when we opened a bottle of champagne." And then they had to convince their colleagues: after presenting the findings at a scientific conference in Florence in October of 1995 he noted that "Some people at the meeting were really intrigued: 'Now we have to look for the reason why we have such a short-period planet.' Other colleagues were looking for arguments: 'It's not a real object, you don't have enough precision.' But we were 100 percent sure of our measurements."

A heavy planet so close to its host star was unexpected. In fact, none of the then existing models for the formation of Jupiter-like planets could explain the formation of what became known as 51 Pegasi b. The find was so unexpected that its discoverers at first wondered if it was in fact a planet or a very small companion that would be more like a star. Such lightweight not-quite-stars are called "brown dwarfs" because they are small and so cool they would only glow in the deepest red and infrared colors. But although brown dwarfs do not have any substantial nuclear fusion in their interiors, and are therefore not true stars, they do emit their own light, which is mostly the after-glow of their formation but mixed for some time with light stemming from the fusion of deuterium. A planet, in contrast, generally only lights up because it reflects the light of the star that it orbits.

No matter whether 51 Pegasi b was a planet or not (later it became clear it really was one), its discovery demonstrated that such small non-stellar objects could be uncovered using the Doppler effect with existing telescopes. So, with the detection of

the first probable exoplanet around an ordinary star, the hunt was on. Just three months later the next two exoplanets were announced: 47 Ursae Majoris b and 70 Virginis b by Geoff Marcy and Paul Butler, based on measurements that they had started in 1987, 8.5 years before the discovery could be announced. In the first few years after that, some two dozen more exoplanets were found each year with this technique. That really changed the attitude of astronomers toward the planet hunters: Marcy recalls how colleagues responded when he started his search in the mid-1990s: "I might as well be looking for little green men, or how the aliens built the pyramids in Egypt, or telekinesis. Even professional astronomers at that time associated planets around other stars with science fiction. It's very hard to imagine that now [with so many colleagues joining the hunt]. It's hard to put yourself in a mindset in which planets were considered the lunatic fringe. But that's what it was." That mindset would change rapidly, however, particularly when the Kepler satellite came into play.

Observing transits

A completely different way to hunt for exoplanets is to look for signatures of transits in the brightness of stars: should a planet's orbit cause it to move in front of its star as seen from the perspective of Earth, the apparent brightness of that star will be somewhat reduced for as long as it takes the planet to move across the face of the star (an example of a transit of Mercury in front of the Sun is shown in Figure 4.5). This could be brief if the projected orbit would take it to just touch one of the poles of the star. It would take longer in the most favorable orientation should the planet cross in front of the stellar equator. For example, for a distant observer looking at Earth crossing in front of the Sun's equator, the dip in the Sun's brightness would last for some 13 hours (which is a little more than one part in 700 of the duration of its orbit). For Jupiter, further from the Sun and moving

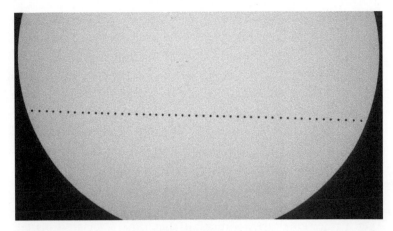

Figure 4.5 A series of images of the Sun was spliced together to make this time-lapse summary of the transit of Mercury that occurred on May 9, 2016. The image was composited from vertical cuts around forty-eight subsequent positions of Mercury that were taken by the HMI instrument onboard NASA's Solar Dynamics Observatory. The transit lasted a little over seven hours in total. A few sunspot groups are visible on the solar surface well to the north of the trajectory of Mercury.

more slowly, the equatorial transit would take 30 hours (which is one part in 3,500 of the duration of its orbit). So, it's not like the stars will be flickering with planetary transits, but rather that an occasional, slight, sustained dimming needs to be looked for.

Measuring these fairly long-lasting transits is very difficult with a ground-based telescope. First of all, the signal is very small. Compared to the stars that they orbit, planets are tiny. Just to illustrate this: in its vast volume, the Sun could hold a million Earths. And even the largest planet in our Solar System is small compared to its host star: Jupiter would fit inside the Sun a thousand times. Because of the way geometry works, the contrasts in transit dimmings are a little more in favor of detectability than the contrasts in volume because dimmings are set by the ratio of surface areas. This ratio tells us what fraction of the starlight is blocked during a planetary transit. It would be one part in about 12,000 for an Earth-sized planet. Jupiter would

take 1 percent of the Sun's light away for a distant observer. The latter is easily detectable with modern-day telescopes and methods.

The ground-based transit method indeed succeeded first for a planet about as large as Jupiter for an exoplanet already known to exist from the Doppler method: HD 209458 b, fourteen times larger than Earth. So this was not so much a discovery as a demonstration of two things: the Doppler signals really did pick up exoplanets, and the dip in the starlight because of an exoplanet transit was in principle measurable. Two teams succeeded independently: one led by David Charbonneau and one by Greg Henry. This is what happened. The Doppler measurements had provided the orbit information so well that if the exoplanet's orbit was such that Earth, exoplanet, and star were lined up, then the timing of this event was already known, at least to Charbonneau's team. So that is what they went looking for, finding the dip in brightness in September of 1999. Henry's team had been monitoring hundreds of stars for three years already. The star HD 209458 was added to their list in May of 1999, and they saw a transit signature in November of that year. The results of the two teams were published back to back in an *Astrophysical Journal* issue of January 2000.

Although this showed that the transit method worked in principle, the teams had done so for a large exoplanet, and one for which they even knew when to expect the transit. For an as yet undiscovered Earth-sized exoplanet, however, the small dip in brightness is very hard to uncover reliably when working with a ground-based telescope: changes in the Earth's atmosphere, including the coming and going of thin clouds, easily exceed the looked-for transit signal, and keeping instruments stable to that level for days on end is hard.

But it is not merely measuring the slight intensity changes that makes this a challenge. If you can only observe for some 8 to 12 hours to see stars in the night sky before the Sun rises again, you would likely see only part of the transit, or—for closer-in

planets—might miss the shorter event altogether. Plus, in order to have a chance of seeing it, you would have to observe every night because the transit happens only once per planetary orbit: once a year for a planet in an Earth-like orbit, and once every 12 years for an orbit like Jupiter's. And then only if the plane of the planetary orbit is properly aligned with the perspective of the telescope, which puts the odds overwhelmingly on the side of just not detecting any transit. Should you be so fortunate as to detect one, you would have to look for at least one repeat to be reasonably convinced that it was not a cloud or instrumental defect, and would really need a third transit to be sure and to obtain an accurate measurement of the orbital period which would give you a clue as to the planet's orbital properties. Now, if we had known, as we do now, that very large planets existed very close to their stars, then astronomers might have been given access to an adequate number of telescopes to actually find them. But that is the clarity of hindsight: before 1995, there was no reason to expect giant planets anywhere closer than where we find them in our own Solar System, and their orbital periods of 12 years and up made them look to be essentially undetectable by transit measurements.

Given these observational challenges, it is no wonder that few such transits were detected until both the odds and the instrumentation were changed: observatories were designed for surveys to measure stellar brightness changes of many stars at the same time, some of which were placed in space where you need not worry about clouds and day–night cycles. Ground-based survey instruments started reporting exoplanet discoveries by the transit method in 2002.

The first instrument in space was the CoRoT satellite, led by the French space agency CNES, launched at the end of 2006. It monitored thousands of stars at a time, for five-month intervals, after each of which the spacecraft had to be reoriented in order to maintain sufficient power on its solar arrays. CoRoT scientists found their first exoplanet in May of 2007, reaching a total of

thirty-two by the mission's end in 2012 that were confirmed by subsequent ground-based measurements to determine the exoplanet masses.

After CoRoT's launch came NASA's Kepler satellite, built to observe 150,000 stars at the same time and for several years without interruption, thus considerably increasing the probability of detecting repeating transits. The Kepler mission to find exoplanets was selected after five designs were submitted in proposal form to NASA over a span of eight years. Eventually, the Kepler satellite was launched in 2009. By 2017, its archive listed over 2,400 confirmed planets, meaning they were seen in at least three successive transits to be certain it was an orbiting object that caused the observed transits. Almost 300 of these have had follow-up mass estimations thus far.

Overall, the space-based missions and the ground-based observations have resulted in over 3,500 exoplanets, of which more than one third have a mass estimation. In addition, the Kepler exoplanet archive contains over 5,000 more possible planet detections that need further confirmation to ensure they are not artifacts of some other process. By the way, once you know the orbit duration of a transiting planet around a known star, the duration of the darkening at the beginning of the transit and that of the brightening at the end of the transit gives direct information on the size of the planet, complementing that obtained from the amount of starlight it takes away during the transit.

The Kepler satellite stared at a single section of the sky for 3.5 years before a second of its reaction wheels failed in 2013 so that pointing to that area of the sky could not be maintained (that did not end the mission, though, as other modes of operation on other areas of the sky were devised). In order to see at least three transits for a given planet in 3.5 years, their orbits need to be at most just over an Earth year long. Hence, planets like our outer giant planets with much longer orbits could not be detected, or at least not as "confirmed planets." Moreover, in order to see transits in the first place, the orbits had to intersect the line of

sight connecting Kepler and their host star. This means that most planetary systems could not be found by the transit method. Several thousand were nonetheless spotted, and subsequent analysis of what fraction of planetary systems is in principle detectable in Kepler observations led astronomers to conclude that almost every star in the sky has a planetary system.

There it is, the quiet revolution, the uncontested transformation in the frame of reference for our place in the Universe, the tipping point in planetary sciences, hiding in a deceptively simple sentence: *Our knowledge shifted from one planetary system to many billions within the Galaxy.* In fact, there appear to be so many planetary systems that their formation is now considered to be inseparable from the formation history of their central stars.

Attributes of exoplanets and exoplanetary systems

Based on the thousands of confirmed and candidate planets we can now begin statistical analyses of exoplanetary systems and of the exoplanets themselves. For example, what was the surprise for 51 Pegasi b turned out to be frequently found in exoplanet systems: multitudes of heavy giant planets orbit close to their host stars. This class of planets quickly became known as "hot Jupiters."

Interestingly, hot Jupiters are not the most common type of planet, despite the fact that many are detected. Their frequency in observations is elevated artificially in both Doppler and transit methods: because they are heavy and large, close-in planets, their large Doppler and transit signals are more easily spotted than the much smaller signals of more distant or smaller planets. Once astronomers looked at the detectability in detail to estimate the biases introduced by ease of detection, the apparent dominance of the hot Jupiters turned out not to be real: the numbers of close-in hot Jupiters and of further-out cooler large planets in reality suggest that there is simply a wide distribution of sizes of

the orbits for these giant planets. Estimates are that maybe one in a hundred of ordinary, mature stars is orbited by a hot Jupiter while very roughly one in six have a "cold Jupiter" such as in our Solar System, with a less-populated domain of "warm Jupiters" in between. If a star has a hot Jupiter then there appear to be fewer smaller planets in its exoplanet retinue, at least with orbital periods between 10 and 100 Earth days.

How heavy can these exo-Jupiters be? It looks like they rarely get to be much heavier than about 3,000 times the mass of Earth. Still-heavier companions to stars are relatively infrequent between 3,000 and 30,000 Earth masses, but then again more frequent for yet heavier companions. The mass range of the relatively rare companions corresponds to masses between 1 percent and 10 percent of the mass of the Sun; that is exactly the mass range of brown dwarfs. Apparently, brown-dwarf companions of stars are much less common than either Jupiter-sized companions or true stellar companions.

There seems to be a second range of less frequently present masses, namely right in the middle of the range of planet masses: the exoplanet population seems to naturally divide into two classes. One class contains planets smaller than four times the size of the Earth at weights of less than 30 Earth masses or 1/10 of a Jupiter mass; the largest of these may not be unlike Neptune. The second class contains exoplanets that are substantially larger and heavier. It seems that typically every second star has at least one of the smaller type of planets in orbits of a year or less. Such planets, often found in compact, multi-planet systems like our own, are particularly interesting from the point of view of extraterrestrial life.

A remarkable contrast between stars that have a small stellar or small exoplanetary companion orbiting them and those with a large Jupiter-like planet was found in the chemical makeup of the central star: giant planets are much more prevalent around stars that, in addition to the hydrogen and helium that make up the bulk of a star, contain a relatively large amount of heavy

elements that dominate our daily lives on Earth. No such contrast is found for more Earth-like planets, nor for small stellar companions. Stars with either other stars as companions or with Earth-like planets can apparently form within any collapsing gas cloud, but there is something special that is conducive to forming heavy Jupiter-like planets: giant planets are seeded more readily when there are abundant heavy elements, although they contain mostly hydrogen and helium once fully formed. Moreover, heavy planets have a preference to be associated with heavier stars: stars of less than about half a solar mass have relatively fewer Jupiter-sized planets, although they have plenty of Neptune-sized planets (Neptune has a mass 1/18 of that of Jupiter). This dependence of the occurrence of giant planets on the heavy-element content and on the mass of their host stars is an insight into the formation of planetary systems that we get to in the next chapter.

Taking in the evidence for all exoplanetary systems together, then about 10 percent of Sun-like stars have a giant planet with a period shorter than a few years, and likely more than half of them have multiple smaller planets with periods shorter than an Earth year. What they have orbiting around them further out remains to be seen, but there is no reason to assume that the close-in ones are all there is—just look at our own Solar System. These more distant exoplanets are just harder to find.

What do we know about the orbits of exoplanets apart from their periods and—with knowledge of the mass of the central star and Newton's law of gravity—the distances to their stars? For those exoplanets found with the Doppler-shift method, astronomers can also determine the eccentricity of the orbits, which is the deviation from circularity. Some early results are emerging. For example, there is a trend for more circular orbits in compact multiplanet systems, which are often those with smaller planets. There is also a tendency for orbits to be more circular when there are more planets. This might tell us something about the opposite of systems with mostly circular orbits: if multiple planets tug on each other irregularly in strongly oval orbits,

this may cause instabilities that could either lead to collisional mergers or to the downright ejection of planets out of their planetary system by close encounters with others. It may well be that there is a natural selection in planetary systems that favors systems with nearly circular orbits.

Another emerging trend is that stars that have giant planets tend to have these planets in more eccentric and wider orbits. Giant planets have orbital eccentricities ranging from pretty much 0 (these are essentially circular orbits) up to 0.9. An eccentricity of 0.9 means that at closest approach to their central star they are ten times closer than at the furthest point in the orbit. Consequently, the amount of starlight warming such planets changes by a factor of one hundred from nearest approach (at periastron) to most distant point (or apastron). To say that that would not be good for stable weather patterns would be quite an understatement. The average eccentricity, as now known, for giant exoplanets is 0.2, several times that of the giant planets in our Solar System, so that their stellar irradiation changes through their typical orbit by more than a factor of two.

Multiple suns

The properties of exoplanet systems discussed so far have been for planetary systems of single stars only. Many of the points of light in the sky that we call stars, however, are in fact close pairs of stars. In such double stars, also known as binaries, the stars have a companion star that they are gravitationally tied to so that they orbit each other. That planets could form and survive in stable orbits while being tugged around by two stars orbiting each other was difficult to imagine. This is the reason why planets were not really expected in the vicinity of double stars. But they exist there, too, as we saw in the example of Kepler-16b depicted in Figure 4.2.

How many stars are binaries depends on the type of star. For relatively heavy stars like our Sun, it appears that somewhat over half (around 57 percent) are double stars. For lightweights

with masses under half that of our Sun, however, maybe only a quarter are members of binaries. There are so many more of these low-mass stars, however, that actually for the whole population of all stars approximately two out of three are single. Yet, the multitude of lightweight stars is also the multitude of faint red ones, so that those that we readily see in the sky are typically the heavier ones: looking at the night sky, slightly more than one in two stars in the sky is in fact a double star that is unresolved in our view. Stars that appear as single dots in the sky are often not just unresolved binary stars but may be triple, quadruple, or quintuple. Particularly those systems with the tightest orbits and shortest orbital periods often turn out to be at least triple systems. How many stars are in fact sets of three or more remains to be determined.

Exoplanets have been found near double stars in fundamentally different orbits. Some orbit one star with the other star far outside of its immediate neighborhood; these orbits are referred as "S type," for "satellite type" because such planets are akin to a natural satellite to these stars. Other exoplanets orbit both stars in the binary, well away from the pair of them, in what are called "P type" orbits, for "planetary type"—not a particularly obvious distinction from S type orbits, but that is what they are called; Kepler-16 (AB)b that was mentioned earlier is an example of this type.

The presence of a fairly distant second star may affect the existence of exoplanets that orbit one star of the binary, but not in a major way: for binaries with separations between the two stars that range from 10 to 1,000 Sun–Earth distances, close-in planets smaller than Neptune appear to be only somewhat less likely to exist than in isolated single stars—something like a 50 percent reduction—while the difference for giant planets appears to be smaller. Whether the same similarity exists between single stars and close binaries with planets orbiting the binary (in P-type orbits) remains uncertain: too few such systems have been found so far to know.

Patterns in orbits of exosystems

We can look at one more property that seems an outstanding attribute of our own Solar System, namely that just about everything except the least massive bodies in the Solar System orbits or spins in the same direction as the Sun rotates. Establishing whether that is true of other planetary systems is a tall order in terms of the observations needed and the precision of the measurements required. We cannot do that for the spin of planets at present, but in some cases we can for the orbital direction of the exoplanets.

One method that can be used to estimate both the tilt of the orbit and the direction of orbital motion relative to the star's rotation works best for large transiting exoplanets. This is based once more on the Doppler effect. During the transit of a planet in front of a star it first blocks light from one side of the star, then from the other side, illustrated in Figure 4.6. One side of the star is rotating toward us, and light from it is slightly shifted toward the blue. The other side of the star spins away from us, and light from there is shifted by the same amount toward the red. So, if a planet first blocks a little of the blue-shifted light and later the red-shifted light we know that it orbits in the same direction as the star is rotating, so the motion is called prograde. If the opposite happens, the orbital motion is opposite, so retrograde. If the orbit is across the rotational equator of the star (orbit 1 in Figure 4.6) the planet will take light from the most blue-shifted to the most red-shifted part of the star's light during its transit, but if the orbit crosses near a rotational pole of the star (orbit 2) mostly unshifted light will be blocked. A strongly tilted orbit (3) may take the same color range as a near-polar orbit (2), but its crossing lasts longer. A planet in a retrograde orbit, no matter how tilted it is or where it crosses the star, will be taking light away first from the more red-shifted part of the star and later from less red-shifted (orbit 4) or even blue-shifted (if it is crossing more to the left across the star than shown here). If it transits from pole to pole, so if the orbital inclination

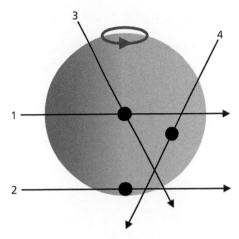

Figure 4.6 Illustration of the Doppler effect during an exoplanet transit of a rotating star. The light from the side of the star that is rotating toward us is shifted toward the blue while that moving away from us is shifted toward the red; the disk illustrates how these color changes are seen across the face of the star. A transiting planet blocks some of the blue-shifted or red-shifted light. How much it blocks depends on the size of the planet. How far toward the blue or red it takes light away depends on its orbit (See Plate 6 for a color version).

is 90 degrees out of the star's equatorial plane, it will take light from the star, but as the star's central meridian is just between stellar parts that are rotating toward us and away from us, no color change is seen. Very careful and precise measurements can reveal any angle in between.

One study using this method looked at seventy transiting planets in different planetary systems. These astronomers found that the orbital tilt can be quite large for close-in giant planets, the hot Jupiters: almost half of these planets show inclinations larger than 25 degrees, with one in six being in essentially pole-to-pole orbits with a nearly 90-degree inclination, and another one in ten orbiting against their star's rotation direction. This is quite a contrast with our Solar System for which no planet orbits with a tilt larger than 7.2 degrees and no planet orbits in a retrograde

direction. There is clearly something to be learned from how these exoplanets came to move in orbits that we do not see in the Solar System.

Characterizing planetary systems in general

Now that we have a view of the general properties of a large number of planetary systems beyond the one that we live in, we should have a look at what matches and what may not in the list of "forensic evidence" from Chapter 3 that summarized ten properties that characterize the Solar System and its planets. But until a sufficient number of exoplanets was assembled, we were not able to know which of these properties would be valid in general and which would apply only to the Solar System, or why that would be the case. But now, with all the new information on exoplanets, we can sort these properties into three categories to see what we are learning: the Solar System properties divide into those that match those of other planetary systems, those that need a modification to include other systems, and those that we cannot yet put into either category based on what we now know.

In the category of properties in which we know our own Solar System to be like most others we only find one:

1. The combined mass in all planets is small compared to the mass of the central star.

Several properties fall into the second category of those properties of our Solar System that are not entirely common properties of exoplanet systems. Something needs to be modified in order for the reformulated property to hold for most planetary systems. Let me do that here, flagging them with an "r" to denote that they were revised:

2a-r. Most planets orbit in roughly the same direction as their star rotates, but heavy, close-in planets (hot Jupiters) frequently have a strongly tilted orbit, with some 10 percent

tilted so far over that they orbit their star in a direction
opposite to the star's rotation.

3-r. Relatively close to a star, planets can be found that are
either composed mostly of rock and metal, or, if heavy
and big, may be mostly composed of volatiles, much of
these in gaseous form. We do not know enough (yet)
about the chemical makeup of exoplanets that are at
least as distant from their star as Jupiter orbits from
the Sun.

4-r. Relatively close to a star, planets can either be in the
mass range from Earth to about a dozen times heavier
if in a multi-planet system, or appear to be lone planets if there is a planet that weighs hundreds of Earth
masses.

5-r. Water (liquid and frozen) is very rare near the Sun and
very common beyond about three Sun–Earth distances.
In exoplanet systems with heavy planets close to their
star, however, water may be abundant in at least those
planets near their stars.

Point (2a) is only part of what was (2) in Chapter 3 because
we do not yet know enough about large moons in exoplanet
systems.

Having only knowledge of numbers, sizes, and orbits of exoplanets as discussed thus far does not suffice to assess whether
the remaining five attributes of the Solar System—relating to the
populations of exo-asteroids, exo-dwarf planets, and exo-comets,
and to the internal thermal state of these and exoplanets—are
valid in general or not.

So there we are: of ten attributes that seemed to stand out as
obvious properties of our Solar System, only one can be demonstrated to hold for planetary systems in general based on exoplanet findings as discussed so far; four attributes needed substantial revision given the knowledge we now have as summarized up
to here, while five await further testing.

Surprises, composure, the sales pitch, and enthusiasm

Among the biggest surprises in exoplanet science was that there are so many giant planets close to their host stars. This category of "hot Jupiters" was not expected, because from the way our own Solar System looked, planetary scientists had inferred that large planets had to form far from the star. That would put their embryos in the cold domain of the Solar System. There, volatile chemicals could readily condense into solids that could stick together. There also, gases would remain longer after the formation of the central star, giving the formed solid clumps time to pull this gas into their gravitational hold. By that scenario, giant planets were thought to form beyond what is known as the "snow line," which is the region where water can exist as a solid without being evaporated by sunlight, or by starlight in other planetary systems. The presence of water in solid form, simply because there is so much of it, adds considerably to the other solids that can coagulate. Then these, with their growing mass, can more efficiently pull in other solids and gases as their gravitational pull increases. Therefore, they can ultimately grab matter far more rapidly than burgeoning planets can do closer to the central star where material exists more as a gas that is hard to hold onto for a beginning planet.

Seeing so many hot Jupiters made scientists realize that their grasp of the Universe was in error: something was playing out that they had not anticipated, largely because the world as they knew it did not reveal everything that could happen—or rather because they had only the Solar System as an example of a planetary system. Now they were forced to rethink things. The most important realization that came out of that was that giant planets may well form by the above scenario, but then they do not simply stay where they form. It seems that they can spiral in toward their star, or they can move outward, or in fact do both given sufficient time, sometimes dragging other planets along or

tossing them about. We think we now know that many if not most hot Jupiters formed much further out from their stars to subsequently spiral into the gravitational pull of their star. But we may still be wrong in that, or not have all the details right. So, astronomers do what all scientists do: they take in the evidence, revise their concepts and models, prove many wrong, and proceed only with the hypotheses, approximations, and theories that are left standing.

In all these findings there are surprises and confirmations, all subject to uncertainties and unknowns. Surprises are how we learn new things, while confirmations establish that what once came as a surprise fits in with the established body of knowledge. In reviewing the scientific literature, it is hard to see the excitement that these surprises have brought: scientific publications are written in such a matter-of-fact style that you have a very hard time gauging whether the authors were excited about what they found or how much of an impact a discovery will have.

One place to quickly look through many publications for the surprises expressed by the authors is this: each such publication is opened with an abstract in which the authors summarize what they did and what they found. This abstract is as much a summary as an advertisement, much like previews are for movies: they serve to attract an audience that needs to select what to view because without doing so scientists would be flooded by too much material to process. Computers can easily search these abstracts, making it easier to find the work that one is looking for. And do we see "surprise" in these "teasers"? The word "surprise" (or any synonym of it, or verb forms based on it) occurs in about one in ninety abstracts of scientific publications about exoplanets. Given how many new things were discovered, astrophysicists must have been surprised more than they admit, but scientific publications do not advertise the excitement of their authors much. Instead, and in contrast to the colorful descriptions in the early scientific literature, they make for pretty dry reading.

Then again, they are supposed to convey new findings concisely and accurately: they are meant to clearly document what was found, and to compare that to what other things have been found before, with little or no emotion showing through. This occurs in all the sciences, not just in astronomy. I have always been struck by the 1953 paper in which James Watson and Francis Crick described that they had pieced together a model for the double-spiral structure of DNA, the chemical code by which all properties are inherited. That was obviously really important and one of the great discoveries of the twentieth century, yet it ends on this dry understatement: "It has not escaped our notice that the specific pairing we have postulated immediately suggests a possible copying mechanism for the genetic material." It had not escaped their notice. Really.

In contrast to the unemotional, cautious, matter-of-fact style of writing in the professional literature, excitement is often over-expressed in the press releases that come from the organizations funding the research. These releases, too, are teasers: by making strong, if not rather dramatic, statements the agencies hope that the media will pick them up for broad publication. And journalists, too, try to attract readership by emphasizing the new or unique aspects. So, we read about the discoveries of the "firsts" or of the extremes: the largest planet, the planet nearest to a star, the extreme water world or lava planet, etc. For example, the finding of CoRot-7b saw headlines like these: "Strange Lava World Is Shriveled Remains of Former Self", " 'Super Earth' May Really Be New Planet Type: Super-Io", "Planet known as HELL IN DEEP SPACE discovered by scientists", "Alien World a Volcanic Nightmare", "Hellish Exoplanet Rains Hot Pebbles, Has Lava Oceans." WASP-12b was characterized as "The mysterious light-eating, super-hot, pitch-black exoplanet" that is "black as fresh asphalt", while elsewhere it is noted that "the newly-discovered exoplanet WASP-12b is pitch-black, reflecting almost no light from its atmosphere. It's also immense, mind-blowingly hot, and being devoured by its parent star at this very instant."

Although life has not been detected anywhere in the Universe outside Earth, the system named TRAPPIST-1 was announced by lines such as these: "Life could jump between TRAPPIST-1's Earth-like planets in decades", "Life may seed from planet to planet in TRAPPIST-1 system, study finds", "TRAPPIST-1 System Ideal For Life Swapping", whereas others were more cautious: "Only One of TRAPPIST-1's Rocky Planets May Be Right for Life", "Is There Life in the TRAPPIST-1 System?" These discoveries and analyses are, of course, worth reporting on, but it is hard to gauge from them just how much any such finding excites the research community or how much it changes our broader understanding.

One way by which we can indirectly see how exciting and fast-moving a research topic is is by looking at how many people work on it or how many scientific publications and conferences it sparks. Counting scientific publications is easy nowadays, with computer-based archives (such as NASA's Astrophysics Data System, or ADS) doing most of the work. And from this we can see that exoplanet studies are exciting quite a few astrophysicists who are dedicating substantial parts of their careers to it. Searching scientific publications from 1995 shows none with the word exoplanet in the abstract. But then the field took off: five years later, over the year 2000, there are five publications, then 71 for 2005, 303 for 2010, and 453 for 2015.

If all this work were simply refining things we mostly knew already or making more accurate measurements of what was estimated before, it would not attract so many of my colleagues. Instead, we can see that exoplanet studies are really teaching us about how the Universe works, about our own place in the Galaxy, about the history and future of the Earth, and also about the possibilities of life elsewhere in the Universe.

Sometimes, of course, scientific publications do articulate the surprise the authors felt about all the things we have learned about exoplanets. This is seen more frequently in what are called review papers than in research papers: authors of review papers

look at dozens, sometimes hundreds, of publications and attempt to integrate what we have learned and identify what is yet to be learned. In review papers, authors are allowed, perhaps even expected, to introduce some of their opinions among the facts. In one such review paper, Joshua Winn and Daniel Fabrycky, having mentioned the late-eighteenth and early-nineteenth century work on the stability and formation of the Solar System by Pierre-Simon, marquis de Laplace, wondered what that scholar might think of our present-day progress: "With so many startling results, it is difficult to guess what would have impressed [Laplace] the most: the high eccentricities, the retrograde planets, the chaotic systems, the circumbinary systems; the ceaseless technological developments that have propelled the field; or the mere fact that we have learned so much about faraway planetary systems on the basis of only the minuscule changes in brightness and color of points of light?"

Sometimes interviews with magazines or newspapers reveal the sentiment of the scientists. Often with a discovery the first reaction is one of skepticism and critique. For example, Michel Mayor, one of the discoverers of the 51 Pegasi b, said: "At the start we were extremely suspicious and looking for different explanations. The first reaction was not to say, 'Oh! We have a planet!' At the start you say, 'Oh, something is wrong.' It's only after weeks or months that you start to be convinced." At other times, when things go well, they show joy. In 2010, William Borucki talked about the first batch of planet discoveries with the Kepler satellite that he guided to become reality: "We never thought we'd have this much this early, it's absolutely wonderful." And that proved to be just the beginning.

Step by step, we have come away from the idea that the Earth is at the center of the Universe and that planets and Sun orbit it. We moved away from the sentiment that the Solar System is the only such planetary system, and we are learning from all the other exoplanetary systems how our own took shape and evolved. Let's see what is being learned...

Notes

1 Exoplanets have a rich diversity of names: some are named after the
   constellation that their star is in as seen from Earth (like 51 Pegasi b
   in the constellation Pegasus), others after the observatory or satellite
   with which they were found (such as CoRoT-7b with the French
   CoRoT satellite or Kepler-186f with the US Kepler satellite or a name
   starting with OGLE from the Optical Gravitational Lensing Experiment), followed by a lower-case letter starting with 'b' for the first-
   found exoplanet and continuing by discovery date to however many
   planets are known ... except if they were come upon at the same time
   in which case the letters are assigned in order of distance from the star.
   If the constellation or observatory name is followed by a capital letter,
   generally just either "A" or "B," this refers to which star in a multiple-
   star system the planet orbits, although in some cases it orbits both
   (such as Kepler-16 (AB)b which means the planet orbits both stars).
   Or the name may start with letters followed by a catalog number;
   the most frequently seen all-sky catalogs in exoplanet science are the
   Henry Draper (HD numbers) catalog (named after an astronomer
   whose widow donated funding for the project to the Harvard College
   Observatory) and the Yale Bright Star Catalogue or its predecessor
   the Harvard Revised Photometry Catalog (HR numbers). Or it may
   be a type of target followed by coordinates, such as a name starting
   with "PSR" for pulsar, or some other characteristic, like "V" for
   variable (from the General Catalogue of Variable Stars) followed by
   a number and constellation. In short, star names can have any of a
   large number of root origins, and may or may not say something
   about the star or its location in the sky. It should not be surprising with so many catalogs around that stars often have multiple
   names. A super-catalog that links all these various names for stars
   and their planets is the SIMBAD database at http://simbad.u-strasbg.
   fr/simbad/. More on naming exoplanets can be found at the IAU's site:
   https://www.iau.org/public/themes/naming_exoplanets/.

5

The Birth of Stars and Planets

A little over a century ago, our understanding of the Universe was—from our present perspective—astonishingly different. Much was being discovered as the nineteenth century ended and the twentieth century began. Take 1905, for example. That was the year in which Albert Einstein (1879–1955) submitted his doctoral dissertation and published his theories about the equivalence of mass and energy, summarized in the famous equation $E = mc^2$; on special relativity, that shows how time and space deform depending on the observer; on the random movement of microscopic particles due to collisions with molecules that enabled the determination of the properties of the individual molecules; and on the photoelectric effect that revealed the ambiguous nature of light that refuses to decide to be either a wave or a particle. Yet, in 1905 no one knew about nuclear fusion and that this is what powers the Sun and the other stars. Its opposite, nuclear fission was just being discovered, but that this process should help keep the deep layers of the Earth liquid and make the continents drift, and keep life going as a consequence of ancient stellar explosions, was nowhere in people's minds. No one knew that magnetism was the cause of sunspots, that these dark areas on the solar surface were not windows into the Sun's solid interior, or that astrophysical magnetism plays a key role in the formation of stars and planetary systems. No satellite had ever left the Earth to view the Moon, the other planets, asteroids, or comets. People could still think that the Solar System was in the center of the Universe. No one even knew that stars were predominantly made of hydrogen mixed with helium. No planet

One of Ten Billion Earths. Karel Schrijver, Oxford University Press (2018). © Karel Schrijver.
DOI: 10.1093/oso/9780198799894.001.0001

outside the Solar System was known and no one knew what white dwarfs were or that these retired stars would be uncovered—a century after their discovery—as the final resting places of some unlucky exoplanets.

And yet, with advancing telescope technologies and with the help of new developments in fundamental physics, the two decades or so after 1905 would come to change our knowledge of the stars tremendously, just as the two decades since the discovery of the first exoplanet around a Sun-like star in 1995 changed our knowledge of planetary systems, including the Solar System. A few of the things discovered pertaining to the lives of stars are these: In 1908, George Ellery Hale (1868–1938) discovered magnetic fields on the Sun, which was the start of understanding the fundamental role of magnetism in the Universe. In 1912, Henrietta Leavitt (1868–1921) published her work on pulsating Cepheid stars that laid the groundwork for measurements of great distances to far beyond our Galaxy. In 1920, Arthur Eddington (1882–1944) suggested that nuclear fusion powers stars. In 1925, Cecilia Payne (1900–1979) demonstrated that the Sun was largely made of hydrogen and helium. In 1929, Edwin Hubble (1889–1953) demonstrated that fuzzy nebulae that had puzzled observers were, in fact, other galaxies like our own Milky Way seen from a tremendous distance, and two years later he showed the Universe to be expanding because these distant galaxies were receding from us and from each other. Around the same time, Georges Lemaître (1894–1966) proposed that the expanding Universe must have started from a compact beginning, leading to what became known as the Big Bang concept. But now I am getting ahead of myself.

The dawn of the beginning

In 1905, with only the beginnings of the tools and methods that we now have at our disposal, scientists were already shaping ideas about the origin of the Sun and the Solar System. They looked

back at earlier concepts and argued them to be only partially correct or fully wrong. But if some of what you think you observe is not, in fact, what is actually there but only in your mind, you can readily be led astray. That happened, for example, in that year, to Forest Ray Moulton (1872–1952), an American astronomer. He published an article in the *Astrophysical Journal*, a periodical that today remains a vibrant professional journal among the preeminent ones for all matters astrophysical. In the article, under the title "On the Evolution of the Solar System," Moulton argues that then existing hypotheses on his theme could not be correct: it was thought that the Solar System and the Sun contracted out of a gigantic cloud of gas (which is still what we think) and that somehow rings of gas were left behind, spun off the contracting cloud, out of which then planets formed (which sounds somewhat like what we still think but is fundamentally different in several ways; more on that in a moment).

After presenting several objections to the prevailing concepts of the time, Moulton then writes "Having given up the ring theory, the problem has been to find, if possible, something more satisfactory." And he proceeds to make a case for another concept, rooted in an idea from his colleague, a geologist by the name of Thomas Chamberlin (1843–1948). In Moulton's words: "It is supposed that our [solar] system has developed from a spiral nebula, perhaps something like those spiral nebulae which [James Edward] Keeler showed are many times more numerous than all other kinds together." That idea had already been floated before the turn of their century, but here the new idea came: he suggested that the spiral nebula out of which the Solar System would subsequently have formed originated because of the passage of another star close to the Sun. He hypothesized that out of the spiraling material that was pulled off both the Sun and off the passing star would later have condensed relatively small bodies that they called planetesimals and protoplanets, that in time collided and merged to form the planets. He argued that this would be consistent with the observations that all the planets orbit and

almost all spin in the same direction because the material pulled off the star during the hypothetical near miss would all move in the same direction. Because in this scenario the gas in the Sun and the matter in the planets would have a different origin, he proposed that it might also account for the fact that most of the rotational energy is contained in the planetary system while most of the mass of the original cloud is contained in the Sun. This was one phenomenon that was hard to comprehend at the time and remains not fully understood even now. But he was to be proven wrong in both of these hypotheses.

Moulton includes a photograph of one of these "spiral nebulae" in his manuscript, namely M51 Canum Venaticorum, shown as seen by the present-day Hubble Space Telescope in Figure 5.1. At the resolution that he had available, something like the inset on the lower left of the figure, it does indeed look very suggestively as if a collision were occurring, with matter being pulled away in a spiraling arc by something passing by. But Moulton had no idea of the true scale of what he was taking as an example of two stars engaged in a near miss. Nor did any of his colleagues at the time. M51 is not a relatively nearby pair of stars involved in a collision, but instead an incredibly distant pair of galaxies— M51a, the Whirlpool Galaxy, and the smaller M51b, which does not have a catchy nickname, each with many billions of stars— that are recovering from a collision that happened perhaps half a billion years ago. At the sharpness of present-day instruments, we can see that the spiral nebula contains a lot of structure with clusters of stars and clouds of gas.

Whenever a concept or hypothesis is shaped, researchers build upon knowledge from others, combined with their own insights. Parts of the outcome of that construction may be right, others wrong. Rarely is everything either fully correct or completely inconsistent with later findings. Science is a collective enterprise in which incorrect ideas are filtered out by series of tests against observations and experiments, dropping incorrect ideas to fall behind the advancing front of knowledge while carrying

Figure 5.1 The Whirlpool Galaxy (Messier 51a, NGC 5194) and, toward the lower right, Messier 51b (NGC 5195) observed by the Hubble Space Telescope. The inset in the lower left is a blurred version representing how telescopes would have seen these galaxies early in the twentieth century. M51b is thought to have passed through M51a half a billion years ago. As the density waves in the spiral arms move through the galaxy, they leave a trail with historical information: star formation is in an initial phase in the dark gas clouds on one side of the spiral arms, is further advanced in star-forming regions that show up in pink next to the dark clouds, and is essentially complete in the clusters of young stars in which the heaviest shine brightest in blue (See Plate 11 for a color version).

Source: NASA/ESA.

forward the ideas that fit. For example, Moulton's concept for the formation of the Solar System through a star–star collision proved to be incorrect, but the concept of it forming out of an initially whirling and contracting cloud—an idea going back at least to marquis de Laplace (1749–1827)—has held up and the concept of planets growing from grains into planetesimals and protoplanets, which are terms that Moulton and Chamberlin already used, is still with us today.

By no means do I intend to criticize Moulton in particular with this illustrative example of what scientists had in mind in 1905. He certainly was not alone in this interpretation. In fact, it was so entrenched in the thinking of the time that historian Agnes Clerke (1842–1907) wrote this in two editions of *The System of the Stars* from 1890 and 1905: "The question whether nebulae are external galaxies hardly any longer needs discussion.... No competent thinker with the whole of the available evidence before him can now, it is safe to say, maintain any single nebula to be a star system of coordinate rank with the Milky Way." Yet, M51 turned out to be pretty much just that. Going against such a deep-rooted idea would take more than opinions and speculations: it would require convincing observations obtained by powerful observatories that, for the first time in history, were from that time onward actually not being built on rooftops or in backyards, but on mountaintops in good climates for optimal clarity of viewing through Earth's atmosphere.

One such example is Lick Observatory, the first permanent mountaintop observatory. It was completed in 1887 on Mount Hamilton on the south side of the San Francisco Bay, well above much of the local weather. One observation on the advantage of that comes from Ludwig Boltzmann (1844–1906), an Austrian physicist famous for developing statistical mechanics which is of key importance to astrophysicists. He visited the observatory in 1905 and experienced the benefit of the high location on his way down to the city of San Jose located at the base of the mountain at the southern end of the San Francisco Bay Area.

That Bay Area often fills with clouds so low that the surrounding mountaintops stick out above them. Boltzmann wrote (translated by Walter Kutschera): "After having seen Mars, large and glaring through the big telescope in the evening almost like the disc of the moon, we returned to the valley. Here, we experienced a dramatic separation from the sky. We had the firmament above and the fog below stretching like the flat surface of the sea. With a jolt the carriage dived into the fog, the stars instantly disappeared; the light of the car lantern could penetrate ahead only a few steps."

Bold ideas

Science is about discovery, and sometimes that goes against the thinking of the time. In the words of Karl Popper, a philosopher, in his "Replies to My Critics" from 1974: "There is a reality behind the world as it appears to us, possibly a many-layered reality, of which the appearances are the outermost layers. What the great scientist does is boldly to guess, daringly to conjecture, what these inner realities are like." "...these are men of bold ideas, but highly critical of their own ideas; they try to find whether they are not perhaps wrong. They work with bold conjectures and severe attempts at refuting their own conjectures....Bold ideas are new, daring, hypotheses or conjectures. And severe attempts at refutations are severe critical discussions and severe empirical tests. When is a conjecture daring and when is it not daring, in the sense here proposed? Answer: it is daring if and only if it takes great risk of being false—if matters could be otherwise, and seem at the time to be otherwise." Popper speaks of individuals here. I would expand that to the scientific community: ideas of any individual can be made into conjectures about how things work by another, and these tested, refuted, or proven as the hypothesis predicted by yet another. Science progresses as a community activity, continually testing and refuting, advancing by eliminating the inconsistent.

What I also want to illustrate by pointing out the example of Moulton and Chamberlin is that we need to realize that although we know much more about the Universe than Moulton did in 1905, we are still bound in our thinking by what we know. Moreover, our imagination remains constrained by what we have seen before, which is a handicap for astronomers who need to somehow visualize what is too small, distant, or shrouded to directly observe. We have known that foundation for this bias in our thinking for a long time, of course. It was astutely captured some 2,500 years ago by Gautama Buddha: "All experience is preceded by mind, led by mind, made by mind" (from the *Dhammapada* translation by Gil Fronsdal).

No matter what we look at, we come to it with preconceptions that may be unconsciously acquired or actively learned. In computer science this is often referred to as the "default option": if not explicitly specified at the time something needs to be evaluated, an option is chosen based on earlier settings. Comedians often utilize people's defaults: the punch line of a joke turns at a point away from the expected into an unanticipated direction or unrecognized association. Something we see at a glance may feel impossible because we do not recognize it in its proper context or orientation until suddenly it changes shape in our minds as something is recognized. Artists may use this in a trompe l'oeil, an intentional creation that is an illusion. And also in paintings in general the constructs in our mind made by past experiences are used over and again as we accept the painter's creation of depth by perspective methods into what clearly is a flat, two-dimensional object. When we see something we do not immediately understand, by default we search for explanations in things that we have encountered in the past. However, it is when things do not fit expectation, when not confirmation but falsification stands before us, that we truly discover something, although we may falter at the first steps to do so.

We need to be mindful that astronomers, too, are subject to defaulting to preconceptions: it may be that much, or even most,

of what we now think to be true about stars and their planetary systems is indeed correct, but details—particularly those that are not subject to direct observation—will almost certainly be proven wrong or at least incomplete. Although certainly not immune to resistance to change, scientists are very aware that "entrenchment" poses a threat. Although in the end erroneous ideas are filtered out, it may take time: entrenched ideas are sometimes hard to dislodge, and for some time scientists may attempt to fit things to what appear to be established concepts until, at some point, the preponderance of evidence forces the error out of our thinking. After the example of "nebulae" above, another example of that is what Cecilia Payne experienced, to whom we return in just a few paragraphs.

The makings of planets and stars

The primary focus of this book is on planets. Planets, in particular Earth-like planets, are formed primarily of relatively heavy elements. For Earth in particular, the top six elements that make up approximately 94 percent of the whole by mass are: 32 percent iron, 30 percent oxygen, 16 percent silicon, 15 percent magnesium, 3 percent sulfur, and 2 percent nickel. Other elements are present at levels of about a percent or less. The material out of which planets like Earth must have formed would have had to contain sufficient amounts of these heavy elements. Present theoretical understanding and observational evidence of the beginning of the Universe in a Big Bang are such that no elements heavier than seven times the mass of the hydrogen atom would exist at detectable levels: the early Universe contained mainly hydrogen, mixed with some helium with four times the mass of atomic hydrogen, and hundreds of millions of times less lithium which is seven times heavier than hydrogen atoms. And yet, the top six elements that make Earth are between sixteen and fifty-six times heavier than a hydrogen atom.

This contrast between chemical compositions of the young Universe and of the Earth presented earlier researchers with a major problem. If we look at it in a little more detail it turns out that there must be at least two steps in resolving this dilemma. First, there is an obvious distinction between the chemical compositions of the bodies in the Solar System. The Sun contains mostly hydrogen and helium that together add up to almost 99 percent of its total mass. The giant planets contain a lot of hydrogen and helium, also, which form the bulk of their dense atmospheric layers, but the terrestrial planets, asteroids, and comets are largely made up of heavier elements. These do exist in the Sun, but in a low concentration of about 2 percent of the total mass for all elements other than hydrogen and helium. So, one part of the problem of the chemical composition in the story of the Solar System is to understand how the material came to be distributed over Sun and planets as it did.

Nuclear power and the ingredients of apple pie

The second part of the problem of the heavy elements is where they came from in the first place. As best we understand the Universe, the heavy elements simply were not there in any appreciable amount early on. They must have somehow been made between the Big Bang some 13.8 billion years ago and before the formation of our solar system about 4.6 billion years ago. Carl Sagan condensed part of the grand challenge of understanding how this came about in his apparently enigmatic pronouncement that "If you wish to make an apple pie from scratch, you must first invent the Universe." If you start with only hydrogen and helium throughout the Universe, how can something like Earth arise which is made mostly of other things?

Nowadays, we have a well-tested theory about how this came about, which by the way also informs us about the age of the Solar System. The solution to the problem of the formation of

the heavy elements lies in the stars and with a process called nucleosynthesis, meaning the placing together of atomic cores: the densities and temperatures deep inside stars are so high that frequent collisions between the fast-moving cores of atoms causes many of them to bond, releasing energy as they do so. This was found early in the twentieth century, stimulated in part because of the equation $E = mc^2$ that Albert Einstein published in 1905 capturing the intriguing equivalence of energy and matter.

A problem that had stumped astronomers for millennia was what powered the Sun. There was, of course, also the problem of what powered the stars, but for the longest time it did not dawn on astronomers that the apparently tiny stars were, in fact, quite similar to the Sun, and that we saw the Sun as we did simply because it was vastly closer to Earth than any other star. Ideas for what powered the Sun ranged from a global fire to the afterglow of the heat released when the gases from a vast cloud fell together into its own gravitational pull. When the Sun was recognized as a star, ideas evolved that considered stars to be cooling spheres of gas, starting hot and blue to eventually drop in temperature to cool and red. Astronomers still refer to blue stars as "early type" and red stars as "late type" although that link with age was proven wrong almost a century ago; once an idea has settled, it is hard to erase all vestiges of it from our thinking and vocabulary. Incidentally, that is no different from knowing about the counterintuitive color scale here for hot things glowing that I just mentioned: we tend to think of blue as cool—as cool as the blue sea—and of red as hot—as the glow of a fire. In reality, however, something glowing blue is much hotter than something glowing red: the temperature scale for stars goes as in the rainbow from hottest blue via yellow and orange to the coolest red.

Concepts to keep the Sun glowing even included the proposal that the Sun was kept hot by meteors continuing to fall

onto it. Around the start of the twentieth century, however, it became clear that none of these concepts could be correct: evidence increasingly showed the Earth to be older than the time span over which any of these hypothesized energy sources could keep the Sun shining. Things started to change in 1920 when Arthur Eddington suggested that nuclear fusion of hydrogen into helium could provide an enormous amount of energy lasting for a long time. How long exactly would depend on how much hydrogen was available in the Sun. At the time, Eddington did not know how much hydrogen there was, but he argued that just 5 percent would suffice. There would prove to be much more hydrogen available than that, however.

Here, we reach Cecilia Payne and run into a second example of the inertia of old ideas. For her PhD thesis, she investigated the abundance of chemical elements in the atmospheres of stars, including the Sun. Her work led her to establish that hydrogen was vastly more abundant in the solar atmosphere than any other element, with helium coming in as a strong second component. But, it seems in large part because she was following the advice of her senior colleagues, that she questioned her own results, writing in her thesis in 1925 that "Although hydrogen and helium are manifestly very abundant in stellar atmospheres, the actual values derived from the estimates of marginal appearance are regarded as spurious," followed some pages further by "The enormous abundance derived for [hydrogen and helium] in the stellar atmosphere is almost certainly not real." Her measurements conflicted with accepted wisdom and, at least temporarily, accepted wisdom was left standing and instead the measurements were questioned. In large part this had to do with the thinking at the time about the origin of the Earth. Just before writing her own work into a fog of doubt she notes that "If, however, the earth originated from the surface layers of the sun, the percentage composition of the whole earth should resemble the composition of the solar (and therefore of a typical stellar) atmosphere." Instead of concluding that her work showed something to be wrong about the idea that

the Earth formed somehow from the Sun's atmosphere without differentiation of elements, the inertia of ideas, her community's peer pressure, and the lack of confidence of a young researcher made her question her own findings rather than the ill-founded preconceptions of others.

In an interview about this recorded in 1968 when she worked at the Harvard College Observatory, she said that "The only thing I remember is saying to [her thesis advisor] Eddington that I was surprised to find how large a proportion of the material of the Universe is hydrogen and he smiled and said, 'Well, that is *on* the stars, but you don't know that it is *in* the stars.'" That was a possibility, of course, although now we know that the Sun is very nearly homogeneous throughout its interior. That understanding came with the years, however, and was not fully developed until helioseismology enabled the probing of the solar interior by sound waves in the last decades of the twentieth century. It also just takes time to shift a collective mindset: as the playwright George Bernard Shaw (1856–1950) phrased it, "New opinions often appear first as jokes and fancies, then as blasphemies and treason, then as questions open to discussion, and finally as established truths." It is now an established truth that stars are made primarily out of hydrogen and helium. In the two decades after Payne's PhD project, scientists worked out the basics of what powers a star, and how that leads to the creation of the heavy elements out of which planets can be formed. For this idea to work there had to be a lot of hydrogen in stars, and the stellar composition had to not be equal to the composition of the Earth. Her colleagues changed their views, and she was recognized to be correct about the dominant chemicals in stars. The old, untenable ideas were thrown out—in part because of her analyses—as the new idea of nuclear fusion in stellar cores and the ensuing creation of the heavy elements was given solid footing in nuclear physics.

The short version of the story of the heavy elements is this. Stars form out of gas floating about in a galaxy, which is

predominantly hydrogen and helium. Then, starting with the hydrogen, they fuse elements into ever heavier forms. The heavier stars reach further into the production of the heavier elements, releasing energy in this chain of reactions up to the formation of iron—with some still heavier elements being formed as side products in the very end phases of stellar lives. When stars run out of nuclear fuel, they eject much of their mass, including the newly generated heavy elements, back into the galaxy—in some cases creating yet heavier elements, including radioactive ones, during the most violent of these expulsions. Then these gases that were hurled off the stars mix with the surrounding interstellar gas, after which this mixture may be captured into the birth phase of the next generation of stars. Some of the gas ejected in a supernova may even be thrown out of a galaxy altogether, particularly out of the smaller ones, later to be captured by another galaxy where it can enter the cycle of star birth and demise along with the other galactic gases.

When Hans Bethe (1906–2005) was awarded the 1967 Nobel Prize in physics for his work on nucleosynthesis, he summarized it like this: "If all this is true, stars have a life cycle much like animals. They get born, they grow, they go through a definite internal development, and finally they die, to give back the material of which they are made so that new stars may live." This process has happened since the beginning of time, and continues today in the multitude of galaxies around us, including our Milky Way. It is why there is more than only hydrogen, helium, and a minute sprinkling of lithium in the Universe to create planets.

This short version is not simply a qualitative narrative. Each step in the process has been quantitatively investigated, tested, improved upon, written about, and then compared to what others found, all guided by increasingly powerful telescopes and computers, all subject to the laws of physics and to the rules of scientific investigation that are enforced by peer review. The above summary also does not convey the scale and complexity of what

transpires. And we have yet to look at what it all means for the formation of planets alongside the creation of their stars.

Veiled nurseries

The first thing to realize about the formation of stars and their planetary systems is that they generally do not form in isolation. At the very beginning of the process, at least in a galaxy like our own, lie gigantic clouds. These temporary areas of increased density form in the interstellar gas that pervades the galaxy as matter slowly swirls around in eddies (Figure 5.2). These clouds, occurring over a total of about 1 percent of the galactic volume, each contain several millions of solar masses, with a hierarchy of structures within them from large to small. Swirling, turbulent motions cause parts of these clouds to become denser, still on huge scales: these "giant molecular clouds" may contain anywhere from a few million solar masses down to ten thousand solar masses. These giant clouds have within them numerous areas much more compact than the average density, still with swirling motions causing structures within them to form and dissolve. It is within these clumps, far smaller than the overall cloud, where stars form in yet smaller and denser cores within the clumps. Molecular clouds are cold, often just a few degrees above the absolute minimum, so that sound moves slowly allowing gas motions within them to often be fast enough to be supersonic, leading to shock waves and swirling chaotic cascades in the gas, all on astronomical scales.

Some of the cores somehow become so compact and heavy that their inward self-gravity wins out over the outward pressure of the gas: the cloud falls onto itself, becoming ever more compact, and eventually turning into a star, often with a planetary system. This involves perhaps only 5 percent to 10 percent of all the gas in the molecular cloud; with time, the bulk of the gas re-disperses into the background interstellar gas.

Figure 5.2 HH 24 inside the star-forming Orion B molecular cloud complex, some 1,350 light years away. At the center of the image, a star is being born. Cloaked invisibly within a cloud, this young star is surrounded by a disk-shaped structure of gas and dust in which planets will likely form. Over the budding star's rotational poles magnetic fields work in a way—yet to be fully understood—to accelerate matter outward in two opposing, narrowly focused jets. The jets ram into interstellar matter, forming shocks that heat the gases and cause them to glow in clumps. Such related phenomena early in a star's life are known collectively as a Herbig–Haro (HH) object. Another such object, still mostly hidden within the molecular cloud, is visible in the upper right corner. See also Figure 5.4 (See Plate 12 for a color version).

Source: Hubble Space Telescope, NASA/ESA.

Figure 5.3 Some of the bright stars in open cluster Pismis 24, in the central region of the nebula NGC 6357, appear to be heavy, with masses up to one hundred times that of the Sun. Toward the left, the bright light from the young star in NGC 6357 is creating a bubble of ionized gas within the cloud (HII region G353.2 + 0.9) from which the star was born (See Plate 13 for a color version).

Source: NASA, ESA and J. M. Apellániz (IAA, Spain).

Collapsing cores often occur in groups of dozens to many hundreds of other cores. These cores are large compared to planetary systems: to collect enough mass to make our Sun, the initial core out of which it formed must have had a diameter of at least some 50,000 Sun–Earth distances. Such a distance reaches out to far beyond the outermost planets and the Kuiper Belt, to at least where nowadays the Oort Cloud is thought be begin. Once stars materialize out of these cores, their grouping makes what are called open clusters (an example is shown in Figure 5.3). They are described as "open" because they are fairly loose formations. Open clusters are made up of stars that are very nearly of the same age, give or take a few million years. Observations reveal that up to 90 percent of stars in the Galaxy form in clusters.

Generally, there is some net swirling motion within the clumps, manifesting itself as a bulk spinning motion once the core

collapses. The more the core contracts, the faster the spinning motion becomes. The analogy with the ice dancer contracting her arms to rev up into a fast pirouette is often called upon at this point: overused, perhaps, but entirely accurate and visually very clear. Even weak swirling motions in the initial "core" in the giant molecular clouds will be greatly amplified by that contraction: by the time the cloud has contracted into a star, it will have compressed by something like a million times in diameter.

At this point we reach a phase in the formation of stars that was recognized shortly after Newton formulated his laws to describe motion and gravitation. Successively working on the challenge, Emanuel Swedenborg (1688–1772), Immanuel Kant (1724–1804), Pierre-Simon marquis de Laplace, and others less well known were the first to formulate in some quantitative terms the concept that the Sun and the planetary system formed out of the self-gravitating collapse of a large nebula. Laplace knew that contraction of a rotating body would make it turn ever faster; so much, in fact, that at some point the centrifugal force of that spinning would exceed gravity. The result, he argued erroneously, would be that rings of gas would be thrown off the contracting sphere out of which he envisioned the planets to be formed. That is not how things are observed to happen nowadays, but then present-day telescopes are very much more powerful than those of centuries past.

The problem of centrifugal forces in the contracting cores on their way to stardom is real, but the Universe has different ways of dealing with it than throwing off rings of gas. Here, magnetism plays a crucial role. The forces that can be carried via a magnetic field that pervades a hot gas can work in fundamentally different ways than the force of gravity. Gravity can only pull things directly toward each other. A magnetic field, in contrast, can either attract or repel, depending on whether opposite or like polarities are closest to each other. Another fundamental property of a magnetic field is that it can carry tension: push on it sideways and it can push back. In other words, magnetic forces

need not be simply pushing or pulling, but can push laterally, thereby applying torque. Plus, there is another type of force that can work when a magnetic field pervades a gas so hot that electrons are being kicked loose and the gas becomes ionized, thus electrically conducting, and in astronomy then referred to as a "plasma." In such a hot gas, a magnetic field can work just like gas pressure, so that even if there is little gas to speak of, a field that is sufficiently strong can resist compression, working against an external force quite efficiently. All of these properties of magnetism are important in the lives of stars and planets.

The present scenario for the formation of stars out of interstellar clouds makes use of gravity, magnetism, and light. Wherever a core happens to form within the giant cloud with sufficient density, gravity wins and starts pulling the cloud onto itself. Once contraction advances over a substantial distance, spinning motions become influential. Spinning is associated with a centrifugal force that works outward from the central axis of rotation. This centrifugal force works against gravity, slowing down the fall of the gas toward the center of the core. But the centrifugal force only works in the direction of the rotation, not along the axis of rotation; in that axial direction gravity is not countered by the spinning. The result is that the collapsing cloud reshapes from roughly spherical to oblate and in due course into a gyrating, vast but wispy pancake made out of interstellar gas that astronomers refer to as a disk. Most forming stars have such disks, which is a good starting point to take the disk as the birthplace of exoplanets given that most stars are now known to have a planetary system.

The compression of the gas also works to warm it up, helping to increase the gas pressure a little, and increasing its electrical conductivity a lot. This is where magnetism comes in: a weak Galactic magnetic field that threaded the initial core will have strengthened because of the collapse, while moreover the revolving star and disk are beginning to work like dynamos, creating their own additional magnetism. The magnetic field is critical in allowing the star formation to proceed: the long-range tension

carried through the magnetic field transports rotational energy outward as gas spirals inward. This transportation engine is also aided by viscosity in the gas as well as by tides within the gravitational field; how much each contributes in what phase and at what distance from the forming star is still being investigated.

Population studies and polychromatic imaging

Powerful telescopes on the ground and in space, working from X-rays through visible to infrared and radio wavelengths, show us some of this process in detail. The formation process of a star is far too slow to follow on a human timescale. But we can look at different star-forming regions in the surrounding Galaxy, thus getting snapshots in different phases of the process. These observations reveal the disks and the forming stars. Interstellar gas is pulled into the gravity of the forming star, migrates inward through the disk, and—probably in a highly variable stream— eventually falls onto the young stellar object. For those systems in which the central star has acquired most of its mass, the disks of gas and dust are estimated to still contain at least one percent of a full stellar mass—plenty to make planets out of, as it turns out.

In order to see the forming star and the evolving disk, astronomers have to use a variety of instruments. Early in the process, once substantial collapse of the cloud has already taken place, the central object that is to be a star remains shrouded inside its own cloud, incubating invisibly in the center of what is reshaping into a disk. Later on, when much of the gas and dust are in the form of a disk, and better still when dust and gas are clearing from it, the surface of the young stellar object can often be seen in light that is visible to the human eye, although it is generally a deep red. The outer atmosphere of such young stars, high above their relatively cool surfaces, is heated by electrical storms in the star's ever-shifting magnetic fields. These layers can reach temperatures of tens of millions of degrees. To see the glow

Figure 5.4 The main image shows the surroundings of the young star HL Tauri in the Taurus Molecular Cloud, some 450 light years from Earth, as observed by the Hubble Space Telescope in light visible to the human eye. The magnified inset is an observation by the Atacama Large Millimeter Array (ALMA) at full resolution. The ALMA image shows the disk around the forming star, not quite a million years old, at wavelengths not visible to the human eye; Hubble could not see HL Tauri very well because visible light is absorbed by the gas clouds surrounding the object, but ALMA, making images with light such as that used to warm things in a microwave, could look deep into the cloud. The gaps in the disk are thought to be formed by the gravitational interaction between the gas and dust in the spinning disk and young planetary bodies forming within it. The dark rings are estimated to lie at 13, 32, 65, and 77 Sun–Earth distances from their central star. Note the object with two straight jets shooting out below and to the right of the center of the image; this is star-forming object HH 30; another such example is HH 24 in Figure 5.2 (See Plate 14 for a color version).

Source: ALMA (ESO/NAOJ/NRAO)/NASA/ESA.

from these layers, X-ray telescopes are needed. The disk gyrating around the star also has a wide range of temperatures, from just a few degrees above absolute zero in the relatively dense mid-plane of the disk where material is shielded from any radiation by the

surrounding gas and dust, to a few thousand degrees very close to the star.

To observe such disks, either by looking at the emitted light or by looking for the absorption of such light, astronomers use X-ray telescopes in space (Chandra and XMM Newton), powerful visible-light and ultraviolet imagers above the Earth's atmosphere (Hubble Space Telescope), icy-cold telescopes in infrared imaging satellites (Spitzer Space Telescope), and multi-kilometer-sized microwave and radio telescopes on the ground (including the ALMA telescope made for microwave radiation in the Chilean Andes and the VLA telescope observing radio wavelengths in New Mexico in the USA). Figure 5.4 illustrates what the combined power of very different telescopes can reveal.

Infrared telescopes are particularly useful to image dust; by filtering different "colors" astronomers can learn about the sizes of the dust particles that range from microns to millimeters. Extended clouds of dust can thus be seen, but when this dust clusters into clumps of matter with sizes from meters to kilometers (or feet to miles) they are actually hard to spot. Only when they grow to large and heavy exoplanets can they be observed in stellar transits or Doppler signals.

Team science

Collecting data from these powerful telescopes is difficult and time-consuming: these costly instruments are in high demand, and obtaining observing time on them is highly competitive. Typically, astronomers have to write research proposals to a funding agency. These agencies have the proposals reviewed and then select a subset from the pool of the most promising ones, typically rewarding from one in five to one in a dozen, depending on the telescope and agency. Then, the lucky few selected astronomers have to have observing time scheduled, which is handled by something often called the time-allocation committee. Such a committee reviews the special requirements

of each observation, taking into account when a source in the sky would be visible, and tries to accommodate scheduling requests so that coordination with other telescopes is possible. But to get time on another telescope, another proposal has to be written, selected, and allocated. Each review cycle easily takes a good part of a year, so many months or even a year or two fly by between proposal and observation.

To increase chances of selection, the astronomers have to demonstrate that they know how to analyze and interpret the observations; this requires special training, so that commonly astronomers need to team up with colleagues that work somewhere else in the country or—as occurs more and more—somewhere else in the world. This is not too different from how high-level sports teams are assembled or how producers and directors of large-budget movies get the best experts to work on their project, except that for scientists it all remains merely a promise until the funding agency makes a favorable decision, and of course there is no profit in this process for them apart from the gain of knowledge, experience, standing in their community, and a publication on their record in the scientific literature.

As a result of this telescope-centric competitive selection process, multi-telescope multi-wavelength observations at any given time of objects in the sky are rare. Fortunately, the evolution of circumstellar disks and their central stars is generally slow, so that sets of observations taken by different instruments at different times can be combined to obtain a more complete picture of the formation of a planetary system out of a contracting gas cloud.

Genesis of a star and its entourage

At some point in the process of the gravitational contraction, the core of the cloud becomes dense enough that it is no longer transparent but transitions to opaque. Now this cloud has the appearance of having a surface, just like clouds in Earth's atmosphere. But for the interior of the star-to-be this has an

important consequence. The gas warms up as it is compressed by being pulled into its own gravity. Here, the energy of the falling motion is by collisions with other gas particles converted into energy of random motion, which is what we know as heat. As the cloud's opacity goes up, it can no longer efficiently cool by radiating its heat away in the form of, mostly, infrared light. From then onward, the temperature will be going up still faster inside the ball of gas and also on its surface. For the inside, this means that with temperature and density both going up, the gas pressure shoots up, which then pushes increasingly against the infall and thus begins to slow the inward motion of the gas. And a magnetic dynamo, much like that still operating in the Sun, but much stronger because it is driven by the fast rotating motion, begins to operate inside the star.

For regions outside the star, the opacity of the center of the cloud means that as its "surface" becomes hotter and brighter, the region around the forming star begins to bask in warming light. The color of this light gradually shifts from infrared to red and— some day, after the disk has long vanished, and provided the star is sufficiently heavy—onward to yellow for stars much like the Sun, or beyond that to blue for the heaviest and hottest stars. Also, the dynamo in the star-like object begins to drive a strong wind blowing into the space around it, analogous to the present-day solar wind that blows against comet tails and against the Earth's planetary magnetic field to cause space weather, but stronger in proportion to the proto-star's stronger dynamo, and aided by the centrifugal force from the rapidly spinning proto-star. It also appears that the gyrating disk itself can drive a magnetic dynamo with its own wind added to that of the central star-to-be. More important, however, is that the irradiation from the hot, magnetically powered atmosphere of the nearby young star is helping to evaporate the gas in the disk from both sides of the central, densest region.

At first, the forming star revolves so fast that it is not far from losing its outer layers due to the centrifugal force—as imagined

by Laplace—although that never quite happens: its own magnetism blows a wind off its surface, which carries away rotational
energy so that the star's spin rate is held in check and eventually
goes down. For a star like our Sun, that process of magnetic braking continues throughout the star's life: the young Sun probably
spun around once every few hours or maybe it took half a day to
complete a turn, but now it takes some 25 days to perform a full
rotation, and it will continue to slow down in the future.

The warming irradiation from the nearly-stellar object in the
center of the disk and the friction-like processes within the disk
itself join with the blowing magnetized wind to push against
remaining materials around the forming star. One early signature
of the magnetism is that of jet-like fountains of gas squirting from
the central region of the disk in directions above and below the
rotational poles (such as in Figures 5.2 and 5.4). Then wind and
heat conspire to clear the tenuous regions above and below the
disk of gas, and ultimately it blows the dense gases clear of what
then soon no longer is a gaseous disk, leaving the beginning of the
planetary system: dust that is coagulating into larger and larger
structures.

By astronomical standards, the clearing-up of gases from what
was the initial circumstellar disk is fast. For the youngest star-
forming clusters (say NGC 2014, the Flame Nebula, with an age of
100,000 years, give or take 50,000) about 86 percent of the forming
stars have disks. By looking at a variety of such clusters, we see
that within some two to three million years after that, at least half
of the observed very young stars have cleared the space around
them of most gas and dust to some 20 Sun–Earth distances. This
is the space where most planets are expected to form; for the
Solar System, only Neptune presently orbits beyond that, at 30
Sun–Earth distances, and even it may have formed much closer
to the Sun. By 10 million years, only some five percent of young
stars still have signatures of their primordial disks. Why do we call
something lasting millions of years "fast"? Because compared to
the overall expected life of a star like the Sun of 10 billion years,

this disk phase lasts a mere 0.03 percent to 0.10 percent of its life time, so the formation phase is over in a flash on stellar timescales (Table 1.2).

We can directly observe that stars are formed with disks, can estimate the durations of different phases, and can more or less measure how much mass is contained within the dusty gaseous disks. This is fortunate, because it is not at all understood exactly how things happen: if we had to predict the outcome of the contraction of an interstellar cloud, we would still have a very hard time not diverging from reality, despite the many studies of recent decades in which astrophysicists hypothesize and then experiment with the help of computers. Many things remain mysterious, or rather ambivalent: there are multiple ideas, or at least uncertainties in the numbers, and observational tests have not yet allowed us to narrow down the field of contending ideas to a single string of processes. Then, of course, there are the simplifying approximations that we have to deal with. These approximations are guided by observations, which is good, but if observations can reveal things only down to the better part of the size of an entire planetary system, it is clear that much occurs that we do not see. In such cases we can have, at best, only indirect information on the processes that are happening. Thus, the above description of the star-forming process is probably roughly what takes place in nature, but many of the processes involved are poorly understood, at least quantitatively and sometimes even qualitatively.

What follows this phase has been studied much longer, and looks like it is on much better footing in terms of basic physics. Inside the forming star, the densities in the core reach first dozens, then over a hundred times the density of water. Within something like 100,000 years, the temperature reaches millions of degrees. When that happens, the particles in the gas are colliding so often and so fast that all electrons are stripped off their nuclei and fly around unbound. Collisions between particles can be so powerful that for a very small fraction of these the colliding

atomic cores actually merge. Merging of nuclei is nuclear fusion. At first, in such forming stars this fuses hydrogen into helium through a series of intermediate steps. At this point the star-like object has become an actual star. It will take some tens of millions of years for this gigantic mass of gas to fully settle, but then a stable balance is achieved so that the fusion reactions in the core will release exactly as much energy as is radiated at the surface, making things outwardly appear unchanging for millions to billions of years, depending on the mass of the star, while the nuclear reactor in the core gradually consumes its available fuel.

Some not-quite stars among the genuine ones

Stars form out of globs within molecular clouds that have a wide range of masses, but with a pronounced preference for lightweights: for every heavy star born there are many more less massive ones. For example, for every three hundred stars within a few percent of a solar mass, there is only one born with ten times that mass. And for every star born with a solar mass, there appear to be some 20 to 50 stars born with only one tenth of that mass. This is less certain than the numbers for heavier stars because low-mass stars are faint and therefore easier to miss compared to their brighter siblings.

As telescopes become ever more powerful, objects have been observed that are like stars, but not really. For star-like objects that form with a mass of less than about 7 percent of a solar mass, or some 25,000 Earth masses, the density in the core never reaches the point where ordinary hydrogen can be fused into helium. There is a heavy form of hydrogen, called deuterium, which carries an extra neutron in the atomic nucleus, that may reach nuclear ignition in the heaviest of these objects, but there is relatively little of it, so it neither generates much energy nor does it do so for very long. Such objects are known as brown dwarfs. Brown dwarfs are not really brown: they glow in the infrared and

a bit in the deepest red or magenta. The first confirmed discovery of such an object was reported in 1995, the same year that the first exoplanet orbiting a Sun-like star was found.

For objects with less than about 1 percent of a solar mass or 4,000 Earth masses, even deuterium will not fuse. Such low-mass bodies by stellar standards never make for a real star: they just glow and cool with time, very slowly radiating away the energy that they were created with as the gases fell onto themselves. This is the point where the substellar domain for which at least nuclear fusion of heavy hydrogen occurs reaches into the exoplanetary heavy-Jupiter domain where no nuclear fusion occurs of any kind.

Until their first unambiguous detection in the mid 1990s, brown dwarfs had been hypothesized about; now we think there may be tens of billions of them in the Galaxy. The numbers of brown dwarfs are such that it seems they are a smooth continuation of the star-forming process on larger scales: they seem to be formed out of yet smaller cores in molecular clouds than those out of which the lightest real stars are formed. Having said that, there are likely fewer than one might expect from a direct extrapolation of the mass distribution for heavier objects, suggesting that at these small masses (for stellar-like objects) we are reaching some change in how these form out of molecular clouds. That, in itself, is no surprise, but it does of course leave us wondering just how small the smallest cores are that actually contract into star-like objects, be they real stars or glowing brown dwarfs. Among the least massive young brown dwarfs is OTS 44, which is still gathering mass, has a disk, and weighs probably only some 1 percent of a solar mass or about 4,000 Earth masses; at the low end of its possible mass range, it may not even be a brown dwarf but in fact a free-floating planetary object. As they have disks, could these also lead to planets, albeit of low masses themselves? We have a lot to learn there, but the first brown dwarfs orbited by planets of about Earth's mass have already been discovered.

Assembling elements

Heavy stars are bright. This means they go through their available fuel faster than lighter stars. So much faster, in fact, that although there is more fuel in a heavy star to begin with, they run out much sooner than lighter stars of the same age. A star like the Sun can maintain nuclear fusion for about 10 billion years. A star ten times heavier than the Sun runs out of its fuel reservoir some 300 times faster, while a star ten times less massive than the Sun does so multiple hundreds of times more slowly than the Sun. That nuclear fuel is initially—and for the longest time—hydrogen that is being fused into helium. In a Sun-like star, this is what happens. First the hydrogen in the core is used, and when that runs out the fusion process gradually moves upward, eating into the hydrogen still available in higher layers. As a result, the outer parts of the star begin to expand, turning into a giant star, glowing red rather than the yellow, white, or blue that it may have been before, depending on how heavy it is. As a giant star with an enormously expanded surface from which to radiate its energy, it is a lot brighter than during its primary adult existence—over a thousand times for a star like the Sun—so that it goes through the little remaining fuel increasingly fast (see Figure 5.5).

Underneath the hydrogen-fusing shell, the stellar core, now no longer generating energy, contracts and becomes denser and hotter. At some point temperature and density increase so much that nuclei with more mass and a larger mutually repelling electrical charge will collide fast enough that a new fuel becomes available: helium can now be fused into heavier products, such as carbon. The ignition of helium, which astronomers call the helium flash, leads to a new state of the star which—if much like the Sun—it can maintain for some 100 million years, or about 1 percent of its primary adult life as a hydrogen-fusing star: it contracts somewhat and its surface becomes hotter again; it is this kind of phase that we see when looking at the central star of the exoplanet V391 Pegasi b. As later the helium also runs out

Figure 5.5 Visualization of the evolution of a star like the Sun, moving in time from the rear to the front of the image. The star forms in the center of a disk out of which a planetary system is likely to form also. When only tens of millions of years old, stellar magnetism is very strong and starspots may be as large as half a hemisphere, often forming at high latitudes if not over the rotational poles. Then the star settles into adult life as a yellow star with frequent, small starspots. After half a dozen billion years, magnetic activity has faded, and few if any starspots occur. When nuclear fuel in the core runs out, the star swells up and shifts from yellow-white to orange and eventually becomes a red giant star. There is little magnetism in the late phases (See Plate 15 for a color version).

in the core, this fusion, too, moves upward in a shell grappling for new fuel. Now there are two shells with fusion: a deep-seated one running on helium and one at shallower depths running on hydrogen. This phase does not lead to a steady state: the star repeatedly gasps for new fuel, contracting and expanding. It can expand so fast that it throws off substantial fractions of its outer layers in what are called thermal pulses. Stars can lose half their mass in a series of such thermal pulses.

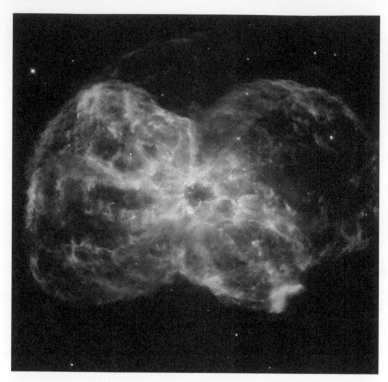

Figure 5.6 Planetary nebula NGC 2440, about 4,000 light years away. Such a nebula has nothing to do with what eighteenth- and nineteenth-century astronomers thought when they named these phenomena. In fact, they form late in the life of stars, as will happen with our Sun, when fuel has run out in the star's interior and the gasps for survival cause the outer layers of the star to be expelled into a giant bubble. The remainder of the star contracts into a white dwarf, the size of the Earth, which in this case is visible as a small bright dot in the center of the image. It glows primarily in the ultraviolet because its surface temperature is some 200,000 degrees centigrade (360,000 degrees Fahrenheit). Over time, it will cool down and fade. The structure of the nebula suggests that the star has ejected its shell in multiple episodes, asymmetrically into the interstellar medium. Different colors in this composite image reveal different chemicals: blue for helium, blue-green for oxygen, red for nitrogen and hydrogen (See Plate 16 for a color version).

Source: NASA, ESA, and K. Noll (STScI); The Hubble Heritage Team (STScI/AURA).

When this last phase is over, after maybe 10 million years, and helium and hydrogen burning have run their course, the last of the gasps throws off a significant amount of gas to form what is called a planetary nebula (an example is shown in Figure 5.6)—which has nothing to do with planets, but that is the name it was given before astronomers knew what they were really looking at. The remnant of the star becomes as compact as the Earth, over 10,000 times smaller in width than at its largest as a giant. It glows white hot at first, which is why they are called white dwarfs, thereafter forever cooling down and progressively, irreversibly dimming over the subsequent billions of years, fading away in perpetuity.

Heavier stars not only run through their fuel faster, they also manage to go further in the production of heavier elements: they fuse together elements like carbon, oxygen, neon, magnesium, silicon, … all the way up to iron, and even produce lesser amounts of some of the heavier elements as a side product in the final phases of their lives. Stars go through these fusion reactions successively, so that the product of a previous step becomes the fuel for the next, also moving upward in the star in shells as time progresses. These stars, too, can eject material as they become pulsing giants. The material in the ejected layers is mixed with fusion products that were dredged up from deeper regions by large-scale overturning flows in the interior, together adding some heavier elements to the interstellar gas once ejected from the star.

Sooner or later, these heavy stars still run out of fuel because once they have converted much of their matter into iron there is nothing left from which energy can be harvested: making heavier elements would indeed consume energy, and the star cannot survive its hunger for energy. Some elements heavier than iron, up to lead, are made in this process when atomic fragments collide with each other randomly, but they do not help the star survive. In fact, forming these heaviest elements by nuclear collision takes precious energy away from the star, only aiding in its demise.

The result is that its inner reaches collapse, with nothing to stop this until the core becomes so dense that quantum-mechanical effects take over and the star changes into a neutron star with a radius of only a dozen kilometers (some eight miles)—or, if a really heavy star goes through this phase, until nothing stops it at all and it changes into a black hole. In the meantime, the outer layers are blown off in an explosion so strong that the brightness of that single explosion can outshine an entire galaxy with billions of stars for a brief time. It is in this explosion, called a supernova (remnants of one are shown in Figure 5.7), that many of the elements heavier than iron are made. All that material is thrown into the interstellar gas, enriching it with elements made during the fusion stages as well as during the supernova explosion itself. A supernova can also be caused in binary stars in which one star overflows onto, and eventually merges with, its partner. It would require lots of details to go through the various scenarios, but the bottom line is that the end effect is the same: a star explodes violently, and many heavy elements are formed in the process. How much is being generated of each element depends on the detailed scenario.

Many of the heavy atomic nuclei formed in the supernova explosion, as well as many of the affected lighter ones, are unstable and decay into lighter elements again. This fission of nuclei is what we call radioactivity. Some radioactive nuclei have very long life-times, some as long as several billions of years. The interstellar gas, enriched with heavy elements, now includes radioactive nuclei that, along with all other gas, will be captured as new stars and their planets form out of the swirling gases.

To summarize, stellar nuclear fusion is the origin of all abundant elements heavier than helium. How was all that put together and verified against observations? One important attribute of stars that helped us advance our knowledge of their life cycles is that stars in the Galaxy form in open clusters, many of them as double, triple, quadruple, or quintuple stars. Another attribute is that some stars never seem to relax.

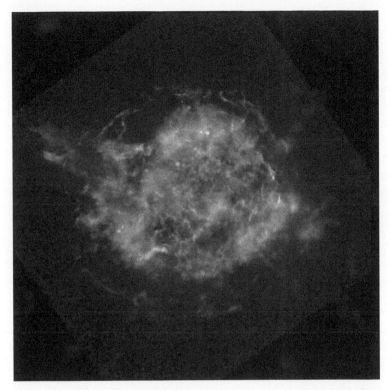

Figure 5.7 Supernova remnant Cassiopeia A. These are the remnants of an exploded star, seen from some 11,000 light years away. The explosion may have taken a little over 300 years to reach the extent now visible. This image is a composite of observations taken in the X-ray range of the light spectrum; none of this is visible to the human eye. The red, yellow, and green colors show X-rays (observed by the space-based Chandra X-ray Observatory) of non-radioactive gases: red for iron, green for silicon and magnesium, and yellow for a wide variety of very hot gases. The blue areas show emission in high-energy X-rays from radioactive gases (specifically here titanium-44), observed with the NuSTAR spacecraft. The structures reveal that the explosion was not equally distributed around the star, but rather was more forceful in some directions than in others (See Plate 17 for a color version).

Source: NASA/JPL-Caltech/CXC/SAO.

Distance markers

Most stars settle into a stable state with their brightness changing substantially only over millions or billions of years, with changes on human timescales limited to fractions of a percent. Some stars are different, however: their brightness pulses regularly, like a heartbeat, going up and down by as much as a factor of two over days to months. By now, we know there are distinct classes of such slowly pulsating stars, reflecting their different chemical compositions and internal structures. But even before this class distinction was made, these stars were—and still are—valuable in establishing distances in the Universe; among them are those called Cepheids after the first one found, the fourth-brightest star in the constellation Cepheus, hence named δ Cephei.

In 1912, Henrietta Leavitt published a relationship she had found by observing Cepheids in a nearby galaxy, the Large Magellanic Cloud. She realized that the pulsation period of what we now call classical Cepheids is longer for brighter stars: those that have a period of about a day are some 200 times brighter than the Sun, while those with periods of 50 days are over 20,000 times brighter than the Sun, while periods and brightness smoothly trend upward together in between. So by looking at Cepheids in other galaxies, and measuring their relative apparent brightness and their pulsation periods, one could actually determine the relative distances of galaxies as seen from Earth. This provided a relative measurement at the time she published the discovery because the actual intrinsic brightness was unknown, only the ratios of brightnesses being known. So, also the distances were relative: one is so many times further away than another. To make that an actual absolute distance one would have to know the absolute distance to at least one Cepheid.

Fortunately, there are Cepheids in our own Galaxy also. Those are very much closer than those in other galaxies, of course, but you still had to measure their distance for them to become useful

in measuring the distance to other galaxies. Over the past century, measurements of even the largest distances in the Universe have become increasingly accurate, still resting on the foundations provided by the measurements to the relatively nearby stars that use parallax and star clusters.

Parallax is "the effect whereby the position or direction of an object appears to differ when viewed from different positions". Measuring the parallax from a known base gives absolute distances. The easiest example that is available to almost all of us is to do is this: stretch out your arm, hold up a finger, and alternatingly close one eye and then the other. The perceived position of the finger shifts relative to what is in the background. If the background is far enough away, and if we measure the distance between our two eyes, the difference in apparent position can be measured to give us the distance between eyes and finger. That is not particularly exciting because you could just use a tape measure, but astronomers have a way of using "eyes" that are really far apart to measure enormous distances into the Galaxy: they record the starry sky, then wait for six months, and record it again, and look for the differences in the positions of the stars in their photographs. The distance between these two recordings is about 300 million kilometers, or 190 million miles, because in the intervening six months the Earth has moved to the other side of the Sun in its orbit. With this method the distance to many relatively nearby stars can be measured. Unfortunately, this does not reach Cepheids because in fact the nearest is too far away for this method to work with measurement techniques currently available.

But there are Cepheids in some of the open clusters that are stellar nurseries, but not many, and not in the nearest ones. Fortunately, there are other variable stars, such as the RR Lyrae variables, of which there are more, and which are closer by. These, too have a relation between brightness and pulsation period. Nevertheless, although there are multiple options for variable stars in open clusters, early-twentieth-century astronomers were still left

with the problem of establishing the distances to enough of such variable stars to know the overall distance scale both within the Galaxy and to other galaxies.

Why do the open clusters come into this? Well, they helped us establish distances to many stars and galaxies, and also in understanding how stars age. By now, we know that there is a simple relationship between a star's color and its luminosity for much of its life, so that in principle measuring both the color and a star's apparent brightness in the sky gives you a good idea of its disance. But no one knew that a century ago. Moreover, that simple relationship is completely messed up in the late phases of a star's life, and also in the earliest phases around the time that planets are still forming. As you cannot know a star's age from its brightness and color astronomers had to find other information.

Ejnar Hertzsprung (1873–1967) and Henry Russell (1877–1957), working in the first decades of the twentieth century, came across an interesting way of summarizing the properties of stars in a diagram, now called the Hertzsprung–Russell (or HR) diagram (Figure 5.8); it is still very much in use among astronomers. If you take a star's luminosity in units of, for example, the Sun's luminosity and pair it against its color in a diagram then most stars crowd into a narrow band that runs diagonally across that diagram from a corner with bright blue stars to the opposite corner with faint red ones, with the moderate yellow ones in between. We now call this band the "main sequence," because that is where stars spend most of their lives and thus most stars fall onto that band in the HR diagram. You can make such diagrams for absolute brightnesses using stars with known distances. You can also do this with the relative brightness for stars in different open clusters, such as the long-known Pleiades cluster, for which parallax distances can be measured. Hertzsprung first, and Russell independently a few years later, noticed that if you did this in separate diagrams for cluster stars something interesting came out: the least-bright stars always ended up forming a common

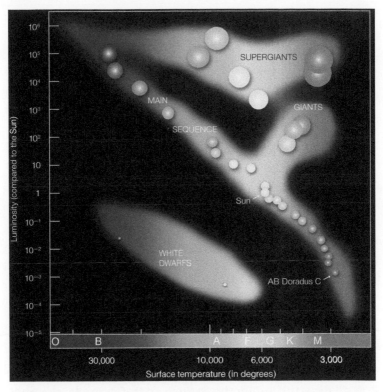

Figure 5.8 Hertzsprung–Russell diagram relating the brightness of stars (vertically) to their temperature (horizontally). For historical reasons, the temperature scale runs backwards. The letters on the temperature bar show the corresponding spectral types. Brightness is shown in powers of ten compared to the Sun's; each tick mark up or down indicates an extra ten times brighter or fainter. The "main sequence" runs across the diagram; there stars remain for most of their lives in an almost fixed place. When stars run low on hydrogen in their cores they swell into giants or supergiants. When they run out of fuel altogether the heaviest stars explode in supernovas. Most stars, however, do not explode but ultimately contract into white dwarfs that start hot and then gradually cool and fade. The star AB Doradus C is among the smallest and dimmest stars possible; yet smaller objects never ignite hydrogen fusion and are either brown dwarfs or, if even smaller, exoplanets. The spheres shown roughly mimic the stellar colors, but the sizes are suppressed in contrast: the future giant Sun will be about 200 times its present size, while heavier supergiants are many hundreds of times larger (See Plate 18 for a color version).

Source: Modified after Press Photo 28c/07 of the European Southern Observatory.

lower part of the main sequence in the diagram. The place in the HR diagram for brighter stars, however, differs from cluster to cluster.

Now what astronomers had to do was to establish the part of the main sequence that was common to other clusters, and then look at how many times brighter or fainter the stars in that were than those in other clusters. That gave them how many times closer or more distant one cluster is relative to another. Parallax distances could be measured to the nearest ones, such as the Pleiades cluster (known also as M45 or the "Seven Sisters") and the Hyades cluster (the nearest open cluster), both of which are visible to the naked eye in the constellation Taurus. The combination of the HR diagram with the parallax distances to the nearest clusters enabled distance determination for any open cluster with enough member stars to outline the common part of the main sequence. Then, with some of the more distant clusters having variable stars within them, the absolute distance to other galaxies could be determined as far out as these variable stars could be seen.

With time, astronomers understood that the differences that were found in the HR diagrams for the brightest of the cluster stars were largely a consequence of the age of the cluster. For all but the youngest clusters where stars have not yet fully settled, the positions of the stars in these diagrams reveal unambiguously how stars evolve. The fact that the brightest cluster stars generally were not on the main sequence showed that these evolved fastest, as we discussed earlier: heavier stars go through their fuel faster and evolve into their giant phases, even as their lighter siblings are still cautiously nibbling away on their smaller hydrogen reserve. The positions of the stars in an HR diagram could thus be used to sort clusters not only by distance, but also by age. Sorting clusters by distance could be done by looking at how many times brighter or fainter the common part of the main sequence was. Sorting by age could be done after scaling to the same distance by looking at where the parts of the main sequence stopped being in common.

Born together, different lives

Along with looking at clusters with numerous stars, the smallest groupings—the binary stars—also helped. This is particularly true for those binaries that are called eclipsing binaries: at some point in the orbit of these stars one star moves in front of another, thus blocking part or all of the light from its companion star behind it. Timing measurements, combined with determination of the velocities of the stars using the Doppler effect and with the law of gravity, will reveal the masses and the sizes of these components. The masses can be determined to just a few percent uncertainty in cases where both stars are observable without one outshining the other entirely. And as they are formed at the same time, such eclipsing binaries provide valuable information on stellar evolution precisely because they are uniquely able to tell us the masses of the stars that are observed. The precision of measurements is so good that they are lending additional information on how common it is that what appears to be a binary star in fact holds three or more companions even if the companions are too faint to be seen directly in the combined light from the stars: they reveal their presence in the Doppler signals from the brightest partner.

It would take computations of internal structure and nuclear fusion rates to translate the information on open clusters and binary stars into an absolute timeline, but with a lot of work having been done over the course of the twentieth century, we now have quite a good understanding of how and how fast stellar evolution proceeds. For the youngest objects that are still approaching truly stellar status as fusion reactors, age determinations based on computer models have to take into account that these stars are not fully settled. Here, too, the inclusion of information from eclipsing binaries is very important.

Such studies have constrained the age of the Pleiades cluster to between 75 and 150 million years, which is well beyond the age at which planets are fully formed and stellar disks have disappeared.

But there are younger clusters as there are older ones: there are over 1,000 other open clusters in the Milky Way alone, with ages that range from around one million years—such as M16, the Eagle Nebula—to almost twice the age of the Solar System at some eight to nine billion years—such as NGC 6791. By establishing timelines for stars in very young open clusters, astronomers also unveiled the timeline for the formation and disappearance of stellar disks and the exoplanets within them. That is how the numbers above came about: statistically speaking, half the disks clear within 2 to 3 million years, almost all the disks are gone by 10 million years. So all the initial formation processes of planets must be completed by then. Or maybe it is better to say that should any substantial accumulation of gas and dust into the planetary system still be ongoing by then, it will be terminated by the disappearance of the building materials. Planet formation may still continue, but now only by collisions leading to mergers and fragmentation of dense bodies already substantial in size.

Ancient worlds

The first generations of stars added their heavy elements to the gases in the Galaxy from which planets could form. Heavy stars are known to evolve fastest, and their supernova explosions release the heavier materials that are critical to planet formation. The slowest-evolving stars that go supernova at the end of their lifetime are about nine solar masses (somewhat dependent on their chemical composition); these live for about 30 million years. So, by the time the Sun was born 4.6 billion years ago, there would have been 300 generations of supernovas of stars at nine solar masses, and more for yet heavier stars. Some 11 billion years ago, there would have been 80 generations of supernovas of stars at eight solar masses, and of course, proportionally more supernovas for yet heavier stars. Why pick that time? That was the time at which the Kepler-444 system that we looked at earlier was formed.

The age of the Kepler-444 system was determined using the Kepler satellite in yet another way by which astronomers nowadays can follow stellar evolution: asteroseismology. This method relies on measuring minute variations in a star's brightness, measuring every few minutes for weeks to months. This reveals waves that run through the star. As for terrestrial seismology, the precise frequencies or tones of these waves map density and temperature in the stellar interior. Because these frequencies change over time with changing fusion-related chemical composition, this method enables a determination of a star's age, even if it is single and not in a cluster of siblings.

The Kepler-444 system shows us that planet formation has been going on for a very long time in the Galaxy, at least going back to the system's age of 11 billion years. This shows us that sufficient heavy elements were already available back then to form dense, rocky protoplanets and from there actual planets. The age of this system and of others that may have formed around the same time makes one wonder just how far life, and perhaps civilizations, may have developed given that they would have had at least six billion years more than we have so far on Earth.

Processes of formation

We are not only learning how widespread planetary systems are, but also about how fast the process of planetary system formation is. For example, among the many stunning achievements of telescopes relating to the formation of stars and their planetary systems is an image of the young stellar object called HL Tauri (see Figure 5.4). The central object, a star-to-be, appears to be less than a million years old, maybe much less than that. Around it, a disk has been imaged in light that is still further away from the visible than infrared. That light is called sub-millimeter radiation and lies almost in the radio range. The image shows a central bright object, surrounded by a disk comprised of alternating bright and dark rings. Astronomers estimate that the central young star

continues to capture mass at an equivalent of a few Earth masses per year.

The innermost region is too small to resolve; there the orbits of the terrestrial planets of the Solar System would easily fit. Further out, dark rings are seen at 13, 32, 65, and 77 Sun–Earth distances. These dark rings indicate a relative absence of bright gases; the question is, what causes the gaps? Hypotheses include that we might be seeing where different volatiles are condensing out, or gravity-related ripples in the disk, or the ordering of disk material by the gravitational pull of young pre-planetary clumps or young planetary bodies that orbit within the gaps or within the bright rings. Different astronomers have proposed different interpretations; the observations and models do not yet allow us to pick one over the others with a lot of certainty. But it is intriguing to note that this disk, probably less than a million years old, just one part in 10,000 of the lifetime of our Sun, is already showing structure on the way to a planetary system if not already with young planets in it.

How can we learn about the formation of planets and planetary systems if we cannot observe individual planets in any detail, let alone the dust, clumps, planetesimals, and protoplanets out of which they would have formed? The answer is: in a way analogous to how we learned about stars, namely observations of populations of planetary systems combined with numerical simulations of as many of the processes involved as our computers can handle. On the observational side this requires interpretation of finger-prints of exoplanets in the light that we observe, with the risks of interpretation bias and ambiguities and uncertainties. On the the-oretical, computational side it requires that simplifying assump-tions are made to enable even the most powerful of computers to do the computations, while hoping that those simplifications do not miss processes or alternatively introduce processes that do not occur in the real world. To minimize erroneous inferences it is essential that multiple groups of scientists work indepen-dently in parallel, not to compete for speed, but to mutually test

and validate their findings by understanding how sensitive their analyses are to how the laws of nature are approximated into computer code.

If some of the networked physical processes that are included in the computer code are put in erroneously, then one of two things can happen: One may find that the end of the model is completely different from the real world, in which case scientists have to hunt through the processes included and find those that were entered in a way that is inconsistent with nature. Or, one may find that the end result is compatible with the real world, in which case either the computer code closely mimics the real world, or the consequences of any erroneously implemented processes are small and do not matter, or nature has a way to rebalance through other pathways so that the outcome is more or less the same anyway. No matter what you find, the model has taught you something, but only if you look in detail at the whole and at all the parts. This requires detailed work by teams who have members who work both in the virtual computer world and in the real world of observatories.

To illustrate the complexity of the network of problems, look at this example of an actual end-to-end test of our understanding. Needed ingredients: formation and subsequent evolution of the material into, through, and out of the gas-plus-dust disk; evolution of gas and dust into precursors of planets, the small planetesimals, including how they collect solids and gases; the gravitational interactions between the planets as they collect more and more material; the gravitational coupling between the hordes of planetesimals and the already largely formed planets, which includes resonances and capture mechanisms leading to gaps or rings; of course, aerodynamic drag between solid bodies and the gas that surrounds them; all of this starting from more or less the right conditions at whatever point the computations are supposed to begin; and ultimately the comparison with observations that have all sorts of selection biases in them owing to the method of observation, the selection of the targets,

interference in signals by stars or clouds of gas near the target or just apparently close to it because they lie somewhere along the direction in which the planetary system is observed, and more.

Of course, there must be enough gas in the stellar disk in the first place to enable the formation of any sizable planets. Or more precisely, there must be enough solid material out of which the first stage of planet formation—dust—can continue toward ever larger objects. Dust can form wherever the star's apparent brightness in the sky, plus any heating that occurs in a disk around a star because of friction or magnetism, are low enough that temperatures of the gases and any forming bodies drop below the condensation point. That condensation point depends on the material in question: refractory materials, which include most metals and rock-like components, can form dust at much higher temperatures than, say, water ice can form, or at which other volatiles form their ices: ammonia, carbon dioxide, carbon monoxide, methane, etc. It also depends on the local gas pressure, but that is pretty low in interplanetary space.

The closest distance to the star where condensation is possible for a given substance is referred to as its "frost line". For water in particular it is often referred as the "snow line" or "ice line". Using the term "line" is, however, rather misleading, as is the suggestion that there is one such fixed distance. Apart from evolving with the brightness of the central star, it also depends on how transparent the disk is, which in turn depends both on how much gas there is in the disk but also how far from the mid-plane you are looking: these disks are densest in the middle, and less so above and below. Dense gas absorbs light efficiently—which is one way by which it can heat—but once absorbed, that light cannot heat things beyond that point: after all, in-home radiators only heat the room they are in because their infrared light can heat but not penetrate the walls. The snow line in a forming planetary system that still has a lot of gas and dust in its disk, which is strongly concentrated toward the mid-plane, is therefore nearest to the star in the mid-plane of the disk. It flares outward away from there above

Figure 5.9 The eight planets of the Solar System to scale, comparing the four terrestrial planets to the four giant planets, shown in order of increasing distance from the Sun. The "snow line" in the early Solar System would have been between the two groups of planets (See Plate 8 for a color version).

Source: Modified after a NASA composite figure.

and below the mid-plane of the disk because there is less matter there and the warming light from the star can reach further before being absorbed, although at some distance from the star the light becomes so faint that radiative cooling of material will always win out over heating. For the young Sun at a time when it still had a pre-planetary disk, for example, the average mid-plane snow line (in near-vacuum, water freezes at a temperature of –130 to –100 degrees centigrade, or –200 to –150 Fahrenheit) was probably somewhere around 1.6 to 1.8 Sun–Earth distances, which is just outside the orbit of Mars, but possibly as far out as 2.7 Sun–Earth distances. Other studies allow a somewhat larger range, illustrating just how dependent the result is on knowing the details of the disk. Frost lines for other volatiles lie still further out, at lower temperatures: for carbon monoxide, for example, this lies where temperatures are as low as –250 degrees centigrade (below –400 Fahrenheit).

The phenomenon of the Solar System's snow line has for decades been considered to be important for the formation of the giant planets, all of which formed beyond it (Figure 5.9). Any

planet begins to form by coagulation of solid particles; so this is the basic process by which a giant planet, such as Jupiter, is thought to start its formation. But then, in order to become a giant, it has to gather a massive atmosphere, too. That can only happen if it first builds up a heavy-element core of half a dozen or a dozen Earth masses, because only then is its gravity strong enough to pull in its extensive hydrogen and helium atmosphere until the planet weighs dozens to hundreds of Earth masses. It is the relatively heavy starter core that enables the growing planet to build up its atmosphere without losing most of it again through evaporation and also to acquire that atmosphere fast enough before the gas in the disk around the planet is blown out.

Water is so abundant among the materials that can form solids in space, that there is about three to four times more solid matter available for the formation of solid bodies beyond the snow line than inside it: within it, water exists as a gas which is hard to capture, while beyond it water is frozen into a solid that can easily stick gravitationally to a growing protoplanet. With more solids around, a larger body can be formed, and it can be formed faster. This appears to have been critical in the formation of the giant planets in our Solar System, and probably would be in any planetary system.

Why is it that astronomers think that planets as large as, or larger than, Jupiter form by first accreting a fairly heavy solid core and then sucking gases into their gravitational well? Why could they not form in the same way as a star does, which supposedly never had a solid core but pulls in gases simply by its own gravity? One piece of evidence, thought of as quite compelling, is that giant exoplanets have been found to form preferentially around stars with high heavy-element concentrations. That supports the idea that they form with help from solids in what is called the "core accretion" process that first forms a solid core that then captures gases. In contrast, star-like objects, including the not-quite-a-real-star brown dwarfs, do not appear to show such a heavy-element dependence even when they are binary companions to a real hydrogen-fusing star, that is, when they are in

positions much like giant exoplanets. This suggests that relatively heavier brown dwarfs form like stars do, by a gravitational collapse of a gaseous cloud without involvement of a solid core and that relatively lighter giant planets form through such an initial solid core.

Orbit migration, the snow line, and time

As soon as the discovery of 51 Pegasi b, the first exoplanet orbiting a Sun-like star, was reported it was recognized that an unanticipated process was needed to explain its history: this giant exoplanet was too close to its parent star to have formed where it was, given the existing hypotheses of giant planet formation, so—unless these hypotheses were to be rejected—51 Pegasi b must have moved in from further out. The process involved had already been discussed around 1980 for clumps in the rings of Jupiter, but now it was obvious that the orbit of an object as large as the planet 51 Pegasi b should have changed over time because where it was, so close to its star, there would not have been sufficient condensed particles to initiate the process of formation for a giant planet, unless the dust made of frozen volatiles itself behaved differently than expected. Following up on the first suggestions for this process, which was recognized to be important for small bodies within a heavy disk, it became clear that orbits of planets as heavy as Jupiter can change over substantial distances through a mutual pulling between the orbiting planet and the material in the disk. Alternatively, planetary orbits might be changed by forces between planets after formation, or perhaps the initial small particles in the disk migrate inward even as planets were forming closer to their star.

Another problem immediately arose with this idea of planet migration through the early disk: if planets can migrate inward in their interaction with a disk of gas or small planetesimal bodies, what stops them from spiraling into their star? Here, it seems, stellar magnetism comes in. The young star rotates rapidly, and

its magnetic field moves along with it. Close to the star this spinning force field whisks away any part of the disk that comes too close. Once planets spiral in to near that cleared zone, their fall toward their star may stop because no disk material is pulling on them from inside their orbit. Later, it became clear that planets can also stop their inward migration for other reasons, and that other conditions can cause outward migrations. These migrations can be very fast, either within thousands of years of the formation of the planets, or much delayed by over 100 million years, depending on what happens. And as it turns out, both these directions of migration appear to be important for understanding why our Solar System, and our Earth in particular, are the way that they are.

Observations reveal that much of the initial molecular-cloud gas condenses into pebbles early on in the disk phase around the forming star, and perhaps has done so already before such a disk is fully formed out of the contracting gas. These pebbles would cluster into ever larger bodies, which would sometimes be destroyed again by collisions with other bodies. After some time, planetesimals would form that would be large enough to survive most collisions with other bodies. This appears to happen within the first million years or so. These planetesimals then fairly quickly collide to form much larger protoplanets, possibly as fast as within 100,000 or even 10,000 years. By and by, these protoplanets merge to form the ultimate planets over tens of millions of years. A planetary system takes shape with exoplanets in orbits that are a consequence of the properties of the initial disk, of quasi-random close interactions between themselves, and over the long haul of slow, mutual deformations of the orbits of all exoplanets in the same system. Much of this is uncovered by comparing numerous exoplanet systems to ever more complex numerical experiments.

With the above in mind, we can grasp why the terrestrial planets have so little hydrogen and helium and also why they are as dry as they are. Of the four, only Earth has a thin surface

veneer of water that is, on average, a mere 0.05 percent of the Earth's diameter. There is likely a roughly comparable amount of water locked up in Earth's mantle, but that still leaves it a very dry body compared to, say, bodies in the outer parts of the asteroid belt that have average percentages for their water contents that are 50 to 100 times higher. First to note in understanding the dryness of the terrestrial planets is that they formed inside the snow line of the Solar System, thus in a region where water ice was scarce, so water could not be captured to begin with. Second, with considerably less solidly frozen material within their vicinity, they grew relatively slowly out of refractory materials, in a phase where gas was already being blown out of the system. They thus never reached the size at which their gravity could capture hydrogen and helium gases while that gas was still around, as the giant planets' protocores did in their orbits much further out. So they never got to grow very much and never had much of an atmosphere in comparison with the giant planets.

There is, obviously, also information on the formation of planetary systems from our own Solar System. The study of decay products from the radioactivity with which the interstellar gas was seeded by supernovas prior to the formation of our Solar System has revealed some interesting timing information. Asteroids seem to have formed some time between one and five million years after the formation of the pre-planetary disk. The radioactivity appears to have been so strong that those bodies that formed early on melted into rather homogeneous clumps, while those that formed a few million years later were lumps of accumulated pebbles. The rocky planets, including Earth, seem to have formed much later, by perhaps somewhere between 40 to 120 million years. These formed by collisional mergers of asteroids that took place well after the gas disk had disappeared.

The relatively strong radioactivity early on in the formation of the Solar System was likely sufficient to keep the early, large bodies in a melted, liquid state for quite some time. The strongest

contributors to this form of heating had a half-life below a million years, so that bodies forming just a few million years later no longer melted. Having said that, melting could also be induced by fast collisions, so that it may well be that these mechanisms contributed differently in different places within the Solar System.

With planetary systems being formed out of spinning disks, it is straightforward to understand why much evidence shows exoplanetary systems to be about as "flat" as the Solar System. Not only is the gas collapsing into a fairly flat system, but the solid dust that formed beyond the frost line corresponding to each particular form of solid is confined to a disk that is flattened yet more by drag as it moves through the gas. This is because the material acts like any sufficiently dense gas and settles into an orbiting disk in which gas pressure helps to create a somewhat denser center than the outlying layers. Dust that is not orbiting entirely within the mid-plane will orbit the young star in an inclined orbit, thus moving up and down relative to the disk. Every time it goes through the dense mid-plane there is a bit more drag on it against the gas, slowing its up-and-down motion through the disk, until at last the dust becomes still more concentrated in the center of the disk than the gas, making it easier to ultimately form a "flat" planetary system. The outliers from "flat" exoplanetary systems, with exoplanets in often mutually inclined orbits, seem to be those systems with close-in giant exoplanets...those that are not supposed to be able to form where they are.

Hardly enough time has passed since the appearance of exoplanet studies as a mainstream science to enable much testing and validating of the proposed formation processes. Nonetheless, trends are emerging. I hope that the trends that I have picked to describe here stand the test of time, but likely some will not, so— as always in science—you should be mindful that all of the findings described above are "preceded by mind, led by mind, made by mind": there may be some "old thoughts" that inadvertently were pulled into the modern interpretation. Identifying any such errors will be a task for the future.

In closing this chapter, we need to update only one point in the list of "forensic evidence" from Chapter 3 now that we know how heavy elements are injected into interstellar clouds out of which planetary systems form:

> 10-r. All bodies in a planetary system are likely to contain limited amounts of radioactive materials in concentrations that depend on details of the system's prehistory. The heat released by radioactive decay supports, at least for some time in the terrestrial planets, tectonic activity.

Whereas we cannot directly establish this to hold for all planetary systems, the evolution of the stars that is essential for the formation of the heavy elements needed to make planets is unavoidably connected with natural radioactivity. So it must be a property common to all systems. How much, however, depends on the environment in which planetary systems form.

6

Drifting Through a Planetary System

The surprise was immediate: the very first exoplanet that was discovered next to a Sun-like star was a giant heavier than all planets in the Solar System except Jupiter, but it orbited its star way too tightly to have formed in the manner that any of the giants in the Solar System were thought to have formed. 51 Pegasi b was, in fact, so close to its star that by the science of 1995 it should not have existed. Where it was, at 1/20 of the distance from the Sun to its innermost planet, Mercury, the standard model of the time showed that nothing could have been cooler than 2,000 degrees centigrade (3,000 degrees Fahrenheit) so there should not have been anywhere near the amount of dust out of which the initial planetary core could have begun to form fast enough to eventually grow into a giant planet. If, somehow, it had been able to start forming, it just could not have collected enough of the gases as these would have evaporated again in the bright, warming light of the nearby young star or would have been whipped away by the spinning star's magnetism. Proposals to address this problem in our understanding were made almost immediately, however, and a decade later a successful idea was left standing, ready for continued testing. That idea was that the orbits of planets can contract or expand, particularly early in the life of a planetary system.

Intuition in science

Advances in the physical sciences are commonly perceived as emerging from seemingly never-ending manipulations of

One of Ten Billion Earths. Karel Schrijver, Oxford University Press (2018). © Karel Schrijver.
DOI: 10.1093/oso/9780198799894.001.0001

complicated equations full of Greek and Latin symbols, supplemented by mathematical shorthand characters that are found on no keyboard, not even on that of the mathematician's computer. Whereas mathematics is doubtlessly an essential ingredient in formulating thoughts both logically and quantitatively, such equation-oriented activities are more often involved in the follow-up exploration of ideas, and certainly in the formulation of a convincing case that is to be made to colleagues in either research publications or proposals. But in my experience it is rare that discoveries in astronomy start with writing down equations. Something precedes that phase, something that convinces the researcher that it is worth going through the tedious process of crunching blends of lengthy equations into an interpretable conclusion. At the beginning, there is some combination of a hunch of a solution spiced with a flavoring of how one might go about tackling that challenge.

In many branches of the natural sciences, advances are made based on expensive, highly technical experiments supported by arrays of specialized measuring devices in minutely controlled laboratory settings. This, too, requires that the researchers embarking on such an experiment have already viewed in their minds the promise of an informative and—even better—exciting outcome. For astrophysicists, however, there is this stumbling block: we cannot experiment with the real world in a physical, tangible laboratory simply because what we deal with is generally too large, too powerful, or too extreme to fit in any laboratory we could afford to build. Consequently, the astrophysicist's laboratory has increasingly become the computer in which ideas can be explored, using equations as foundations, and vast amounts of numbers as the links to the reality that presents itself around us in the Galaxy when viewed through powerful telescopes. Nonetheless, in the end, as Arthur Eddington put it in 1938 in *The Philosophy of Physical Science*, "For the truth of the conclusions of physical science, observation is the supreme Court of Appeal." But the computational experiment,

supported by mathematical theory, helps us guide our thoughts toward the experiment, because, as Eddington continued, "Every item of physical knowledge must...be an assertion of what has been or would be the result of carrying out a specified observational procedure."

Regardless of whether we address a challenge with equations in pure mathematical terms or implement it on computers, we have to start somewhere. We have to grasp not only the problem in the beginning, but also the promise of an outcome and the pathway that leads us from one to the other. In other words, we have to imagine how we do something first, long before we can do it or before anyone decides to fund us in doing it.

No matter how wild our imagination is, we start from what we know, from what we have experience with. For astronomers at the beginning of their careers this means that they start with a store of everyday experiences—just as we all do in life—combined with what they are taught. What happens if there is something out there in the Universe that we have not seen in everyday life, which for an astronomer, in truth, is just about everything except for the most basic processes? What happens when something has been only very recently discovered, so obviously has not been taught about in classes either at all or at least not enough because we do not know what is going on? In that case we have to rely on a gut feeling of being onto something; in other words, on our intuition.

The crucial importance of intuition is not often emphasized by researchers, be it in their professional writings or in how they communicate with the general public. I am not sure why that is. Perhaps it is a feeling that the initial searching for a solution to a scientific riddle is not interesting or not particularly supposed to go with the image of a purely rational scientist. Perhaps it is because that initial intuitive process is forgotten, or thought of as an unnecessary digression from the ultimate findings. But I am convinced that the creative process of science is an inextricable interplay between logic and intuition.

In the form in which I mean it, intuition is not some pre-existing instinctive view of the Universe that astronomers are born with. One way of defining it that resonated with me was formulated by Raymond Louis Wilder (1896–1982) when he was writing in 1967 on the role of intuition in mathematics. Mathematics and astrophysics are not all that different in my mind: both deal with mostly imagined worlds ruled by the logic of equations. Wilder equated intuition to "an accumulation of attitudes (including beliefs and opinions) derived from experience, both individual and cultural. It is closely associated with ... knowledge, which forms the basis of intuition. This knowledge contributes to the growth of intuition and is in turn increased by new conceptual materials suggested by intuition."

He immediately acknowledged that "Before one can have a really intuitive feeling about ... problems, one must have worked on them." For scientists, "their choice of the problems on which they work is guided by what the collective intuition deems the most fruitful direction for research." Hence, "The major role of intuition is to provide a conceptual foundation that suggests the directions which new research should take. [...] The role of intuition in research is to provide the 'educated guess,' which may prove to be true or false; but in either case, progress cannot be made without it and even a false guess may lead to progress." He then notes that this individual intuition is eventually shared within a science, becoming a collective intuition. This also influences how the really big problems evolve: they are often handed over from one generation of researcher to the next. Wilder says: "I believe that what happens here is that the collective intuition in the field of a particular problem continues to grow, being passed on by the older workers to the younger. Ultimately, due to a combination of a more mature collective intuition (which has been growing unnoticed), new methods, and individual genius, someone (usually a younger [scientist], relatively new in the field, and possessing a fresh individual intuition) is able to solve the problem."

In practical terms, on the theme of this book, this is how these words can be interpreted: No one had seen any other planetary system than our own. Thus the thinking of scientists was guided largely, and tested exclusively, against the properties of this one system that we knew. Processes that were not obviously ongoing, or that were not clearly required to explain present-day properties of the Solar System, were not given much attention or were not even imagined. And there were quite a few of these. But then both inquisitiveness and intuition were marshaled to the frontier of science by the discovery of exoplanets in unexpected places.

The changing orbits of planets

Within months of the announcement of 51 Pegasi b, an exoplanet in a system so different from our own that it rattled our thinking, there was a publication that proposed a solution: the close-in giant planet 51 Pegasi b was to have formed much further out, well beyond the snow line that lay several dozens of times more distant from its sun than its present orbit, and subsequently to have spiraled inward. Such orbit migration had been discussed for the Solar System. Initially, it was considered to be relevant only for the small bodies, the planetesimal–asteroid–comet population, but early in the 1980s discussion of the consequences for young, forming planets in the Solar System appeared, mostly in relation to planetary rings. But publications on the subject came in at a tiny trickle of one or two per year, not reaching the mainstream of research. In studies that have appeared since 1995, orbit migration of planets is mentioned in hundreds of research publications.

By 1995, the conceptual leap occurred that the orbits of giant exoplanets can contract over large distances during and after initial planet formation would allow the hypothesized formation process of giant planets as formulated for the Solar System to survive. Instead of a major revision of the formation process, that leap was based on the realization that exoplanets and their progenitors, small exo-planetesimals and larger exo-protoplanets,

literally did not orbit in a vacuum. They were embedded in the circumstellar disk full of gas and dust and larger clumps of matter that would initially have contained quite a bit of mass compared to the mass of the early exoplanets. Such a relatively heavy disk would not leave the young exoplanet alone: gravitational tidal forces would work in such a way that the planets' orbits would change, typically causing the planets to spiral in toward the star. Scientists later realized that some disk properties could also cause outward spirals.

Initially, when it was proposed, this was a helpful idea for the formation of our Solar System: as a planet formed, its gravity would pull in the pebbles, dust, and maybe gas near its orbit, but once it had cleared this zone out, how could it grow further? The idea of letting the young planets spiral in was just great: by moving the planet into another orbit, it would come into new feeding grounds filled with more disk material, all orbiting the central star, that it could suck up, then move on in further, suck up more, and so on until the entire disk was essentially cleared and the orbital migration would stop. Or until the migrating exoplanets reached the inner edge of the disk, where the star's magnetic field creates a cavity beyond which no inward tidal forces from the disk would work. Whereas this concept would allow Jupiter and the other Solar-System giant planets to grow efficiently, it did require rather special conditions to prevent them from falling too far toward the Sun when they did so.

The idea that planet migrations would somehow be stopped by emptying the disks just in time to leave so many planetary systems with heavy giant planets around in the Universe seemed to rely too much on an improbably fortuitous timing of events. However, it did open up new ways of thinking: at least the thinking now shifted to a paradigm in which (exo)planets did not necessarily form where they were found.

Around the time the concept of planet migration was developed for exoplanets, the growing collective intuition was beginning to guide research and research funding. Experiments were

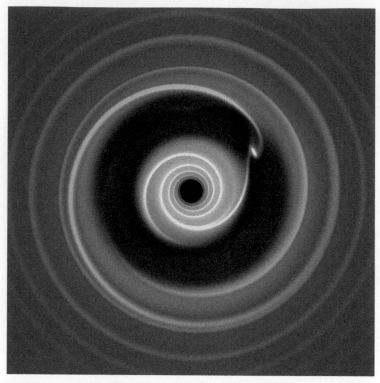

Figure 6.1 Numerical experiment of a protoplanetary disk. In this computer model, a model planet (in the center of the bright patch in the dark clearing) of three Jupiter masses is placed at 5 Sun–Earth distances within a gaseous disk that gyrates around a Sun-like star; the star (not shown) is in the center of the image, surrounded by an opening in the disk which in real life would be a region whipped clear by the magnetic field of the spinning star. The gravitational interactions in this rotating disk lead to a spiral-shaped wave that orbits the star in an almost fixed pattern along with the planet, changing only very slowly as the orbit of the planet changes due to the gravitational forces between the planet and its wave. The brightness scale goes from black for near vacuum to white for the most massive regions in the gas disk.

Source: Simulation by Phil Armitage.

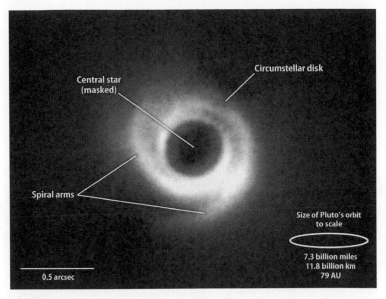

Figure 6.2 Spiral waves in an object named SAO 206462 that is an exoplanetary system in the making. This system is a distant target that appears very small in the sky (the scale of 0.5 arcsec is equivalent to a U.S. 10-cent coin seen at eight kilometers or five miles), but it has the size of an entire planetary system as indicated by the scale in the lower right (one AU or astronomical unit equals the Sun–Earth distance). The very young, bright star in the center of its circumstellar gas-dust disk has been masked out in order to see the structure of the disk around it. The disk shows two spiral arms, reaching so far out that on the scale of the Solar System they would reach as far as the orbit of Pluto.

*Source:*NAOJ/Subaru, after a study by Muto and colleagues (2012).

set up in the astronomer's primary laboratory, the computer. Much work has been done this way regarding planetary migration over the subsequent two decades. With these computations it became clear that planets could set up spiral waves in the gas that would predominantly let them spiral inward through their large-scale, long-range, unbalanced forces (one such example in the virtual world is shown in Figure 6.1 and a real-world example

seems to be found in an object known as SAO 206462, shown in Figure 6.2).

Then, as the question arose of how such spiraling motions could be stopped, resonant couplings were noticed as important. Whenever pairs of planets had orbital durations that always pulled in the same phase, orbital migrations could be stopped altogether; this would be the case when periods had ratios forming whole numbers, for example 1:2, 2:3, and 2:4. Moreover, if some process were capable of creating ring-like density enhancements (such as seen in the case of HL Tauri in Figure 5.4) then this would influence where planets form and where they migrate to.

The problem with Mars

Problems with planet formation were not limited to exoplanetary systems, however. Within the Solar System, multiple studies over the preceding decades had shown how, starting from dust, via planetesimals and heavy planetary embryos, at least the inner terrestrial planets Venus and Earth would have been able to form roughly where they did and in the models could have orbits and masses roughly as they do in real life. But Mars was a problem. Most models put it too far out from the Sun, and could not reproduce the real lightweight planet that Mars is in the real world: coming in at only a tenth of the mass of the Earth, models up to a decade ago could only make a Mars-like planet by assuming special, and seemingly artificial, conditions. Without those, Mars would typically outgrow Earth in the computer experiments. One way to make Mars stop growing early in its life, thereby stunting its development, would be to remove most of the planetesimals out of which it would have to grow at least from somewhere between the orbits of Earth and Mars and from there quite a way outwards. But that seemed, for just a couple of years, an odd assumption to make as there appeared little evidence to support it. Scientists were not happy with that discrepancy.

But by 2011 Kevin Walsh and colleagues came up with a plausibility test for this truncated planetesimal disk through a concept that had been proposed already a decade earlier: could they gather convincing evidence for the argument that had been made that Jupiter had gradually fallen toward the Sun close enough to gobble up or kick out much of the mass that would have otherwise formed a heavier Mars? Could it then later somehow have been pulled outward again from the Sun's gravitational hold to where it is now, moving from about 1.5 to 5 Sun–Earth distances? Well, by then astronomers were able to grasp the idea of planets spiraling in, because that had started to come into mainstream thinking fifteen years earlier. But how do you move something the size of Jupiter—at 318 Earth masses—outward by over 500 million kilometers or 300 million miles against the Sun's gravity? Moreover, if this process could have happened, and did take away growth material for Mars over a wide range of distances in the Solar System, then how come there are so many asteroids at 2 to 3.2 Sun–Earth distances, right in the zone through which Jupiter would have had to travel twice in what is now called "the Grand Tack"? That name has a nautical origin that visually works as an analogy, although physically it does not: sailing ships cannot sail directly into the wind, but can move against the wind up to a certain angle on either side of the direction from which the wind comes, and the "tack" is the turn from going against the wind on one side to the other, so this movement bears a remote resemblance to the back and forth that Jupiter is envisioned to have done in its Grand Tack.

The solution to the conundrum of how Jupiter could have moved into and out of the inner Solar System was found in pairing it up with another planet. The early work on planet migration focused on the orbital movement of a single planet through the disk. In the computer experiments, astronomers noticed that for planets of different masses, the behavior was different. For a body that is up to a few Earth masses, the interaction with the surrounding disk sets up spiral density waves in the disk that

effectively take rotational energy from that body, causing it to spiral inward fairly fast, gradually falling toward the central star in a spiraling orbit. A heavy body, like a fully grown Jupiter, clears the orbits near its own by its gravitational pull, which also leads to an inward spiral, but at a rate that is set not by gravity but by how fast material from outside the cleared gap migrates into it. Both of these processes develop on timescales of tens to hundreds of thousands of years. For planetary masses between those of Earth and Jupiter something happens that can cause the body to spiral into its star in a matter of decades. In the literature, these processes are subdivided into Type I, Type II, and Type III migration, but we need not go into that distinction here. Just note that all of these processes are scarily fast for planets that are in the midst of formation processes that span multiple millions of years!

Once it was established that bodies of different masses migrated at different speeds, planetary scientists realized that this meant that if there were multiple such bodies going around, a diversity of things could happen. Specifically, Jupiter and Saturn would make a very interesting pair: the heavier Jupiter would spiral in more slowly than a lighter Saturn. As Saturn would likely have started growing somewhat slower in the less-dense settings further out in the primordial disk, the faster-spiraling proto-Saturn would play catch-up with its bigger sibling. But another process would intervene to keep the early Saturn from coming too close to the early Jupiter. That process is called mean-motion resonance, or orbital resonance. What is that about?

Resonant clocks and planetary orbits

Much in nature is based on cyclic repetitions: sound, light, clocks, tides, seasons, even our heartbeat and respiration. In many cases, there are several things that cycle that are not entirely independent of each other. Possibly the first to notice that such cycling objects could couple into a joint movement was Christiaan Huygens, a Dutch scientist. After being bedridden because of a "small

indisposition" he wrote a letter, dated February 27, 1665, in French as was common those days, to Robert Moray (1609–1673), an influential member of the Royal Society of London. He writes that he noticed that two pendulum clocks, which he had invented in 1656, hanging a few feet from each other on the same wooden beam had their pendulums swing in remarkable synchronicity. If he stopped and restarted one of these off the beat of the other, they soon recovered their synchronous behavior, both running at the same rate. When he moved them to opposite sides of the room, they lost their synchronicity, with one losing several seconds per day relative to the other. He was puzzled about what it could be that coupled them, pondering whether somehow tiny vibrations in the air were the mediating agents. Understanding of coupled oscillators still remains difficult in many circumstances. In an experiment in 2015, 350 years later, Jonatan Peña Ramirez and colleagues finally demonstrated that the synchronization of Huygens's pendulums was in fact most likely caused by the flexing of the wood upon which they were mounted. But even this recently the authors noted that, prior to their experiment with two large pendulum clocks mounted on a common wooden structure, "Huygens synchronization is still an open problem. This claim may be surprising, specially if one considers the fact that the behavior associated to pairs of coupled oscillators has been extensively and exhaustively studied."

Huygens performed, in effect, an early laboratory experiment of what is now found in many processes, from mechanical resonances in cars to biological rhythms in nature. Coupled oscillators influence each other by whatever force exchanges energy between them. When the natural frequencies of the oscillators is essentially the same, as was the case for Huygens's pendulums, perfect synchronicity can result, but more often than not oscillators have different frequencies and their coupling does not lead to synchronicity at all. In general, the coupling transfers energy from one oscillator to another. This can lead two oscillators to run together, or to run in some fixed multiple of each other's

motions, or to run in a complex and seemingly erratic pattern by transferring energy back and forth from one to the other, thereby continually changing the amplitudes of the oscillations. The latter occurs, for example, when two quite different pendulums are suspended from a common string.

The natural period of a pendulum is determined by its mass and by its length. If two are made that are identical, then these two, if coupled, can swing back and forth in perfect synchrony, either fully in step or moving in opposite directions all the time. If they are made so that they differ slightly, however, Huygens's experiment can be recreated: if not coupled, they swing at slightly different rates, but when coupled they will synchronize so that neither swings entirely at its natural frequency, but both at the same. If the pendulums have natural frequencies that differ by essentially a simple ratio of whole numbers, such as 1 to 2 or 3 to 5, then the coupling will tend to make them swing exactly at that ratio of whole numbers. The state of two or more oscillators at periods that are ratios of whole numbers is called a resonance.

That all this has something to do with the Solar System was worked out by Laplace some 120 years after Huygens's first observation. He was intrigued by whether a planetary system with as many planets as the Solar System was intrinsically stable or not. One thing he worked out, with the help of Newton's laws, was what happened between Jupiter and Saturn. Two orbits of Saturn take very nearly as long as five of Jupiter. That is far from the nearly identical oscillators that Huygens observed, of course, but nonetheless, Laplace noted that this movement in this repeating pattern was able to influence the planetary orbits into having clear non-circularities. He also began work that revealed a similar resonance in the largest moons of Jupiter: he argued that this resonance was important to the orbital periods of Io, Europa, and Ganymede whose periods are 1.77, 3.55, and 7.16 days, respectively, with ratios of very nearly 1:2:4. Note that if the orbits of bodies are moved into a resonance, that means that they are moved away from orbits where they are not in near-perfect period ratios; that

leads to zones of avoidance in disks such as many of the gaps in Saturn's rings (Figure 6.4).

Such orbital resonances are now recognized throughout planetary systems. In our own, they govern the clustering of asteroids leading and following Jupiter (the Trojans in Figure 3.2), the orbits of Pluto and Neptune, and induced gaps in the planetary rings and in the asteroid and Kuiper belts. Several exoplanetary systems have pairs, triplets, and even quadruplets of planets in resonant orbits.

When two planets are put in a state of an orbital resonance they cause their orbits to become more eccentric and then coupled. This happens most clearly for resonances involving period ratios of low numbers, such as 1 to 2 or 2 to 3. This kind of resonance can even cause the orbital planes to be torqued away from each other into a non-flat planetary system, something that is important for the inclined, eccentric orbits in some exosystems, such as exoplanet HD 80606 b as discussed in Chapter 9.

The initial hypothesis of pairing of Jupiter and Saturn in an orbital resonance was found to have unanticipated impact potential for the Solar System. Researchers found that this powerful pair of giant planets could have reshaped the disk of gas, dust, and planetesimals in several ways. Once paired in resonance, not only could they stop spiraling toward the Sun, they could even move outward again. How exactly they did this is hard to formulate in words alone, but it can be visualized now based on a variety of computer experiments: the planet pair creates a big gap in the disk by both steering matter toward the Sun and outwards, in huge density spirals (an example of a computer simulation of that is shown in Figure 6.3), as well as by throwing other smaller bodies off their orbits. Although the effect is like pushing things away, gravity can only pull; the clearing of a gap is actually the result of the coupling of several effects, some even working against others. There are the gravitational tides caused by the collective of all moving bodies. There are near-collisions between bodies that are both like a viscosity of a fluid and the pressure of a gas: groups

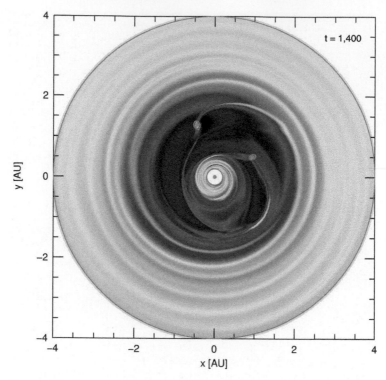

Figure 6.3 Computer experiment for a young planetary disk gyrating about a forming star in the center with a clearing in the disk immediately around it. Orbiting in the model disk are two Jupiter-mass planets with orbital periods of one twice that of the other. The model planets (the two dots away from the center, smaller than the ovals shown around them) interact primarily with gas in the disk on the side opposite from the other planet. Everywhere but in the innermost white region and outside of four Sun-Earth distances (or AU) where the model assumes a vacuum, the intensity shows the mass contained in the disk, from near vacuum in black to densest in white: the resonant pair of planets has sculpted a region of low density in the disk—a gap—through the phenomenon of spiral density waves both inside and outside of the gap.

Source: Kley and Nelson (2012).

Figure 6.4 Daphnis orbiting in a gap in Saturn's rings. This small moon is only eight kilometers (five miles) across. The image, taken by the Cassini spacecraft, shows how the gravity of this moonlet shapes waves on the edge of the gap that it has cleared in the rings.

Source: NASA/JPL-Caltech/Space Science Institute.

of interacting bodies can scatter each other about to gradually spread out to fill a larger space, so that some of them can, for example, migrate into a gap cleared by a growing planet. And there is the overall conservation of momentum of orbital motion that means that if one thing slows down in an orbit, other things must speed up. Altogether, these effects slow down matter inside the orbits—causing it to move closer to the central star—and speed up matter outside of the orbits—causing it to move away from the star—thereby opening up a gap in the disk.

As the pair of planets interacts in a mutual gravitational pull with the density spiral, and also exchanges orbital energy with bodies that they scatter about, they themselves—remaining locked in their orbital resonance—can move through the planetary system. Their direction of motion depends on whether they primarily cause other matter to spiral inward—in which case they themselves will move outward—or the other way around—with the opposite result. That, in turn, depends on how much other matter there is just outside and inside their orbits. Thus, the direction of motion depends on what is referred to as the density gradient: the change in density with increasing distance from the central star.

This led to many different avenues of research as well as possibilities for the evolution of a forming planetary system. How much matter is where in the disk at different distances from the star depends critically on how fast mass is fed into the disk by the collapsing cloud early on. It also depends on how fast matter moves through the disk to fall onto the star, and once the star is bright enough it also depends on how fast gases are being evaporated from the disk into the disk winds.

The forming planets also create gaps in the disk as they collect matter from around them, while moreover a heavy sibling can affect the surroundings of the lesser partner or partners, which will change their rate of growth as well as the rate, and even direction, of their subsequent migration. Any substantial density jump to higher densities outside a region of lower density can stop planet migration. Thus, such a location becomes what has become known as a planet trap. These planet traps shift in time, however. One reason for that is because the frost lines move: at these lines, different for different substances, solids would coagulate out of the gas which helps build the planets. These frost lines depend on the stellar brightness and on the transparency of the disk, both of which are changing fairly dramatically in the early phases of a planetary system as the young star forms and as the surrounding gases are cleared away.

Ultimately, closest to the star, there is a point where the magnetic field of the rapidly spinning star sweeps material clear away, forming an inner edge to the disk; that would be the final point at which an inward migrating planet can become trapped. Such traps will disappear once the disk disappears, but when the disk disappears, the pulls between planet and disk also disappear, so that the last trap location may be where the planet remains. Then again, if a planet migrates too fast, the traps may not be able to capture it, or the capture into an optimal orbital resonance may fail to take hold, so that such planets may fall into less stable resonances, or they may ultimately spiral so close to their star that they are demolished, shredded by the star's gravity and evaporated in its heat.

Testing the scenarios and Earth's water content

If all this sounds like there are many things that influence the evolution of a planetary system, then you are putting your finger on a sensitive spot: computer experiments have revealed a great diversity of possible planet configurations in hypothetical planetary systems in which different processes were included somewhat differently. Plus there is the element of chance because the planets ultimately form by collisions between bodies moving in a swarm: running the same experiment but with slightly different initial positions of the players can, and frequently does, result in a different outcome, as also expected in the real world.

On the one hand, it is comforting that there is sufficient freedom in creating simulated planetary systems that mimic the great diversity of observed planetary systems. On the other hand, it is important to make sure that one does not simply invent freedoms that nature does not have. After all, the processes involved are all subject to the laws of physics—the dynamics and composition of the initial cloud, the resulting circumstellar disk, the formation of the first planetesimals, and the ultimate planetary system that forms. Computer experiments are also subject to

the laws of physics, albeit necessarily formulated as simplified approximations. Moreover, they are started from some initial configuration of disk matter and early planetary bodies as input provided by the researcher; these may or may not be consistent with each other, depending on whether the computation could be started at the very beginning of it all. Computer models suggest that the planets would migrate far faster than seems to be the case in the real world. Ways to prevent that from happening in the virtual world are being experimented with, but the computer models would have to include much more detail to test these ideas out properly. Unfortunately, our computers are not, or at least not yet, powerful enough to do so, despite all their present-day capabilities and speed. So what do you do to prove that, at a minimum, you have not taken liberties with the laws of the Universe?

This brings us back to Walsh and colleagues. They did not actually work through how the inward-then-outward migration of Jupiter and Saturn came about, but strictly focused on what the consequences would have been on what we can still, today, observe. Namely: would the terrestrial planets still have formed as they did, would Mars have suffered severely stunted growth, and would the asteroids have managed to get to where we see them today despite Jupiter coming through their range twice? If all this could be answered satisfactorily, we would be able to learn more about when and how far this "Grand Tack" occurred, and perhaps other parts of the history of the Solar System might be unveiled.

In their experiment they used the properties of the asteroid belt as the primary test of the consistency of their model with the real world. For their computer model, they had to make a series of choices: they had to choose where to put the proto-giant planets, how fast and far to let them migrate, how much and at what rate to acquire mass from the surrounding disk, and how much mass was where in the planetesimal asteroid population. They experimented with various choices, settling on the following. They started with a plausible gas disk based on work by colleagues, and chose representative sizes for the population

of planetesimals so that they could model the drag forces acting on the model planetesimals as they moved through the model gas. Their model Jupiter is started fully formed at 3.5 Sun–Earth distances, which is just outside the snow line where growth of a giant planet is thought to most readily occur. Their model planets Saturn, Uranus, and Neptune are still growing at the start of the simulation, positioned at 4.5, 6, and 8 Sun–Earth distances, respectively, which is much closer to the Sun than they are now; Neptune, for example, is currently at 30 Sun–Earth distances. After some time Saturn is sent migrating inward to catch up with Jupiter, which is also sent migrating inward, albeit more slowly, when Saturn begins to catch up.

None of this is what is being tested, of course: these are the prescribed properties and processes. What they tested was what the consequence was for the many thousands of small objects, for which they chose 100 kilometers or 60 miles as the typical size, that were distributed throughout the disk from close to the Sun out to 13 Sun–Earth distances. Then they tracked where these all ended up. Why did they think this was an informative test? Because it would enable them to see not only whether an asteroid belt would exist after the Grand Tack of Jupiter, but also where the asteroids in it might have come from. This goes back to the discussion of the volatile-poor S-type (silicate rich) asteroids versus the volatile-rich C-type (lots of carbon, and water) asteroids in Chapter 3. S-type asteroids are found mainly within about 2.8 Sun–Earth distances, thus largely within the snow line, while C-type asteroids dominate beyond that distance. In their experiment, Walsh and colleagues did not have to specify the chemical composition; they merely needed to trace where each model asteroid went from beginning to end of the simulated evolution. They simply flagged two different categories of asteroids: let us call all those simulated planetesimals that started within their model Jupiter orbit mS-type, and all others mC-type.

From their model run, several things were inferred. First, the inward migration of the giant planets displaced many mS-type

model asteroids inward through orbital resonances, moving them closer to the Sun, where they would be available for the later formation of the terrestrial planets. After the two migratory passes of Jupiter and Saturn—inward, then outward—a little over 1 percent of the model planetesimals where the real asteroid belt is now found themselves first scattered outward, and then back inward, to match the real-world S-type asteroids. Many of the model planetesimals that started further out, the mC type, were scattered back from well beyond the snow line inward to also populate the region where the real asteroid belt resides. The ultimate orbits of this population at the end of the computer experiment overlap somewhat with the mS-type ones, but average a little further out, just like the real C-type asteroids that lie on average beyond the real S-type asteroids. Some of the mC-type planetesimals were scattered even further inward, into the domain of the infant terrestrial planets.

So, not only did their model show that an asteroid belt could survive (well, actually reform after) the double trip of Jupiter through its domain, but also that it could mostly maintain the mS-type/mC-type separation by distance across the snow line. Plus, it implanted supposedly water-rich mC-type asteroids inward of the snow line: water could not condense there in the real world, but inside an asteroid body that had initially formed further out in the colder regions of the Solar System it could survive without evaporating. This is fascinating, because it may well be why there is as much water on Earth as there is: C-type asteroids, shepherded there from much further out by the migrating Jupiter, made it available in the finishing phases of the formation of the terrestrial planets. Plus, as a result of Jupiter's pushing and pulling, there would be less asteroid material left for Mars to form out of, but still enough to make it the smaller brother of Earth at 11 percent of Earth's mass.

Walsh and colleagues also looked at what the computer suggested would happen if Jupiter had started at a very different location or migrated to a different distance. This turned out to be less

successful in matching the real-world asteroid belt, suggesting that Jupiter may indeed have formed much closer in than it is now, a little outside the snow line, as theoretical arguments would suggest.

A few years after the experiment by Walsh and colleagues was published, it was discovered that merely by growing into giant planets far from the Sun, even in fixed orbits, giant planets can scatter nearby asteroids into orbits closer to the star in the center of the emerging planetary system. These scattered asteroids would have formed beyond the snow line, so would contain a substantial amount of water, such as C-type asteroids do. On one hand one might now say that there are two scenarios to nudge water-rich asteroids from far from the Sun onto collision courses with the terrestrial planets closer to the Sun, so that the test by Walsh and colleagues leaves us with an ambiguous outcome: the overlap of S-type and C-type asteroids may or may not be a consequence of orbit scattering by migrating giant planets. Both scenarios may work, but the asteroid populations leave it unclear whether one of them or both actually occurred in the young Solar System. But, as described above, the "Grand Tack" does provide an appealing picture as to why Mars is so small. A rather intriguing conclusion emerges from these studies from the perspective of exoplanet habitability and extraterrestrial life namely that water could be delivered to many of the exoplanets that orbit fairly close to their stars simply by the formation of giant planets further out, without the need for these giants to perform just the right "Grand Tack" to avoid destroying these further-in planets, but still shower them with water…well, actually shower them with ice-laden asteroids. It seems to me that this all provides an appealing possibility: any planetary system in which giant planets form fairly far out from their host star would have seen delivery of the volatile water onto inner planets by way of ice-carrying exo-asteroids nudged toward their star by the giants' growth, while migration of giant exoplanets can also do that and moreover influence the sizes and orbits of Earth-like exoplanets closer to their star.

With this scenario of growing and moving giant planets that shepherd and scatter smaller bodies around one can take the next step: what does the model suggest should happen in the zone of the future terrestrial planets? Planetesimals are expected to run into each other, throwing fragments back into different orbits, or merging in full or in part to form larger planetesimals, that—once large enough—are called planetary embryos. The embryos will continue to gather planetesimal mass, and occasionally run into one another. It is a battle for dominance, a survival of the fittest, but initially defined fully by chance: eventually, the largest few will be left orbiting the Sun. To study what might take place, one can run multiple simulations with slightly different initial conditions and look at the results.

One such multiple-run model experiment showed that typically three or four planets would emerge out of the collision sequences, and that the outermost one—the virtual sibling of the real Mars—would come out much smaller than the others. Unfortunately, this work did leave the formation of a light Mercury somewhat of a mystery because none of the runs resulted in one. Is that by chance, or does it mean we should perhaps also wonder about what Mercury's low mass means or whether it is special that there is no planet orbiting inside of it? A later study ran over a hundred models for the formation of the terrestrial planets and only in roughly one in ten did something with a mass and orbit like the real Mercury result, in contrast to planets like Venus and Earth which formed in most of these experiments. So, maybe having Mercury there is a matter of chance for our Solar System.

When running such simulations, it is also possible to investigate what these many collisions do to the nascent planets. Some of the collisions involve such large bodies and occur at such high velocities that the energies involved are enormous. As a result of the heat involved in such poundings, large parts of impacted bodies would melt, to later solidify again by cooling, only to be hit again, or to be impacted already while still liquid. If the resulting magma seas are deep enough for long enough,

then chemicals may differentiate, with heavy ones sinking below the lighter ones. Thus, an iron-dominated core can form below a silicate-rich mantle, such as happened to the real-world Earth and its sibling terrestrial planets but likely also to many of the smaller bodies out of which the full-sized planets eventually formed.

When such a differentiated "bullet" planetesimal hits another or a larger planetary embryo, its iron core may well shoot right into the core of the target, there again leading to melting, and a more differentiated, enlarged protoplanet. There is likely considerable mixing by incomplete penetration, or some remixing as the impactor drags some of the target's mantle down into its core upon impact. But over time, in these young forming planets, particularly once really big and mostly melted in their interiors, chemical differentiation will occur, as it did in the Earth. Such simulations appear to be quite successful in making model planets that resemble Earth, Venus, and Mars. They also suggest that most of the Earth's water was added in impacts after the planet was roughly three quarters complete.

The orbits of these model terrestrial planets forming close to the Sun do not evolve very much in the computer experiments. But it has been argued that it is possible things did change in the real world: maybe some planets did form very close to the Sun, or maybe some spiraled in very early on to near the Sun. The movement of Jupiter in and out of the inner Solar System could have led to instabilities that could have pushed these innermost early precursor planets into the Sun. There is no remaining evidence of such a scenario in our own Solar System, but the high frequency of close-in exoplanets—well inside of where Mercury is in our Solar System, such as the Kepler-11 system in Figure 4.1— would leave that possibility open: we may not have any such close-in planets in the Solar System simply because the migratory movements of Jupiter and Saturn further out early in the history of the Solar System caused them to disappear by shepherding their demise into the Sun.

The above life stories of planetary systems, many experimented with in computer simulations, lead to an impressive variety of evolutionary pathways. We have seen scenarios in which a giant planet can spiral toward its star after forming far from it. This spiraling motion could stop at the inner edge of the disk created by the star's magnetism, or—if the timing is just right—when the innermost parts of the disk are evaporated and blown away. If the motion of such planets is stopped near the star, perhaps in one of the "planet traps" mentioned above, this would create an exoplanetary system with a close-in giant planet, that is a "hot Jupiter" or "warm Jupiter". If a slightly less massive second planet were to catch up with the spiraling giant, the pair could be caught in an orbital resonance that, in the right configuration, could migrate outward again as has been argued for our Solar System. This kind of tacking motion would cause a cleared inner region of the disk where the terrestrial planets now orbit. That would open up the possibility of the formation of lightweight terrestrial planets at intermediate distances from the star, such as Mars, and also seeding the innermost, dry parts of the forming planetary system with water that froze onto planetesimals initially outside the snow line. Any such inward motion, regardless of whether an outward migration were to follow, would shepherd material well inside the giants' orbits toward the central star. This adds to the available mass to form planets, of course, but it may also lead to orbital instabilities that could destroy and push existing close-in planets into their star, or could destabilize the orbits of multiple existing close-in planets to then collide to form larger bodies, like the super-Earths found in other planetary systems. In all of this, we need to remember that stars come in a great variety of masses, as do their parent disks and forerunner molecular clouds, so that we should expect different dominant evolutions for massive stars with perhaps multiple planets by far outweighing Jupiter and for low-mass stars where there may not be enough mass in the disk to form even a single giant planet.

It is conceivable that all of these scenarios play out somewhere in the many planetary systems throughout the Galaxy, and as such contribute to the great diversity of exoplanet systems. That said, perhaps they do not all work as astronomers nowadays envision. What is clear, however, is that upon realizing the possibility of inward or outward migration of planetary orbits many new paths opened up for our thinking about planetary system evolution and for the research into it.

Instabilities and bombardments

One of the processes mentioned above is that of orbital instabilities. To look at one example that is being investigated, let us go back to our own Solar System, and pick it up just after the proposed Grand Tack by the Jupiter–Saturn pair. There is some evidence that there have been other rearrangements among the giant planets subsequent to this movement.

As long as there is a lot of gas in the disk, the orbits of growing planets appear to be well insulated from couplings that upset the system: if movements were to develop that are substantially off circular paths, these will be efficiently damped by drag against the gas and collisions with solids, pushing everything back to circular orbits. But once the disk clears, this is no longer the case. Through resonances, orbital eccentricities can grow which may ultimately lead to orbit crossings. With enough time, and thus with enough movements of one planet across the orbit of another, this will lead to interactions. Rarely will full collisions occur, because that requires a more or less direct line of flight between the two objects. But close encounters, with strong gravitational attractions, will happen. This may cause planets to be thrown out of the plane of the original disk, or one of the interacting planets may be ejected from the system altogether.

Terrestrial planets well inside the orbits of interacting gas giants are not shielded from the effects of such interactions and may be bombarded by large bodies thrown around by the far-out

Plate 1 Earth and Moon from a million miles away (see Figure 1.2 on page 7)

Plate 2 Artist's impression of an icy exoplanet near a cool dwarf star (see Figure 1.4 on page 18)

Plate 3 Earthrise over the Smyth's Sea region on the near side of the Moon (see Figure 3.1 on page 65)

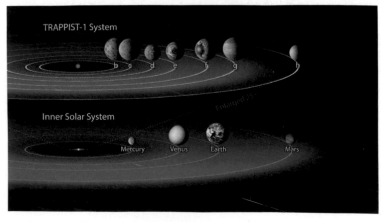

Plate 4 Visualization of TRAPPIST-1 (see Figure 4.4 on page 137)

Plate 5 Sun, Jupiter, Earth, and Ceres to scale (see Figure 3.3 on page 75)

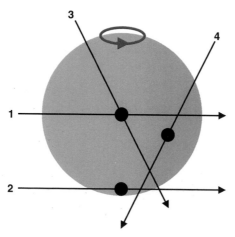

Plate 6 Doppler effect during an exoplanet transit of a rotating star (see Figure 4.6 on page 158)

Plate 7 Moons of the Solar System (see Figure 3.4 on page 77)

Plate 8 Solar System planets to scale (see Figure 5.9 on page 212)

Plate 9 Apollo 17 Lunar Roving Vehicle at Shorty Crater (see Figure 3.7 on page 86)

Plate 10 Comet McNaught (see Figure 3.12 on page 108)

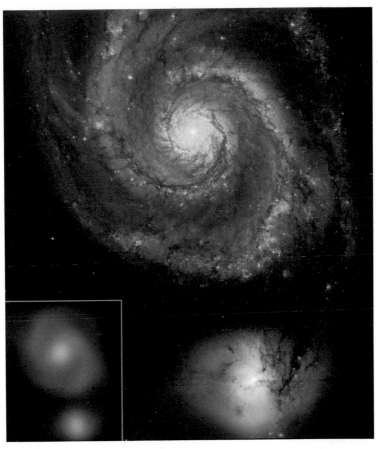

Plate 11 Galaxies Messier 51A and B (see Figure 5.1 on page 171)

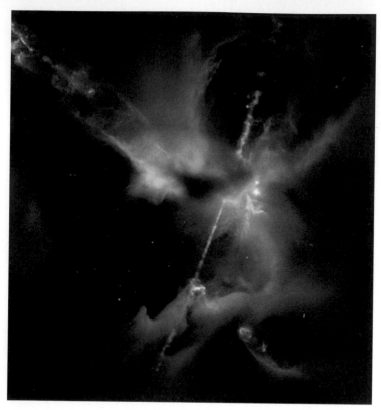

Plate 12 Forming star HH 24 inside the Orion B molecular cloud complex (see Figure 5.2 on page 182)

Plate 13 Open cluster Pismis 24 (see Figure 5.3 on page 183)

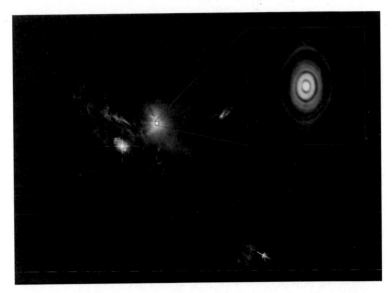

Plate 14 Stellar disk with gaps around HL Tauri (see Figure 5.4 on page 187)

Plate 15 Evolution of the Sun (see Figure 5.5 on page 196)

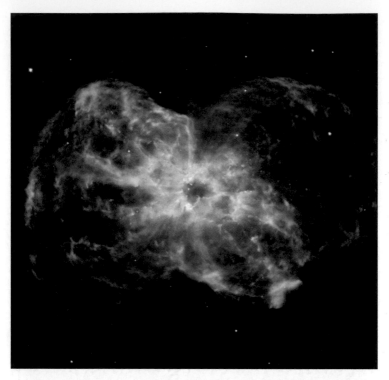

Plate 16 Planetary nebula NGC 2440 (see Figure 5.6 on page 197)

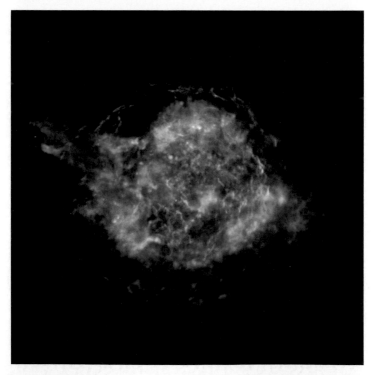

Plate 17 Supernova remnant Cassiopeia A (see Figure 5.7 on page 200)

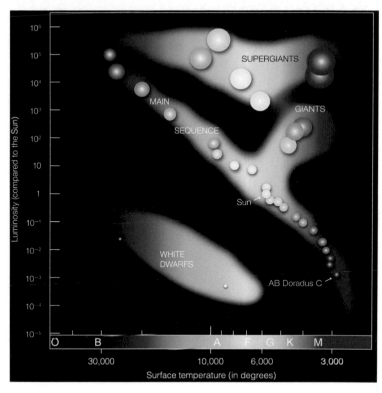

Plate 18 Hertzsprung–Russell diagram (see Figure 5.8 on page 204)

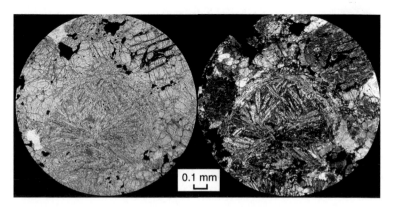

Plate 19 Chondrules in a fragment of the Chelyabinsk meteorite (see Figure 7.1 on page 257)

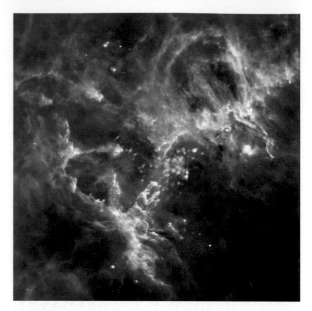

Plate 20 A segment of the Eagle Nebula (M16) (see Figure 7.3 on page 265)

Plate 21 Gravitational lensing (see Figure 7.4 on page 273)

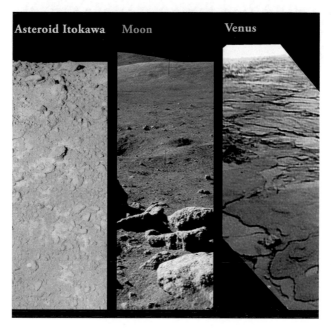

Plate 22 Surfaces in the Solar System: Itokawa, Moon, Venus (see Figure 9.1 on page 306)

Plate 23 Aurora seen from space (see Figure 10.5 on page 362)

Plate 24 Surfaces in the Solar System: Mars, Titan, Earth (see Figure 9.2 on page 307)

Plate 25 A volcanic eruption on Jupiter's moon Io (see Figure 12.2 on page 394)

Plate 26 Saturn's moon Titan (see Figure 10.1 on page 350)

Plate 27 Sojourner and Mars Pathfinder on Mars (see Figure 10.2 on page 355)

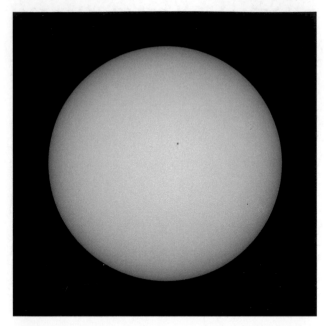

Plate 28 The Sun in visible light (see Figure 10.3 on page 358)

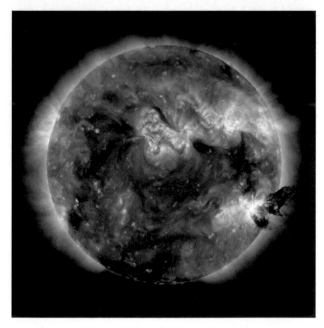

Plate 29 The Sun in extreme ultraviolet light (see Figure 10.4 on page 359)

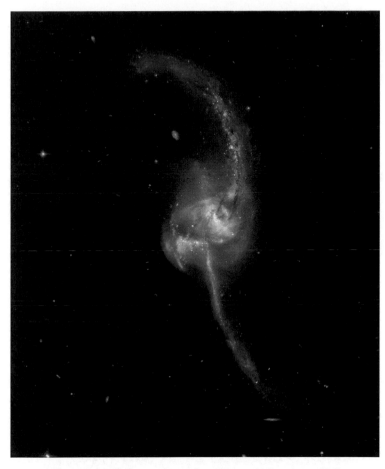

Plate 30 Colliding galaxies (see Figure 11.1 on page 384)

giant planets. The detailed tests of the consequences of Jupiter–Saturn migration on the rest of the Solar System described above were carried out several years after Alessandro Morbidelli and colleagues proposed another event in the history of the Solar System to explain the Late Heavy Bombardment. This supposedly intense storm of asteroid impacts was, at the time, considered highly likely to have befallen the Moon well after its formation. More recent work suggests that the phase of lunar bombardment may have been more protracted than originally argued for, but even if the bombardment turns out to have been less cataclysmic than argued at first, the concept of it did help spur innovative research. This work took place before the Grand Tack concept was tested, but a decade after 51 Pegasi b was discovered. Because Morbidelli was working at the observatory on the Côte d'Azur in Nice, this became known as the "Nice model"—it is indeed a nice model, though that's not the origin of the name. This concept resulted from computer experiments in which a large and heavy disk-shaped population of planetesimals totaling dozens of Earth masses interacted through gravity with a set of giant planets. The researchers noticed that this interaction could cause the orbits of the giant planets to be pulled out of the mutual orbital resonances that their earlier evolution would have put them in. These orbital changes could culminate in a major rearrangement in the orbits of all the planets and asteroids throughout the Solar System. The giant planets, in particular Uranus and Neptune, could move outward. These two could even exchange places in their ordering from the Sun. This transition would scatter vast numbers of planetesimals about, many of which would either be thrown in toward the Sun or end up in wide orbits distant from the Sun. Some might even be ejected from the Solar System altogether, evicted from their family, doomed to a solitary, dark, and cold existence.

Although by no means proven to have happened in real life, this concept could explain several properties of the Solar System that can be observed today. For example, it can explain why

the giant planets Uranus and Neptune are in their current relatively distant orbits from the Sun, whereas they would have been expected to form more easily in a denser part of the disk, closer to the Sun, albeit located outside the snow line to ensure that they would grow fast enough early on. The population of asteroids thrown inward from the initially distant regions may well be related to a phase of heavy bombardment by asteroids that is thought to have come to pass somewhere between 400 and 700 million years after the formation of the Solar System began. The multiple scatterings by the giant planets would throw asteroids into orbits outside the plane of the planets, which is consistent with the asteroid belt and scattered Kuiper Belt being far from flat, with orbital inclinations up to some 40 degrees out of the average plane of the planets, much more than the planets which orbit within just a few degrees of the average plane. Moreover, the random scattering of many planetesimals may explain why relatively many of the giant planets' moons orbit against the dominant direction of the Solar System. Such moons are members of the population of so-called irregular moons: the retrograde moons, but likely also many of the prograde moons with strongly inclined orbits, may well be captured scattered asteroids that would have encountered the planets from random directions in their eccentric orbits around the Sun.

By the way, I mentioned HR 8799 back in Chapter 4, which is a system with four directly imaged giant planets that are so young that they still glow with the heat of their formation. These four appear to have nearly the same masses in contrast to the four quite distinct giant planets in our Solar System, but otherwise there is a remarkable analogy in their arrangement: if scaled for the fact that their central star is brighter and heavier than the Sun, they are in an arrangement very similar to that of Jupiter, Saturn, Uranus, and Neptune in terms of their relative distances to their "Sun", HR 8799, and the irradiation that they receive from that star is quite similar to that of our giant planets from our Sun. It is not an identical system to our own, clearly, but it is comparable

in that it shows (at least until now for HR 8799) multiple giant planets surviving the anticipated spiraling migration in a phase where a dense and dusty debris disk still exists.

The full history of all of the major planets and moons in the Solar System will take more study to put together. That is hardly surprising given than 4.5 billion years have gone by since their formation and no one has been around to record their good times and their bad. So far, explorations of the orbital evolution of exoplanet systems leave the modelers of the Solar System in the position of prehistoric archeologists: they know where objects were found in the dig, but how and why they came to be where they are is all subject to interpretation. In the case of Solar-System evolution models, there is no uncertainty due to sociology or psychology as would be the case for archeologists. But there are other uncertainties: in order to model the history in a computer, the laws of physics need to be reformulated into much-simplified descriptions and such reformulations may or may not leave all the processes as they would occur in nature. Nonetheless, the formation of planets, their subsequent migrations, orbital resonance locking, and then chaotic perturbations as the coupling with the disappearing disk also vanishes leaves an attractive and—as far as we know—consistent picture in our minds of the history of our Solar System. It also leaves enough room for entirely different exoplanetary systems to form by the same processes requiring only differences in timings, mass flows, stellar brightness, and, of course, the consequences of the randomness of the collisional processes that are essential to the formation of the planets.

This brings us back to our collective and individual intuition and research priorities: researchers working in this area are retuning their minds by adding knowledge of these processes, and they are transferring that knowledge to the next generation in the classes they teach and the research projects they direct. Despite uncertainties in the details of these processes, the ideas that grew out of recent findings have put us onto new paths of intuitive exploration, no longer solely guided by Solar-System expertise.

With new observations and more powerful computers, yet more lines of research are pursued.

For example, the various ways in which orbits can deform in multi-planet systems through instabilities can have major effects even on large planets. Experiments show that, particularly for tightly packed, close-in sets of planets, the consequences of orbital rearrangement could be enormous. It is possible for planets to be literally ejected from the planetary system, but they can also be thrown out of the plane of the original disk. The latter would be in line with the observation that there is a population of such systems in which the orbits have substantial inclinations relative to each other and to the equatorial plane of the stellar rotation. Some can be scattered so much that they move against the star's rotation, thus becoming retrograde orbiters.

Where does all this leave our own Solar System? It seems that the orbits of the set of four giant planets will be stable for the duration of the Sun's lifetime as a mature, stable fusion reactor, and that the orbits of the terrestrial planets have only a small likelihood of instability. So, we should be safe in that regard. Asteroids can still come our way, though, and some are big enough to be more than a headache for all life on Earth. That is why all the near-Earth objects are being monitored. Thus far, none pose a near-term threat.

Characteristics of planetary systems

At the end of this chapter, it is time to revisit the list of "forensic evidence" distilled after reviewing our own Solar System and revised after learning about exoplanetary systems and the details of stellar and planetary formation. Now, we can revise it once more, having learned about the evolution of orbits in planetary systems. When viewing the entire list, six of the ten points are identified as likely related to the phenomena discussed in this chapter which are consequences of changes in orbital diameters, or orbit migration, which is a very important and exciting

development in the science of planetary systems. They are identified by adding an "R" to their identifier:

1. The combined mass in all planets is small compared to the mass of the central star.

2-R. Most planets orbit in roughly the same direction as their star rotates, but heavy, close-in planets (hot Jupiters) frequently have a strongly tilted orbit, while some 10 percent have had their orbits tipped over so far that they move a direction opposite to the star's rotation. About half of the small and distant moons of the large planets in the Solar System orbit in the opposite direction; knowledge of exoplanet moons is, as yet, too limited to know their orbital statistics. Orbit migration, orbital destabilization, and planetary scattering are likely causes for the inclined or retrograde orbits.

3-R. Relatively close to a star, planets can be found that are either composed mostly of rock and metal, or, if heavy and big, may be mostly composed of volatiles, much of these in gaseous form. Orbit migration likely plays an important role in this. We do not know enough (yet) about the chemical makeup of exoplanets that are at least as distant from their star as Jupiter orbits from the Sun.

4-R. Relatively close to a star, planets can either be in the mass range from Earth to about a dozen times heavier if in a multi-planet system, or appear to be lone planets if there is a planet that weighs hundreds of Earth masses. Orbit migration likely plays an important role in this.

5-R. Water (liquid and frozen) is very rare near the Sun and very common beyond about three Sun–Earth distances. In exoplanet systems with heavy planets close to their star, however, water may be abundant. Orbit migration and related scattering of planetesimals appear to be important to the distribution of water to bodies in different locations in a planetary system.

6-R. Between Earth and Jupiter lie surprisingly light bodies, including the small planet Mars, the dwarf planet Ceres, and the multitude of small asteroids that all together contain but a small fraction of the mass of a planet. Whether similar asteroid belts exist in extrasolar systems is as yet unknown, but orbit migration by two or more giant planets may lead to similar conditions.

7-R. Beyond the planets, we find multiple dwarf planets and many comets. Orbit migration of giant planets may well play a role in shaping the population and their orbits for many planetary systems.

8. Dwarf planets and planets were essentially entirely liquid some time in their lives, allowing for gravitational separation of chemical compounds in their interiors, letting heavy materials sink toward the center and lighter ones float to the surface. For terrestrial planets this led to iron cores and rocky mantles, while for smaller moons this created rocky interiors and icy (or watery) shells.

9. Metal and rock deep inside in several planets and moons in the Solar System remain liquid, despite billions of years of cooling down. Water remains liquid in some moons of the giant planets despite the cold of space so far from the Sun.

10-r. All bodies in a planetary system are likely to contain limited amounts of radioactive materials in concentrations that depend on details of the system's prehistory. The heat released by radioactive decay supports, at least for some time in the terrestrial planets, tectonic activity.

7

Lone Rovers

Stars often have a nearby companion star or, less frequently, even two or more, forming systems in which they orbit so close to each other that they look like a single star as seen from Earth. Apart from these lifelong companions, however, distances between most stars in the galaxy are vast. Such large separations are beneficial from the viewpoint of long-term stability of planetary systems: if other stars came too close, this would upset the orbits of everything from planets to asteroids. That would at least lead to catastrophic impacts on planets, and could even spell disaster for the entire planetary system.

A nearby supernova?

Whereas most stars are quite isolated for most of their lives, star formation is most commonly observed to occur in clusters of hundreds to many thousands of stars that, by astronomical standards, are tightly packed together. This makes us wonder: was our Sun always alone or did something happen for it to shake its company? Bearing on that question is the puzzle of the radioactive compounds with half-lives of at most a few millions of years. As we shall see in a moment, these compounds appear to have existed throughout the emergent Solar System around the time that the gases were condensing into solids. These compounds, called short-lived radionuclides, cannot have formed long before the Solar System started to coalesce because otherwise there would be no evidence of them having once existed. They certainly did not form afterwards inside the young solid bodies. So they must

One of Ten Billion Earths. Karel Schrijver, Oxford University Press (2018). © Karel Schrijver.
DOI: 10.1093/oso/9780198799894.001.0001

have formed just about at the time that the pre-solar nebular core began to contract. Some of the radionuclides can be released from stars in multiple ways so their origin is ambiguous, but one in particular—iron-60—is thought to form only in a supernova. How likely is it that there was a supernova near the young Sun if it were as truly isolated as it is now? Or did this happen while the Sun was in a densely populated star cluster in an environment similar to that in which most stars are born? But then, where are its erstwhile close neighbors now?

For the sake of argument, and to illustrate the magnitude of the odds involved, let me make a few assumptions, which will then reveal themselves to be incompatible with the real Universe. This is merely a thought experiment, but such experiments have proven to be useful tools to explore the logic of arguments. Such a "Gedankenexperiment" is often used to demonstrate that what is being argued cannot be correct, leaving the opposing alternative as a more likely condition. Let me assume that stars are scattered evenly throughout the Galaxy and that new stars are being formed randomly all over the place. In such a hypothetical world, aged, heavy stars would also occasionally be exploding in supernovas also uniformly around the Galaxy. In the actual present-day Milky Way Galaxy a supernova occurs about once in 50 years. Over a period of, say, two million years—a couple of half-lives of a short-lived radionuclide—there would thus be about 40,000 supernovas that might generate this radioactive material. That sounds like a lot, but given that there are over 100 billion stars in the Galaxy, that means that for any given star to have a nearby supernova occur within that two-million-year period, it would have to involve one of no less than 2,500,000 neighbors. In order for the supernova materials to reach and mix with the pre-solar material within the decay time of the short-lived radionuclide that should be there, the supernova would have to occur within some five to a dozen light years from the nascent Sun. There are now only about thirty stars within a dozen light years from the Sun. Because the distance unit "light year" is used repeatedly

throughout this chapter it is good to keep this in mind: the characteristic size of the Solar System, as given by the full width of the orbit of the outermost planet Neptune, is very nearly 1/1000 of a light year.

The ratio of the number of stars that are now within a dozen light years from the Sun to the number needed—statistically speaking—to have one explode in a two-million-year interval would be 30 to 2,500,000. Thus the odds of such a supernova happening close enough in time and space in a randomly filled Galaxy are 1 in 83,000. Although not entirely impossible, this large ratio suggests that it is very unlikely that the assumptions that I just made are consistent with what is actually the case. If, in contrast, we were to fill that volume of five to ten light years with a few thousand stars instead of a few dozen, the probability of a timely, nearby supernova would go up correspondingly, but it would still only be of the order of 1 in 100. Astronomical observations have shown us an easy way to increase the probability for a supernova near the forming Sun by not only packing stars close together but moreover by increasing the probability of supernovas around the time that many stars are still forming: because most stars are born at roughly the same time in clusters that contain thousands of stars within some dozens of light years of each other, and because the heaviest of these stars go through their fuel supply very fast, some will go supernova even as nearby less-massive stars and their planetary systems have not even completed their formation.

Signatures of a primeval explosion

Now, let us go through that step by step. First, let us have a look at the evidence for a nearby supernova early on. After all, if a radioactive isotope existed at the formation of the Solar System with a half-life, say, of the order of a million years, and we were to look at meteorites now, 4.6 billion years after they formed, no measurable amount of that radioactive material would be

left. How can you measure what is no longer there? Clearly you cannot, but you can look for what is left in its stead. It turns out that even if nothing of the original is left, you can still perform chronology or dating of sequences of events.

Geologists have developed this method of "radiometric dating" over the past century, ever since the discovery of radioactivity in 1896, starting with Henri Becquerel (1852–1908), his doctoral student Marie Curie (1867–1934) and her husband, Pierre Curie (1859–1906). The most widely known of this type of work is called carbon dating. Carbon comes in a few different forms that behave identically chemically, but that have slightly different masses of the atoms because they have different numbers of neutrons in their nucleus. These neutrons act as a glue to keep the protons together whose number sets the chemical properties of the elements. These different forms of the same chemical element with different nuclear masses are known as isotopes. The most stable form of carbon is known as carbon-12. Another form that occurs in nature with two more neutrons, known as the isotope carbon-14, is radioactive with a half-life of 5,730 years, which means that after that time only half of the initial amount of carbon-14 is left; after another such period half of that is left, and so forth.

If you know how much carbon-14 exists in something compared to carbon-12, and if you know the ratio of carbon-14 to carbon-12 when that something was made, you can determine its age. Well, except that you need to meet two conditions: first, you need to know the starting ratio in the past, and, second, you can only do this for things no older than a handful of half-lives because otherwise too little carbon-14 is left to measure accurately.

Fortunately, we know the carbon-14 to carbon-12 ratio in all living things quite well, because there is a steady source that makes new carbon-14 even as existing carbon-14 decays away. The entire Solar System is continually bombarded from all sides with atomic particles that travel through the Galaxy at nearly the speed of light. Earth is largely shielded from these particles

because the magnetized solar wind keeps most of them at bay. Moreover, we on Earth's surface are shielded from them quite effectively by Earth's magnetic field and atmosphere. But the upper atmosphere is exposed to a ceaseless shower of these cosmic rays. When these particles collide with atoms in the atmosphere, they shatter their cores leading to a cascade of similar but ever-less-energetic collisions until energies no longer suffice to break atoms. Some of the pieces of the atoms are captured into other atoms; one such capture, of a neutron into a nitrogen atom, leads to the formation of carbon-14.

The cosmic rays from far outside the Solar System have been coming in in a fairly steady stream for a very long time, which led to the creation of a comparatively steady number of carbon-14 atoms per year. We know that rate pretty well nowadays. This process has been going on so long that a balance has been reached between new carbon-14 being made and existing carbon-14 decaying away as it transforms into nitrogen-14. With a balance between input and decay, there is a very nearly fixed carbon-14 to carbon-12 ratio over many centuries in the past. Not any more, by the way, because society is putting a lot of additional carbon-12 into the atmosphere by burning fossil fuels, but that is only since the beginning of the industrial revolution around the middle of the nineteenth century. People also messed with the carbon-14 content of the atmosphere via a series of nuclear explosions in the second half of the twentieth century. Before these human influences, however, carbon-14 and carbon-12 were ingested in a fixed ratio into the metabolism of anything living on Earth. That includes trees, for example, which—like all plants—use atmospheric carbon dioxide to grow. Once a tree dies—or, more precisely, once a tree ring is left dead behind the living outer shell underneath the bark of the growing tree—nothing more is ingested into that ring. From that moment onward, carbon-14 gradually decays away. Measuring what is left at any time later sets the time at which the wood was formed. That is the basis of all carbon dating.

Ancient meteorites

Carbon dating as applied by archeologists cannot be used as such in the hands of astronomers, because they have a few very big problems facing them. First, they generally do not know the initial ratio of different isotopes when the Solar System formed. And second, many of the useful radioactive isotopes have half-lives so short compared to the age of the Solar System that they simply no longer show up in measurements. There is a way, however, by which one can address both of these problems simultaneously, provided two things are true for an object: first, all parts of the object must have been made at just about the same time from a very well-mixed gas, and second, that gas must have condensed into a solid quickly enough for anything inside it to be locked up without possibility of escape. The first requirement does not mean that the entire object needs to be chemically uniform. Fortunately not, because that would be an unworkable requirement: as dust solidified out of the gas from the pre-solar molecular cloud that eventually formed the Solar System, all sorts of differentiations happened, depending on condensation temperatures and how chemicals settled into crystalline struc-tures or settled subject to gravity in liquid interiors of asteroids. What that first requirement means is that the material must be isotopically uniform. The requirement that different isotopes of any of the chemical elements are present in the same relative numbers everywhere is quite a good approximation throughout the Solar System for any atom that is fairly heavy; it would not be true for hydrogen and its heavier chemical twin deuterium, for example, but these are not radioactive and are not used for dating. In astronomical settings the second requirement, also referred to as "closure" of the object, means that it must have solidified fast compared to the half-life; for radionuclides with half-lives of hundreds of thousands to a few million years, that requirement is met for chondrules (Figure 7.1), the pebbles in meteorites, that formed in the early Solar System.

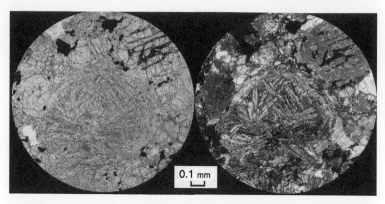

Figure 7.1 Chondrules in a fragment of the Chelyabinsk meteorite seen under a microscope using plain (left) and polarized (right) light to bring out the crystalline structures. The Chelyabinsk meteor occurred on February 15, 2013, over Chelyabinsk Oblast in the Russian Ural Mountains on the border with Kazakhstan (See Plate 19 for a color version).

Inside a chondrule, there is a great diversity of tiny structures, bits of material with different chemical compositions, that at some time early in the Solar System's history clumped together. Each of these bits may have a different mix of chemicals, but is assumed to have the same mix of isotopes for each element. How can one use a 4.6 billion-year-old chondrule to establish if there were ever radioisotopes in it with half-lives of about a million years? Imagine the fragments of the chondrule just formed. Radioactive isotopes are decaying and disappearing. Each isotope atom that decays transform into an isotope atom of another element. Let us make this specific, so that it is easier to follow: There is a radioactive isotope of aluminum (Al for short), called Al-26, that decays into an isotope of magnesium (Mg for short), called Mg-26. With time, there is therefore less Al-26 and more Mg-26. Eventually, there is no measurable amount of Al-26 at all, leaving only Mg-26. That new Mg-26 has now been added to any Mg-26 that was already there when the granule within the chondrule was formed. There is also another isotope of magnesium, Mg-24,

that is neither radioactive nor formed as decay product from another radioactive isotope, so that this Mg-24 has always been there in the same amount. Consequently, in the fragment of the chondrule, the ratio of Mg-26—which was in part newly added—to Mg-24—which has stayed the same—is now different from the original isotope ratio. How much more Mg-26 there now is depends on how much Al-26 there was originally. That will differ depending on what fragment we are looking at; we cannot measure it because it is no longer there. We can measure a stable form of aluminum, though, Al-27. Now all we need to do is measure two ratios for several chemically different fragments of the meteorite pebble: Mg-26 to Mg-24 to know something about the decay product (Mg-26), and Al-27 to Mg-24 to know the original chemical makeup of the fragments. Each granule within the chrondrule may have different amounts of aluminum and magnesium. The combination of measurements of several such granules now tells us how much Al-26 there was relative to Al-27 when the chondrule fragments "closed," that is when they solidified.

Discoveries from the Lunatic Asylum

Looking at isotope ratios is what Typhoon Lee (1948–) and his colleagues did, working in the Lunatic Asylum. Yes, really, they list their affiliation as the "Lunatic Asylum" at the Division of Geological and Planetary Sciences of the California Institute of Technology, located in Pasadena. They were making a word play there: their laboratory had developed a very precise mass spectrometer to measure isotope ratios in rocks to be brought back by the NASA Apollo missions. These lunar rocks would be measured on that fully automatic device, called "Lunatic 1" (a contraction of lunar and automatic), and from there the step to their laboratory's name is a small additional joke. They took their work very seriously, though. In 1976, they published results of the analysis of fragments of the Allende Meteorite, which fell in Chihuahua, Mexico, on February 8, 1969. They demonstrated that Al-26 had

been present when the meteorite formed, and as it has a half-life of only 0.7 million years, something must have created that radioactive isotope within at most a few million years prior to the solidification of the parts that make the Allende Meteorite.

In 1960, John Reynolds (1923–2000)—working on the Richardton Meteorite that fell on June 19, 1918—had realized that another radioactive isotope had been present, namely iodine-129, that decays into xenon-129, with a half-life of 16 million years. That was the first indication of some form of enhanced radioactivity early on in the Solar System, but the much shorter half-life of Al-26 narrowed the window down from a few tens of millions of years to just a few million. Since then, signatures have been found for a dozen short-lived nuclides that taken together suggest that, within about two million years before the Solar System began its formation and possibly with the Sun already a protostar, a star exploded nearby enough that it added radioactivity to the cloud out of which the Sun was settling down as a star, right in the middle of the formation of materials that would make the earliest asteroid-like objects.

After years of research on the various radionuclides and their histories, it seems that not only a supernova, but in fact several sources contributed to the pool of radioactive materials in the cloud that preceded the Solar System. One of these was possibly a supernova, or at least a star at a stage in which the outer layers, enriched with radioactive products from stellar nuclear fusion, would be explosively ejected. This is most strongly supported by the iron-60 isotope which is essentially impossible to make in nature by other means than in a supernova, but it is not clear just how much iron-60 was really there at the beginning of the Solar System, nor how close a supernova could be expected to occur adjacent to a forming planetary system: if the exploding precursor star were too close, its light might have evaporated an embryonic planetary system even before it properly took shape; it probably means that the supernova was not very close in space or time to the forming Solar System.

Sources other than one or more supernovas, including stellar winds and cosmic rays, likely contributed to the diverse populations of short-lived radionuclides. All of these other sources require fair proximity of other stars in order to yield enough material that can impact the nebula around the time the disk formed out of which the Solar System grew to explain the abundances of the short-lived radionuclides. All these studies and the comparison of short-lived with long-lived radionuclides also enable the determination of the age of the oldest solidified bodies in the Solar System: 4.57 billion years.

This early radioactivity was so strong that astronomers consider it likely that the heat this generated was sufficient to melt the earliest asteroids, thus explaining, although perhaps only in part, the difference between the achondrite and chondrite meteorites: the achondrites stem from slightly older asteroids in which Al-26 was still sufficiently present to heat and melt the body and to allow it to mix, if not gravitationally settle. Those forming just a few million years later had much less Al-26 left. This could consequently no longer melt them, leaving the constituent chondrules as separate neighbors found in many meteorites.

It may be that the nearby stellar explosion itself influenced the early formation phase of the Solar System not only by fueling the melting of the first asteroids, but even earlier by creating a shock wave or flows in the molecular cloud triggering the formation of the Solar System. Whether this was indeed the case remains under investigation. The very fact that there was such an explosion near to the Solar System carries a clear message, however: the Sun was not alone when it formed. Analyses have suggested that this explosion would have had to happen most likely within about five light years from the early Sun, perhaps up to some ten light years.

The disintegration of the nursery

Supernovas are rare events in a galaxy. First, they only occur for stars heavier than about nine solar masses, which themselves

are not formed all that abundantly compared to less massive stars: estimates suggest that only about one in 400 stars are heavy enough to go supernova. Second, these heavy stars do not live for very long; no longer than about 30 million years for the least massive ones that can go supernova, and less for the heavier ones. Other processes that result in short-lived radionuclides within the birth environment and time frame of the Solar System also lead to insights into how many stars should, statistically, have been fairly close to the forming Sun and its planets. Combining all these arguments quantitatively, the conclusion is that the proto-Sun should have been formed in a very young star-forming cluster with a total mass of between about 500 and 3,000 solar masses, distributed over some 2,000 stars, contained within distances of 5 to 12 light years. Such numbers are of course quite uncertain, however, so we should read this to mean that the environment of the young Sun may have contained between 1,000 and 10,000 sibling stars, with 2,000 being a reasonable working assumption.

Star clusters form out of a molecular cloud that is weakly bound by gravity, and consequently the swarm of stars that forms out of that cloud is also only weakly bound. The stars hover in each other's neighborhood for a while, but then "evaporate" from their nursery by moving away from the cluster if their own motion is fast enough to escape the gravitational pull of the swarm. Alternatively, they may be pulled away by the tidal forces working from the many stars further away in the Galaxy. The typical timescale on which this happens for a cluster like the one thought to be the birth environment of the Solar System is 100 million to 300 million years. Therefore, many of the members of the dispersing birth cluster were still around throughout the formative years of the planets, including during the migratory movements of the giant planets. Less-tightly packed groupings, often called embedded clusters, fall apart on a timescale of order ten million years. This means that most stars and their planetary systems have mostly fully formed, and have been subjected to

nearby supernovas, before they commence their solitary migration around the Galaxy.

Just as moons orbit their planet and the planets orbit the center
of mass of their planetary system, the stars orbit the center of mass
of the Galaxy, all spinning about in a mostly disk-like structure
with spiral arms (Figure 7.2). In the 4.3 to 4.5 billion years subsequent to its birth cluster falling apart, the Sun has orbited the
center of the Galaxy some twenty-seven times. All of its siblings
will have done the same, more or less, but will have dispersed
along a trajectory leading or trailing, and also somewhat sideways
from the solar orbit (just like comet nucleus fragments behave
within the Solar System, such as the example of Figure 3.14).
Although candidates are being considered, no sibling of the Sun
has been unambiguously identified thus far; as measurements of
stellar motions and distances improve (such as with the GAIA
satellite of the European Space Agency) this may change, aided
also by comparisons with the detailed chemical makeup of the
Sun which should be quite similar for its siblings. It will be a
challenge, though: it is estimated that perhaps several tens of
the Sun's original siblings still reside within 300 light years from
the Sun, sprinkled among some 300,000 other stars in that same
volume.

The environment within the Sun's birth cluster was very, very
different from the present surroundings of the Solar System.
Instead of some 2,000 stars, nowadays there are three stars within
five light years. All three are actually contained in a single stellar
triplet, jointly called Alpha Centauri, one of which is Proxima
Centauri, the nearest known star hosting a planetary system, and
the third-brightest star in the night sky. A dozen stars lie within
12 light years. No star heavy enough to ever become a supernova
is anywhere nearer than several hundreds of light years. Today,
you have to go out to a distance of 50 light years from the Sun
to encompass a number of about 2,000 stars. Of these, by the way,
only about 10 percent are visible to the naked eye, the rest being
too faint to spot.

Figure 7.2 A composite view of three galaxies illustrating the flat, pancake-like structure of galaxies like our own: *top:* edge-on view of the Needle Galaxy, or NGC 4565, photographed by the Galaxy Evolution Explorer; *center:* ultraviolet image of the Andromeda galaxy, or M31, also by the Galaxy Evolution Explorer; *bottom:* artist's concept of the shape of our own Milky Way Galaxy. Stars seen surrounding the galaxies in the top two images are foreground stars within our own Galaxy that are all a hundred times or more closer to Earth than the galaxies behind them. The images are scaled to the same apparent width, but the Andromeda galaxy—some 260,000 light years across—is estimated to be twice as heavy and about twice as large as our Milky Way Galaxy; the Needle Galaxy is roughly 100,000 light years across.

Source: NASA/JPL-Caltech, and R. Hurt (SSC/Caltech).

With so many more stars close by early on, the night sky around the very young planetary system would be filled with many hundreds more stars visible to the naked eye...which, of course, did not exist at the time as life had yet to develop. Moreover, these stars would only become really visible a few million years into the formation process of the Solar System when the cloud out of which it was born became transparent. The typical separation between stars packed this tightly in a star-forming cluster, for example, is about one half to one light year, or of order 30,000 to 60,000 Sun–Earth distances. That is safe enough to be well away from the forming Solar System, which itself is no wider than about 50 to 100 Sun–Earth distances. But these stars are all moving about and close encounters would definitely be possible. The proximity of so many other stars also had consequences for the state of the gas in the young Solar System.

The effects of stellar neighbors

One possible interaction that has been proposed and is still being researched is that one or more of the cluster stars passed by the young Sun close enough to upset the orbits of objects in the Kuiper Belt. This is a possible alternative to the "Grand Tack" scenario in which the movement of the giant planets in and out of the inner Solar System would eject asteroids into a widened distribution as seen in the Kuiper Belt. Others argue that both these processes actually happened, so that consequently there may have been multiple periods of fairly heavy bombardments of the young planets in the inner parts of the Solar System by gravitationally deflected minor planets starting from further away.

There is, however, a coupling other than by gravity. Researchers have argued that the environment in which Earth originally formed would have been like what we now see in star-forming regions of sizes somewhere between the Eagle Nebula (Figure 7.3, also known as M16, with some 8,000 stars in several star-forming regions) and the Orion Nebula (M41, with only about 700 stars).

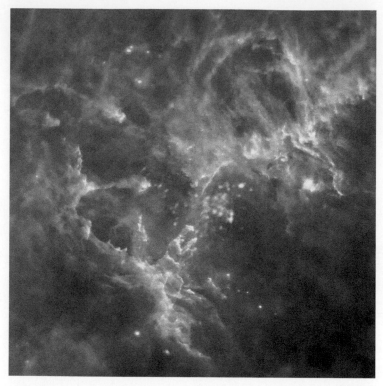

Figure 7.3 Composite image of a segment of the Eagle Nebula (M16) transformed from colors invisible to the human eye into colors that we can see: X-rays (by the XMM/Newton satellite) and far-infrared (by the Herschel spacecraft). This image focuses on star-forming region NGC 6611. The gas shows up in the infrared, while many of the young and magnetically active stars—only one to two million years old—in the open cluster show up in X-rays in the center of the image, looking somewhat fuzzy because the X-ray image is less crisp. Just below and to the left of the most densely populated field of stars in the center is a cloud structure known as the "pillars of creation," a region of ongoing star formation (See Plate 20 for a color version).

Source: ESA/Herschel/PACS/SPIRE/Hill, Motte, HOBYS Key Programme Consortium.

In such dense regions, some quite heavy stars will form. Not only will these explode as supernovas when they run out of nuclear fuel after a few million years, but they stir up the neighborhood during their phase as steady fusion-reactors. Stars that are this heavy are bright and hot. Their surfaces are so hot, in fact, that much of their light is radiated in the visible blue domain (so we perceive them as blue) and in the ultraviolet (that is invisible to the human eye and is mostly stopped in the Earth's atmosphere).

The ultraviolet radiation from these hot stars, which are classified as O- and B-type stars (see Figure 5.8), is so energetic that it can break the electronic bonds of the hydrogen molecule. It can actually knock the electrons out of their orbits within the hydrogen atom. Thus, the molecular material around such stars is transformed into an ionized gas, or plasma, of free-flying electrons and atomic cores. Ultraviolet radiation of the hot stars disrupts molecular hydrogen, and that, through a series of intermediate absorptions and emissions of light, results in heating of the gases surrounding these hot stars. The temperature of the gas rises from a little above absolute zero to of order ten thousand degrees. And with the rise in temperature, the gas pressure increases in step, so that the ionized gas begins to rapidly expand pushing against the surrounding cold molecular gas. The speed and range of the expansion of these bubbles of ionized gas depends on the density of the medium around the stars, of course, but is for a long time so fast that it is supersonic: a shock front is formed that sweeps up surrounding gases in front of it—working almost like a snowplow—as it propagates outward from the hot stars, disturbing if not disrupting the molecular cloud within the nascent star-forming cluster.

The full effects of these shock waves, the ionization bubbles, and the supernovas are still being investigated. It is possible that these shocks can push surrounding gases together enough that this triggers a collapse that enables the formation of other stars. With enough bright stars and their effects working long enough, however, it may disperse material and contribute to the end of star

formation within the open cluster. So it may be that some stars form aided by the bubbles blown out from around the heaviest stars in the cluster, while ultimately the same effects keep most of the gas in giant molecular clouds from forming stars at all; estimates put the average efficiency of capturing material in stars compared to material escaping from a molecular cloud at about 8 percent, with a large range depending on the details of a given cloud. The influence of the newly formed heavy, bright stars reaches well beyond the open clusters themselves: the bubbles created by starlight and supernovas aid even in the disruption of the embedding giant molecular clouds, thus limiting their life-times to typically from a few to 10–20 million years, with heavier ones living longer. But then, the redispersed gas reconfigures to assemble other, new molecular clouds and the processes start over from scratch.

What does all this do to the cloud cores in which an exoplan-etary system is forming? The shocks generated by the early gen-eration of heavy stars run ahead of where ionization and heating are happening. So a dense core may be pushed into collapse by the shock, or by other motions in the giant molecular cloud. But regardless what initiates the collapse, there will be a time that the gases surrounding the core become heated and rush outward from the nearby hot star or, more likely, multiple such stars. For us viewing from Earth, this leaves a dark blob of dense gas, in an otherwise largely cleared and transparent setting. These blobs are now in a race against time to form a planetary system because that growth needs the gas that is now being heated and evaporating off the outer edges of the blobs. These blobs are known as evaporating gaseous globules, or EGGs. If no protostar has formed deep inside it, the EGG will simply evaporate entirely. If there is a forming star and planetary system in its center that can hold on to its matter, the outlying cool gases around them will eventually evaporate, so that then even the young planetary disk is now being externally heated and thereby subjected to evap-oration. The ionized protoplanetary disk is known—through the

combination of fragments of syllables in this description—as a proplyd. Almost two hundred of these have been found in the nearby Orion nebula alone, for example.

This external evaporation process adds to the internal one related to the brightening and warming star taking shape in the center of the planetary disk and the disk's own heat engine. How the material in such disks behaves, and how much time is available for planets to form inside them, thus depends both on the population of stars in the extended environment of a forming planetary system and on the properties of the core out of which it formed, which are related to the type of star it will end up with in its interior. Sorting out the consequences for the properties of exoplanet systems of exposure to internal and external ultraviolet radiation during planet formation, orbital migration, and the extent of the Kuiper Belt is subject to much ongoing research.

For our Solar System, it seems there must have been an interesting combination of at least one heavy nearby star that exploded in a supernova long enough before the Solar System was fully formed that the material that was thrown out mixed with the pre-solar cloud but close enough to the system's formation to leave short-lived radioactivity mixed into the gas. Any other heavy stars within the Sun's birth cluster must have been distant enough not to disrupt the pre-planetary disk by either other supernovas or evaporation by ultraviolet light so fast that the planets would have had no time to form. Even the size of the open cluster in which the Sun formed is important: just large enough that at least one nearby supernova occurred, but not so large that many hot, bright stars of type O and B would have formed nearby to blow too much of the gases away.

This having been said, it seems plausible that somehow the solar disk was truncated somewhere between 30 to 100 Sun–Earth distances, somewhere around the range of the outer edge of the Kuiper Belt. The energy input from a hot star could be responsible for this. This hypothesis leads to a consistent overall picture about how much mass would have been in the initial disk and for how

long compared to how much there now is in all the planets after also allowing for the concept of the giant-planet migrations, including a switch in orbital distances for Uranus and Neptune. In that picture, mass in the disk may even have been flowing outward beyond where Jupiter formed, partly compensating for the evaporating gas from the outer edge of the disk. Although this implies an inward flow of gas onto the young Sun and an outward flow toward the evaporation edge, the net effect seems to have been that the disk could survive longer, thus giving the major planets the time of order 10 million years they are thought to have needed to form. And, finally, it suggests a mass distribution in the original disk comparable to what is estimated from some other observed disks.

Exoplanets floating between the stars

Even though the relative isolation of planetary systems for most of their lives is beneficial to their long-term stability and survival, this does not necessarily mean that planetary systems are generally safe places for planets: in the past few years, evidence has been mounting that planet-sized objects are not just found in planetary systems, but that many are floating between the stars, apparently having been ejected from their birthplaces.

Substantiation for this has been accumulating increasingly since the last decade of the twentieth century. In part, it has come from ever more detailed observations of young star-forming regions. The study of such clusters offers two key advantages: there are many objects within a small area, and these have all been formed within a few million years of each other. Both of these help enormously in translating the observed brightness and color of an object into what its mass might be. Looking, for example, into the Orion Nebula Cloud, about 160 objects were found with masses between 4,000 Earth masses (above which objects are classified as brown dwarfs) and 1,500 Earth masses (which is as small and faint an object as the instruments could

spot) that appeared isolated, not bound to a star. In addition to these planet-mass objects there were also somewhat heavier brown dwarfs floating among the nearly one thousand yet heavier proper stars in the field of view: there were almost five times as many brown dwarfs as there were free-floating planet-mass objects. Other young clusters have also been found to contain a substantial number of planet-mass objects floating about between the stars.

In young star clusters, these objects, in the mass range of heavy Jupiters, are so young that they are still glowing from the heat of their formation, and are still fairly large in this phase prior to contracting into a state of a mature planet. Consequently, they are fairly bright at infrared colors. Lower-mass free-floating planets also exist, but these cannot be imaged in the same way as young planets because they would have cooled down and shrunk and thus be too faint to detect with today's heat-detecting instruments. There is, however, a way to see them, namely if they transit in front of a background star. However, contrary to the darkenings of transit observations that the Kepler satellite relied on, one should instead look for transient brightenings of the background star.

Now we enter a territory that we have not yet explored in this book: gravitational lensing. We have to answer a question not dissimilar to the one we started the radionuclide puzzle with. There the question was "How can you measure what is no longer there?" Here the question is "How can you detect a dark object floating in a dark space?" For exoplanets that are heavy enough and close enough to their star one can use the Doppler effect, but for exoplanets with less mass or a wider orbit the gravitational pull on their star is weaker—decreasing in proportion to their mass and to the square of the size of their orbit—and so is the resulting Doppler effect. Transit measurements can succeed for exoplanets that are quite distant from their star, but here the problem becomes one of proving that an observed dimming is indeed an exoplanetary transit. First the transit needs to be seen

twice to prove that it is not some artifact, and then at least a third time to demonstrate that the transits are periodic, as expected for an exoplanet orbiting its star. For a planet like Jupiter around the Sun, with an orbital period of almost 12 years, this would mean a wait of 24 years from the first detected transit. For a planet in a Neptune-like orbit it would take 330 years! That is why transit measurements are not useful for exoplanets orbiting far from their star.

How, then, could you go about proving the existence of exoplanets that orbit very far from their stars or that float unbound between stars? The answer here is: you use the background stars and the bending of light by gravity. This brings us back to Albert Einstein in the first decades of the twentieth century: his work on general relativity showed that light bends as it travels close by a heavy object. Well, after he published his energy–mass equivalence ($E = mc^2$), light, traveling about in tiny packages of electromagnetic energy called photons, was expected to bend even in the way that Newton described gravity: something with mass would be deflected from a straight line of flight by the gravity of any mass that it passed by. Einstein, however, described gravity in such a way that it distorted space itself; the result was that in Einstein's equations the deflection of light was double that of Newtonian equations.

This idea was put to the test—literally testing the concepts of Einstein's general relativity—by looking at how light from distant stars was bent by the nearest really heavy object: the Sun. But to see the stars just outside the Sun's disk, where the magnitude of the bending was measurable, the Sun's light had to be blocked. The then Astronomer Royal, Frank Watson Dyson (1868–1939), realized that the 1919 eclipse would pass in front of a densely populated star field—the young open cluster known as the Hyades—and would offer a great opportunity to test Einstein's concept. Arthur Eddington led an expedition to that total solar eclipse to measure the effect; another, led by Andrew Crommelin (1865–1939), was sent in case the weather would not cooperate in one

place. Eddington's team went to Príncipe, an island off the West African coast, and Crommelin's to Sobral in Brazil. The eclipse of May 29, 1919 was successfully observed by both teams, although through clouds over Príncipe, and the outcome proved Einstein right. The news made newspaper headlines all over the world: once the scientific analyses had been done and checked, the *New York Times* of November 10, 1919 reported: "Light is all askew in the heavens. Men of science more or less agog over results of eclipse observations. Einstein theory triumphs."

Here, interesting footnotes to history can be made. One footnote is about Albert Einstein's work: In an earlier computation, Einstein had realized that light had mass and therefore would be bent by the Sun's gravity, but in a way such that Newton's and Einstein's predictions would have been identical. Alas, the weather was bad for a 1912 eclipse expedition that had planned to measure the bending of light, and then the First World War made expeditions impossible to the next one in 1914. By the time the 1919 eclipse came about, Einstein had worked out his general theory of relativity, which says that gravity works to distort space. His updated prediction was made in 1915 with his new theory of general relativity, and because his revised prediction was double that from the Newtonian case, the confluence of unfortunate events led to a fortunate outcome: the measurements by Eddington and Crommelin confirmed his theory of general relativity! Another footnote is that Einstein presented his correct prediction to the Prussian Academy of Science in 1915. But as Germany was at war with England, where Eddington lived, no direct communication occurred. This happened via the country in the middle, the Netherlands, which was neutral in the war so that information could pass through there indirectly, first going through the hands of Willem de Sitter (1872–1934)—later a famed cosmologist himself—who forwarded it to land on Eddington's desk as secretary of the Royal Astronomical Society at the time.

Gravitational lensing is about light being bent around a heavy object. Light from far-away galaxies can be bent around nearer

galaxies, and create an oddly distorted ring-like image of the distant galaxy (Figure 7.4). But for the study of dark but massive planets within the Milky Way Galaxy another manifestation is looked for because the ring-like distortion formed by a lensing

Figure 7.4 This image of the galaxy cluster SDSS J1038+4849 was taken with the Hubble Space Telescope. The arcs of light—segments of what are known as Einstein rings—are created by light from a distant galaxy somewhere around the center of the rings that has been bent around by another galaxy nearer along the line between that galaxy and Earth. A similar bending of light is used to find nomad planets: these bodies bend light from stars behind them, but they are not heavy enough to create true rings that can be imaged separated from the star's light itself. Instead, transiting nomads bend enough starlight around that there is a temporary brightening of the star as seen from Earth for as long as the nomad planet is in front of it (See Plate 21 for a color version).

Source: NASA.

exoplanet is too small to resolve: instead, astronomers hunt for an apparent temporary brightening of a star. This is a consequence of a relatively nearby object moving across a more distant star. In doing so, its gravity focuses light a bit more toward Earth than without that foreground object, similar to how a magnifying lens can be used to start a fire by focusing the bright light of the Sun. Of course, there are many other reasons a star could brighten temporarily, including large stellar flares or the disappearance of starspots followed by the appearance of others. Fortunately, the way light changes with time because of lensing has a very particular, well-known form. This signature of transient brightenings of stars by distant planets passing in front of them as seen from Earth is called microlensing. The name may sound a bit odd because, obviously, a planet is nothing "microscopic" to us, but compared to an entire galaxy, for which gravitational lensing was first used, it clearly is miniscule.

The sheer size and general emptiness of the Galaxy means that exoplanetary microlensing events will be rare, even if there are many free-floating planets and despite the vast number of background stars. However, as early as 1991, Shude Mao and Bohdan Paczyński realized that it was, in principle, possible to find exoplanets this way, pointing out that "A massive search for microlensing of the Galactic bulge stars [a direction in the sky where there are many stars to observe] may lead to a discovery of the first extrasolar planetary systems." They wrote this a year before Aleksander Wolszczan found the first exoplanet around a pulsar and four years before Michel Mayor and Didier Queloz announced the discovery of 51 Pegasi b. Mao and Paczyński immediately realized that it would be an observational challenge, however, cautioning that only a few exoplanet microlensing events might be seen even if a sample of ten million stars were to be continuously monitored for a year. What they were thinking of, however, was a microlensing event in which the light from a distant star would be lensed by a foreground star crossing in front of it, while additional lensing spikes could then occur because that

foreground star might be orbited by an exoplanet; the first such detection was made in 2004. When they assessed the possibility of finding exoplanets by microlensing, however, Mao and Paczyński were not thinking at all about lone exoplanets drifting in front of distant stars without being accompanied by a star. It took twenty years before Takahiro Sumi and colleagues could report that they had detected a class of microlensing events for a population of exoplanets that favored the idea that the "planetary-mass population that [they] have identified here may have formed in protoplanetary disks at much smaller separations and then been scattered into unbound or very distant orbits." Many of these exoplanets do indeed appear to float freely in space, far from any star with which they would have formed originally.

In order to find such lone planets, surveys of star fields are required that can cover tens of millions of stars whose brightness needs to be accurately measured a few times an hour for months or years on end. Modern-day dedicated telescopes with multi-million-pixel camera systems can do this. However, these monitoring surveys detect many transient brightenings caused by many different processes. In order to extract those events related to exoplanet microlensing, first all of those need to be identified that are associated with non-planetary-mass objects such as brown dwarfs, certain types of variable stars, and even distant supernovas, black holes, and neutron stars, or image artifacts like cosmic rays hitting the camera. After sorting and sifting for cause, only one in a hundred survive as true microlensing events caused by exoplanets.

For observations made in the direction of the Galactic Bulge, the most densely populated central area of the Galaxy, astronomers calculated that for every star there are about two microlensing objects that are very roughly as heavy as a few hundred Earth masses; lighter ones are not detectable this way while much heavier ones would be brown dwarfs or small stars. The process of down-selection from observed brightenings to those attributed to planetary-mass objects may miss things it

should have eliminated or may eliminate events it should not have. Consequently, there are considerable uncertainties in the resulting findings, and the resulting frequencies of such objects could be off the true value by a large number. Moreover, it remains ambiguous whether the inferred planet population causing microlensing is associated with objects that are either orbiting their star in a wide orbit—typically 10 or more Sun–Earth distances from a star—or with objects that are floating freely between the stars. The fundamental difference between these two types of exoplanets is that we expect all planets to form in orbits around stars, so if there are planets that are not orbiting stars at all, then this would be quite surprising.

Separating distant orbiting planets from unbound planets is complicated and uncertain, and must be done by statistical means because the sample of directly observed distant planets is fairly small. This means that frequency distributions of properties of exoplanet systems need to be extrapolated from the exoplanets that were detected generally rather close to their star into an unknown range of possible distant planets. Moreover, assumptions need to be made about the masses of the lensing objects. This kind of statistical argument, often unavoidable when working so close to the limits of what is observable, suggests that there are about as many free-floating planets as there are stars, but with a large range of allowed values within the uncertainties. For now, we might simply conclude there are many free-floating planets out there, likely numbering in the billions for the entire Galaxy.

How would such free-floating planets—also known as rogue planets, orphan planets, or nomads—come to exist? The names given to these objects suggest that they formed as planets around a star, but then somehow came to escape its gravitational pull. But the heaviest of these objects have masses that overlap with the lower tail end of the range for brown dwarfs, which themselves are often thought of as forming like stars, in the center of a gravitationally collapsing cloud. In fact, both of these may happen, although perhaps one mechanism might be much more

important than another. Possibly, the balance even depends on the general environment in which they form, namely on the properties of the molecular cloud and the star cluster that forms out of it. And likely the balance will depend on the mass of the object.

Perhaps some of the free-floating planets form as stars do, in the center of a cloud that collapses under its gravity. As far as we can see, brown dwarfs generally start with a disk around them that readily lives as long as the disk around a heavier star. Masses contained in these disks for stars and brown dwarfs are of order 1 percent of the mass of the central object. Although for a brown dwarf this means that Jupiter-mass planets cannot form for lack of building material, multiple smaller planets could still form around them. We simply do not know as yet how far down in mass this process continues. Future observatories, particularly those working in the infrared where these objects may be seen, or by surveys for gravitational lensing, will have to provide us with information.

Thrown out of the system

Computer experiments are showing that the approach of stars in a dense star cluster can perturb planetary orbits. This can be by tilting of orbits out of the plane of the original pre-planetary disk, or by direct ejection of one or more planets into interstellar space. Yet others are subject to an apparently mild destabilization of the orbits of the planets, but then some millions of years later this may still end in the ejection of a planet from the system. Not only can planets be ejected from a planetary system by a passing star, but there could even be a planet exchange between such stars so that possibly some planets orbiting stars may have been born around another. Even more indirect effects need to be considered: for the many stars in wide binaries, passing isolated stars or the combined tidal effect of multitudes of distant ones can perturb the orbit of the companion star, which thereby can perturb the

planetary system with consequences similar to those if a traveling star had passed by directly.

The computer experiments suggest that as many as one in ten to one in five planetary systems in open clusters will lose a planet by the gravitational interactions between sibling stars. They also show that planetary systems are more likely to lose a planet by internal interactions between planets than by the passage of relatively more distant stars. Apart from, and likely often prior to, planet ejection there is a phase of orbital distortion. As we saw in Chapter 4, something like one in four of the close-in hot Jupiters found have orbits that are strongly tilted away from the direction where the disk would have been in which planets originally formed. A fraction of these inclined hot Jupiters have their orbits upset so much that they are now retrograde planets, orbiting against their original direction. An alternative to such scenarios is that a free-floating planet may come close to a planetary system and be captured, so that either it enters in an unusually inclined orbit or it causes others already there to change their orbits. It remains to be seen how many such orbits can be attributed to any of the proposed mechanisms mentioned above. Both observational and theoretical studies show that these are not highly unlikely mechanisms but that they possibly affect tens of percent of planetary systems.

Thus far, observations of unbound planets are, not surprisingly, very difficult and the low frequency of detections leaves very large uncertainties even for unbound planets with masses as large or larger than Jupiter at 318 Earth masses. If we take the Solar System as a guide, it is clear that there is a very much larger number of much smaller bodies orbiting the Sun in the asteroid belt, the Kuiper Belt, and the Oort Cloud. Computer experiments have shown that the lightest object in a close encounter within a planetary system stands the largest chance of being ejected. They also show that systems with giant planets are far more likely to eject bodies, even if these are as large as Mercury. In fact, for a Solar System like ours, it may be that dozens of such

bodies, and many more smaller ones, were ejected. Moreover, the Oort Cloud objects are very nearly unbound to the Sun, and are readily influenced by distant stellar encounters and tides from the surrounding Galaxy.

Some have estimated that there may be up to 100,000 objects for each star that are as heavy or heavier than Pluto, at a little more than 1/500 of an Earth mass; that may seem small, but Pluto still has a diameter of 2,400 kilometers (or 1,500 miles). Such estimates are made with several assumptions that make the result highly uncertain, but imagine that it is correct. Even if the true number turned out to be 100 or 1,000 times too large, the number of free-floating bodies with masses between those of the giant Jupiter and the dwarf Pluto is simply astounding.

The number of smaller bodies is expected to be much larger still. Most of such objects are so small that they are hard if not impossible to spot with telescopes, unless they should come relatively close to Earth. In 2017, the first such small nomad was discovered: the oblong, reddish object known as 1I/2017 U1 or 'Oumuamua (after the Hawaiian word for scout), measuring roughly 240 by 40 meters (or 800 by 100 feet), passed through the solar system, approaching the Sun to within a quarter of a Sun–Earth distance, before racing away again at such a high speed that it is clear that 'Oumuamua is not bound to the solar system but a truly interstellar traveler.

What is it like on a nomad planet?

One thing to think about is what these nomads would be like. As we see increasing corroboration that massive planets can migrate long distances and their orbits can be lifted way out of the disk plane in which they formed, it seems possible that not just aster-oids or dwarf planets can be ejected but also that full-sized planets end up being thrown out of their system. This could happen well after planets are fully formed, and be caused either by orbital instabilities or by passing stars. Could this happen to a planet like

Earth, or even to Earth itself? Not in the Solar System as it is now, but it could have long ago, as in any young planetary system. The consequence of such an ejection would be that it would go very dark and very, very cold on the ejected planet, with the oceans and perhaps the atmosphere itself freezing over. That would be the end of life on the surface as we know it. But there would still be geothermal heat. Some scientists are wondering if life could persist, or even form, on such an apparently inhospitable world. Dimitar Sasselov (1961–), director of the Harvard Origins Project that studies the origin and evolution of life, noted in a discussion with other scientists that "If you imagine the Earth as it is today becoming a nomad planet—it gets expelled and then it's flung into interstellar space. Life on Earth is not going to cease. That we know. It's not even speculation at this point. People who study [life in the most extreme conditions on Earth], in particular in the deep biosphere in the crust of the Earth, already have identified a large number of microbes and even two types of nematodes that survive entirely on the heat that comes from inside the Earth. And because the internal heat of the Earth is going to continue at this level for at least another five billion years, this entire deep biosphere is going to be completely uninhibited by the Earth becoming a nomad planet." If a Jupiter-sized planet were to be expelled and kept its moons another potential to support life could be carried by nomads: tidal forces maintain deep zones of liquid water beneath the surface ice of several of the moons of the giant planets.

For now, however, we are going to leave habitability and the potential of life, until the final chapters. While reading those chapters, which focus on terrestrial planets and moons within planetary systems, it is worth keeping in mind that there may be more in the way of living on planets and moons than in the sunlit or starlit worlds. Just imagine emerging onto the frozen surface of a moon—perhaps like Europa that orbits Jupiter—through a thick layer of ice to see a pitch-black sky full of stars, but with one circular patch blocked out by the nearby giant nomad

planet, with only the faint auroral glow over your moon and over the planet's polar regions, powered by the flow of interstellar gas around the planet and by the orbital motion of the moon that you are on ... unless, that is, you happened to be on a rogue planet that was not merely ejected from its planetary system but in fact thrown clear of its galaxy. In that case the faint, fuzzy, star-like light sources in the sky would actually be distant galaxies, and you would be very, very much alone.

8

Aged Stars and Disrupted Exosystems

In retrospect, the first evidence for exoplanets came from extinguished stars and demolished planets: fragments of planets, moons, and asteriods having crashed into dead stars called white dwarfs had contaminated the atmospheres of what once were their stars, and the tainted glow from these white dwarfs signaled the existence of the remains of such unfortunate planets. But although astronomers saw the signature of the contamination in these stellar atmospheres, it took them almost a full century to figure out what caused it and what we could learn from it.

The tool of light

One way to find out which chemicals are contained in a material is to inject shavings of it into a flame so that it heats up and glows, and then to unravel its light into its constituent colors. If sufficiently spread out, such a color spectrum shows sets of well-defined colors. This started to become a tool for astronomers to study stars around the middle of the nineteenth century through the work of Léon Foucault (1819–1868) and Anders Ångström (1814–1874). As they studied the colors emitted by a flame with a heated element in it, they noted something interesting when they compared that light to light passing through that flame from a lamp behind it: the particular colors emitted by the flame viewed against a dark background were identical to those absorbed by it

One of Ten Billion Earths. Karel Schrijver, Oxford University Press (2018). © Karel Schrijver.
DOI: 10.1093/oso/9780198799894.001.0001

when a lamp was placed behind it. They concluded that atoms of any element could glow at particular colors by themselves and also could absorb light from an external source at those same colors, all depending on the chemical makeup, the temperature, and the density of the gas.

Early in the 1800s the first observations of the Sun's color spectrum were made by William Wollaston (1766–1828) and Joseph Ritter von Fraunhofer (1787–1826). They noted that sets of particular colors were largely missing from the broad rainbow continuum emitted by the Sun—many of the most pronounced absentee colors are still today referred to as "Fraunhofer lines," and we still use the same letters that he used as their identifiers. The work of Foucault and Ångström provided a beginning of understanding why this happened. By 1859, Gustav Kirchhoff (1824–1887) and Robert Bunsen (1811–1899) had made a further step as they demonstrated that each element has its own characteristic colors, making its color spectrum into a unique fingerprint. That turned the study of light into a broadly applied method in physics by which to analyze the composition of chemicals. In fact, in the early years of its development, it was a tool to discover hitherto unknown elements, including the Sun's second-most-common element, helium. It was named after Helios, the Greek personification of the Sun, because the element was spotted in the Sun's spectrum but was unknown to exist on Earth at the time.

Stellar astronomers now analyze the colored fingerprints hidden in starlight for much of their work: light from deeper layers in a stellar atmosphere passes through higher layers where elements both absorb light at their specific colors and also add their own glow. The strength of the absorption or emission provides us with clues as to how much of each element there is and what its temperature is. Over time, the detailed analysis of the spread of colors in starlight developed into the astronomer's most important and complicated observational tool: spectroscopy, the study of details of light and how it is generated and interacts with matter. After

the work by Pieter Zeeman (1865–1943) in the late nineteenth century it even became possible for astronomers to use spectroscopy to measure the strength, and under some conditions also the direction, of magnetic fields permeating hot glowing gases throughout the Universe. Nowadays, spectroscopy is the primary tool by which astronomers learn about compositions, temperatures, and densities of anything that shines, as well as of any gas that absorbs or scatters light from another source.

Although already a useful tool for decades by then, spectroscopy was only given its theoretical underpinning in 1913 when Niels Bohr (1885–1962) proposed the theory of the atom that explains the sets of colors emitted or absorbed by the diverse elements. Astrophysical spectroscopy and the theory of what happens to light as it is emitted by, and travels through, gases is a key class taught to aspiring astronomers, and it is one of the most difficult. You have to understand what individual molecules, atoms, or ions do with photons of light, how their motion modifies that, and how it works when there is a magnetic field present. Then you have to work through what happens for multitudes of such particles that form a gas of different temperatures and densities. You have to learn to think about what astronomers call "radiative transfer" in terms of "optical depth," which quantifies how far each color of light penetrates before being absorbed or scattered, and "source functions," which describe how much of the light that you see originates from where along a line of sight. It is one of the hardest classes to go through, but as light is the astronomer's primary tool there is no skipping these courses and their tough homework exercises. I was taught this topic by the professor who later became my PhD advisor, Kees Zwaan (1928–1999), sitting on the century-old benches of the old observatory in Utrecht, the Netherlands. Those classes were harder than any others, and seemed very abstract at first, but later I realized that thinking about light in terms of his equations was essential in order to interpret the fascinating stories hidden in starlight.

The contamination of white dwarfs

What does the principle of spectroscopy for chemical analysis have to do with exoplanets? We already saw that it comes into play when using the Doppler effect to measure the back and forth of a star subject to the pull of one or more planets around it; that very method relies on the presence of element-specific color fingerprints in the spectrum of a star where these serve as reference markers for precise measurements of color (or wavelength) shifts that are the Doppler consequence of planets orbiting a star, first successfully applied to discover exoplanets in 1995. Another timeline by which that question can be answered starts in 1910. Henry Russell was working on early versions of the Hertzsprung–Russell (HR) diagram (see Figure 5.8) in which the intrinsic (distance-corrected) brightness of stars is plotted against the color of stars. Most stars lie in a narrow band for most of their lives; this strip in the HR diagram is called the main sequence. Russell found the first major outlier to such a diagram in which a white-hot star was far less bright than others, placing it well below the main sequence. That first outlier was 40 Eridani B. In 1913, Adriaan van Maanen (1884–1946) came across a second such star, now known as Van Maanen 2.

It was a third one, Sirius B, that triggered the discovery of what these stars really were: because of the Doppler effect that it created as it pulled on its binary companion, Sirius A, astronomers could estimate its mass. This turned out to be very nearly the same as that of the Sun. But the surprise was the combination of this mass with its size, which in 1915 was deduced from the combination of the surface temperature, observed brightness, and measured distance: this object packed a mass like that of the Sun into the volume of something like the Earth, which is a million times smaller than the Sun.

These compact, white-hot objects are nowadays known as white dwarfs. We have identified them as the most common end of the evolution of a star: this is how all isolated stars end their

lives if starting out weighing less than roughly nine times as much as the Sun. Heavier stars end their lives in explosive supernovas leading to black holes and neutron stars. But over 99 percent of all stars weigh less than that. Those end their lives through much more gradual evolutions that result in a never-ending cooling phase where much of the original star's mass is compressed into something that is commonly about as small as the Earth. Because it is so small and heavy, the surface gravity is absolutely crushing at about 100,000 times that at the surface of the Earth.

White dwarfs are quite common: there are probably over ten billion of them throughout the Galaxy. They are faint, however, because they are so small. And therefore they are hard to spot. Among the earliest white dwarfs found was one chanced upon in 1917 by Adriaan van Maanen among other objects for which he was measuring distances. By 1920 he had realized that this particular object was very faint compared to other stars. More-over, through spectroscopy he had noticed that there appeared to be a lot of heavy chemical elements in the atmosphere. Not having been included in earlier catalogs, it had no official name. He decided to refer to it as "Anonymous 1." Two years later, a colleague, Willem Luyten (1899–1994), dubbed it "Van Maanen 2" and introduced the term white dwarf. In retrospect, the obser-vations made by van Maanen provide the earliest evidence for extrasolar planetary systems, but no one realized it for a long time: the mystery of the unusual amount of heavy chemicals in its atmosphere, and in the atmospheres of other white dwarfs, would remain unresolved for another 90 years.

The heavily compressed gas that makes the atmosphere of a white dwarf is strongly affected by the high gravity: the heavier elements sink out of the atmosphere into the dwarf's interior so fast that the mixing processes that would homogenize the chem-istry again under less crushing gravity do not succeed—unless the temperature is really high. White dwarfs start in a truly hot and bright state, with a surface temperature of some 100,000 degrees centigrade (180,000 degrees Fahrenheit), and then gradually cool

because there is no heat source that can replenish the internal energy once that is radiated from the bright surface. When conditions are initially that hot, heavy elements in essence can be levitated into the atmosphere by the light from deeper within the white dwarf which pushes outward on the atoms. When the dwarf cools below about 20,000 degrees centigrade (36,000 degrees Fahrenheit) and truly becomes white hot rather than shining ultraviolet or blue, this no longer works, and for most of these white dwarfs nothing stops the fall of heavy elements. The timescale on which this segregation of elements occurs is substantially shorter than the cooling timescale for white dwarfs. The color spectra of such aged white dwarfs should consequently display only the elemental signatures of hydrogen and maybe some helium. That is exactly what most of them do.

Some 25 percent to 30 percent of the relatively cool white dwarfs do not live up to this expectation, however. The members of this minority population have significant amounts of elements in their atmospheres that are heavier than carbon. Signals have been reported of at least eighteen of these chemicals. These eighteen elements can be grouped into several classes of behavior depending on the chemical and thermal settings inside planets, moons, and asteroids throughout a planetary system: the so-called chalcophile elements sulfur and copper that form compounds that float near the surface of a molten planet; the rock-loving (lithophile) elements oxygen, sodium, magnesium, silicon, phosphorus, scandium, vanadium, and strontium, including also the refractory ones (that solidify at high temperatures) aluminum, calcium, and titanium; the rock-and-iron-loving elements chromium, manganese, iron, and cobalt; and the purely iron-loving (siderophile) elements cobalt and nickel. This separation into different types of elements reflects how they preferentially bond into compound chemicals: lithophiles readily bind with oxygen forming relatively light compounds, while siderophiles readily mix with molten iron and then this heavy mixture sinks below lighter compounds when subject to gravity

in a hot, molten state inside planets. But we were talking about remnants of stars, not planets (at least, not yet.)

Astronomers realized that the presence of heavy elements in white-dwarf atmospheres required their having been captured from somewhere outside the deceased star. Moreover, they realized that this capture would have had to occur quite recently, in astronomical terms, or might even be ongoing. After all, with more time the heavy elements would sink into the interior and would not be detectable in the surface layers. Through much of the twentieth century, astronomers commonly assumed that these heavy elements might be somehow captured out of the interstellar medium through which the white dwarfs traveled. There were clear problems in making that work, but no obvious solutions in astronomers' common thinking. As recently as 2003, a scientific review paper on knowledge of white dwarfs noted that the "heavy elements are believed to be accreted [captured by gravity] from the interstellar medium when the star occasionally encounters a cloud or high gas density region. After some time passes, gravitational diffusion removes these elements from visibility." Only a few suggestions were made that it could be the result of accretion of sizable solid objects, for example comets that would impact on white dwarfs originating from an equivalent of the Sun's Oort Cloud. That, too, was shown to be problematic, however. Although this hypothesis could in principle explain why hydrogen was not the dominant atmospheric element, comets cannot explain the observed chemical mix in these white dwarfs: comets are rich in carbon (often bound in carbon dioxide) and oxygen (much of it chemically locked in water molecules), which later were shown rather lacking in white dwarf atmospheres.

The composition of shredded planets

In the first years of the twenty-first century, knowing how many planetary systems actually existed in the Galaxy, the

interpretation swung about entirely, in a reorientation as drastic as the one a century earlier that followed Cecilia Payne's discovery that hydrogen dominates in stars (as described in Chapter 5): heavy elements in white-dwarf atmospheres are now seen as contamination by fragments of disintegrating minor planets that crashed into the white dwarf's atmosphere. And that is where flame spectroscopy and white dwarfs come together: the heavy elements seen in the otherwise pristine hydrogen or helium atmosphere are equivalent to shavings thrown into a laboratory flame, leading to the spectroscopic fingerprints that tell us about the composition of the planetary fragments that fall into the white-dwarf atmosphere.

What does the white-dwarf contamination tell us about the chemical makeup of the bodies that impact these objects? In the Solar System, four elements add up to over 90 percent of the mass in chondrite meteorites and Earth itself: the rock-loving oxygen, magnesium, and silicon, plus iron. For contaminated white dwarfs overall these same four add up to over 85 percent of the mass in the contaminants. Ratios of the iron-loving elements manganese, chromium, and nickel to iron suggest that exo-asteroids ending up in white dwarfs have undergone iron-core formation just as the larger asteroids and terrestrial planets have in the Solar System. The separation of elements into a rock-dominated exoplanet mantle and iron-dominated exoplanet core is also consistent with the white-dwarf observations.

Moreover, in both the Solar System and in the extrasolar set-tings around white dwarfs, an interesting absenteeism is displayed by carbon: whereas it adds up to one third of the mass of oxygen in the Sun and in the interstellar gas, it is deficient by a factor of ten in the inner Solar System bodies and also in contaminated white dwarfs. In white-dwarf contamination, we may thus be seeing evidence that the lack of carbon in Earth's overall composition is reflected also in many other exoplanetary system bodies, at least in those whose fragments eventually come to fall into white dwarfs. Water presents a more complex story, but observations do

suggest that water is as rare in exo-asteroids as in Earth where it amounts to only about 1 percent of the whole, being abundant only in a thin surface layer.

How can exoplanet remains fall into a white dwarf?

Just a minute, though...well, actually we need to embark on a path that takes us through a century of research to appreciate what is happening. In order to become a white dwarf, a star first goes through a giant phase, which for the Sun is as large as Earth's orbit. Any planet orbiting within the peak size in this phase will be enveloped by the star and destroyed. And then the giant star shrinks into a relatively tiny white dwarf, which should have nothing orbiting it in its wide surroundings. With the neighborhood cleared of planets and asteroids during the giant phase, how does anything substantial end up crashing into the tiny white dwarf that forms out of the giant? The question is even more pregnant with consequences if we realize that for white dwarfs with an essentially pure hydrogen atmosphere any heavy-element contamination in its atmosphere disappears within weeks, so that if such contamination is observed it cannot be something that happened in the distant past but must be essentially ongoing.

This brings us back to stellar evolution, visually summarized in Figure 8.1. As stars exhaust the hydrogen in their cores through nuclear fusion, they reach a phase of expansion. Exactly why that happens is beyond the scope of this book, but the essence is this: stars at the end of their lives increase dramatically in size, more than once, and some can undergo an instability in which they pulse from large and bright to cooler and compacted multiple times with a cycle of some thousands to tens of thousands of years. During this phase, the stars also lose much of their mass, either in a dense wind or as layers thrown off during the pulses of expansion, or both. A star like the Sun is expected to lose

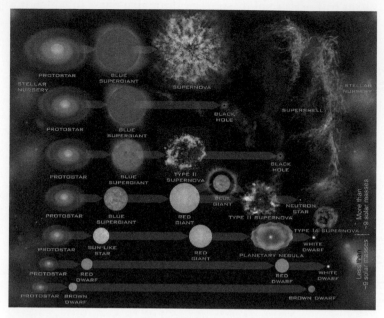

Figure 8.1 Representation of stellar evolution for stars of different masses, showing stellar timelines from birth in stellar nurseries on the left to demise as much of their material is ejected back into the interstellar medium where it can contribute to a new generation of stars in new stellar nurseries. The heaviest stars, shown at the top, have the shortest lives and end their existence in supernova explosions that blow vast bubbles whose edges are called supershells if formed by multiple nearby supernovas. Stars like the Sun with intermediate masses end their existence as white dwarfs after phases as red giant stars and ultimate expulsion of their outer layers to form what are called "planetary nebulae." Everything that is ejected in any of the explosions at the end of the lives of stars is enriched in the elements that make planets, is mixed with the interplanetary gas, and may then be used in the next generation of stars and their planetary systems as these are formed in the stellar nurseries of open clusters. Pseudo-stars that are too light to ever have proper hydrogen fusion are called brown dwarfs; these just glow red at best and will be forever gradually cooling down.

Source: NASA/JPL.

about half of its total mass, to go through a phase in which it is over 4,000 times brighter than it currently is, and to expand as a giant to just beyond the size of the current orbit of the Earth. A star twice as heavy loses some 70 percent of its mass, reaches a maximum brightness of almost 10,000 times the Sun's present-day brightness, and swells to well beyond the equivalent of Mars's orbit. If it starts at five solar masses, the star will lose 80 percent of its mass, blinds at almost 50,000 solar luminosities, and grows to engulf Jupiter's orbit. Any star starting with more than about nine solar masses also goes through a giant phase, but it ends in a supernova, in which most of the outer layers are ejected in a gigantic explosion, leaving in the center a 10-km (6-mile) neutron star or an essentially infinitely compact gravitational anomaly known as a black hole. In short, being anywhere near a dying star is not good for a planet, even if the star ends up not exploding as the heaviest ones do.

But let's stay with the stars that end up as white dwarfs after mass-losing phases as giant stars, which is the case for any star starting with less than about nine solar masses (somewhat dependent on the star's elemental composition). A white dwarf is an object composed mostly of carbon and oxygen in its interior, enveloped by a thin "atmosphere" that is predominantly hydrogen, although in some that also contains significant amounts of helium. We know they all have masses much like the Sun. After all, lower-mass stars evolve relatively slowly so that the Universe is simply not old enough for any star with a mass below about 80 percent of that of the Sun to have existed long enough to have become a white dwarf, while stars substantially heavier than the Sun are less frequent. Therefore, most white dwarfs that now exist in the Galaxy originate from stars with masses between roughly 1.5 and 2.5 solar masses.

Incidentally, when a star has a nearby stellar companion, forming a binary star, the evolutionary options multiply, because each component evolves at a rate set by its mass and the masses are generally different. In binaries, stars may envelop their

companion star in their giant phase, may spiral in toward each other as a result of stellar winds, may spill over onto their companion as they evolve if already close to each other, may be impacted by a supernova from their dying sibling, or may end their lives forever existing as a pair of stellar end-products in the form of a white-dwarf binary, or a neutron-star binary, or a white-dwarf–neutron-star binary. Although something like half of the stars we see in the sky are binaries, triplets, quadruplets, or beyond, their stories are too complex and diverse to pursue here. In the context of the fate of exoplanet systems like our own Solar System I shall stick to the story of the single stars.

A preview of the future

Whereas the mass loss during the giant phase is fast by stellar time scales, any planetary system body will be orbiting the star many, many times during this phase. Consequently, such bodies have time to adjust their orbits smoothly to the changing mass of their central star: with less mass there to attract them, the gravitational pull is reduced so that the centrifugal force in the orbit causes them to gradually spiral outward into wider orbits where a new balance with the reduced gravity is found. The orbits therefore expand in a way that is inversely proportional to the stellar mass.

For the Solar System, this is what computer experiments suggest will happen: In another 4.5 billion years, the Sun will expand to become a giant star. Its radius is likely to grow to about the current orbit of the Earth. But because the Sun is losing mass in this phase by as much as 40 percent, the orbit of the Earth will have expanded to about 1.7 current Sun–Earth distances, so that it may remain beyond the reach of the Sun's atmosphere. Whether it really will, depends on another effect: for planets fairly close to their star—as many become simply because their star swells up more in relative terms than it loses mass—gravitational tides that couple the rotation of the star to the orbital motion

of the planet also need to be incorporated in predicting what will happen.

The process works roughly like this. If the planet is relatively close to the stellar surface, then its gravity pulls that surface up into a bulge, just like the high tide in Earth's oceans subjected to the gravity of the Moon and the Sun. That bulge in turn interacts with the planet through its gravity. If the planet moves faster in its orbit than the star spins, the planet pulls the bulge along with it, and the bulge in turn works on pulling along the entire star. With time, the star will thereby gradually increase its speed of rotation. There is an equal but opposite pull of that tidal bulge on the star that works to slow the planet. As the planet loses energy of motion, it descends into a lower orbit. This tidal coupling will only cease to change stellar spin and planetary orbit when the periods match and the planet appears to hover over a fixed spot on the rotating star. That spin–orbit interaction is what is expected to happen to the Earth: the vastly expanded, old Sun will spin much more slowly than the Earth's orbital movement, so that the Earth should gradually spiral inward when the Sun is at its largest.

Now we are faced with two opposing effects on the Earth during the Sun's far-future giant phase: the Earth's orbit should widen as the giant Sun loses mass, yet tighten owing to the gravitational tides. Quantitatively determining which one will win out remains a difficult problem, so that the ultimate fate of the Earth in orbit around the aging Sun remains unclear. Present-day studies lean toward Earth not quite surviving the Sun's giant phase because of fairly strong tidal effects that make the spiraling inward win. Not that it matters much: by then, some eight billion years into the future, the Earth's atmosphere if it has not entirely evaporated, will no longer support plant life, while the habitability zone will have moved past it billions of years before then because it moves outward with the brightening of the Sun; all of that is considered in the final chapters.

The fate of any exoplanet depends on the same two processes: the magnitude of the tidal forces in which the masses of the planet

and of the star are important determinants, and the size and mass loss of the evolving star, both of which are determined by the star's mass. For close-in exoplanets, the hot Jupiters, the situation is particularly precarious, of course. For example, if orbiting a star of 1.5 solar masses, a Jupiter-mass exoplanet would not survive the first giant phase of the star even if the planet started out in an orbit at two Sun–Earth distances: despite the fact that the star would not grow beyond about 0.9 Sun–Earth distances in that first phase of expansion, the tidal forces would enforce a death spiral into the star. In general, even those exoplanets that at first survive the expansion of their star may still not survive the ultimate phase of its thermal pulses because of the vast amount of material that is then blown off the star.

The giant phase of a star thus clears out a region in its inner planetary system where planets, asteroids, and other bodies are engulfed, either directly by the expansion of the star or indirectly by tidal forces that cause these bodies to spiral in toward the expanded star. The gap that is cleared by tides is larger for heavier planets, because the tidal forces are larger and have a longer reach for such planets. However, if there are smaller planets orbiting inside the orbit of a larger one, then the latter's spiraling in may well capture the smaller planets or perturb their orbits, spelling doom for the smaller of the company before the larger one is itself pulled to pieces ahead of it terminating its own inward spiral into the star's atmosphere.

Apart from the gravitational tidal forces acting on exoplanets too close to their star, there is also the effect of the star's brightness: with a star growing thousands of times brighter as an aged giant than as a long-lived adult star, the evaporation and sublimation of entire bodies that will occur throughout the planetary system is important. Large exoplanets may survive this phase, at least partially, because it may take longer for them to evaporate than the giant phase of the star persists, but they will lose at least part of their atmospheres. For the Solar System, it is thought nothing smaller than the large planets will survive

the Sun's giant phase, even those objects that orbit as far out as the distant Kuiper Belt: many of the distant ice-heavy bodies will evaporate fully after a phase in which most will already have been disrupted by the loss of ices, depending on their internal makeup.

Large planets orbiting sufficiently far from their star may survive, in whole or in part, the giant phase. However, if there are multiple large planets, their survival through their star's giant phase becomes far more uncertain: even if as only planets they might survive the giant phase of the star, the presence of other planets subject to the changing mass and radius of the star is likely to destabilize the remnants of exoplanetary systems left over after the giant phase. Such destabilizations are similar to those that occur as a planetary system forms, which we discussed earlier: some will have nearly immediate consequences, others may not develop severe symptoms until a few hundred million years later. So, such instabilities may be initiated during the giant phase but may not result in major orbital rearrangements until well into the star's white-dwarf phase.

The study of what can happen to orbits in multi-planet systems as stars evolve is currently only feasible via computer experiments. These reveal that there are many different pathways for a planetary system, depending on the star's evolution, and the orbits and masses of the planets. It also depends on chance processes determined by just how close planets get as they either scatter around each other or fully collide: starting with the same virtual planets and orbits with only a slight change in initial positions can lead to very different outcomes for the computer experiment. One extensive study explored sets of four to eight planets with masses either like those of the giant planets in the Solar System or like the terrestrial planets, all in different combinations of their relative masses. It appears that throughout the evolution of the star from adult on the main sequence, through a giant phase with substantial mass loss, to ultimately a cooling white dwarf, a planetary system can remain stable or may destabilize, can eject planets or see them merge in collisions, can lose planets through

capture into the stellar envelope in the giant phase, or may lose them when they are pulled apart by coming too close to the white dwarf. Any of these can happen in any phase of evolution, and planet loss through collision, engulfment, or ejection can happen to any number of the planets in the system.

These numerical experiments show that a changed orbital arrangement can appear to be stable, or at least not lead to catastrophic loss of planets, but that destabilization may take hundreds of millions of years to develop. A small fraction of these destabilizations can shift orbits so much that one or more of the planets can eventually come to orbit closer to the white dwarf than the maximum extent of the giant phase just before. This state would misleadingly suggest that exoplanets survived an evolutionary phase of their star during which they should have been engulfed. Ejection from the planetary system is more commonly the case for interactions between heavy planets, whereas terrestrial-mass planets more commonly end up in collisions; intermediate-mass planets suffer collision or ejection in about equal numbers. The more planets a system contains, the likelier it is that strongly elliptical orbits will develop for one or more of them, thus increasing the probability of coming too close to the white dwarf to survive at some point in such orbits.

There are at least three scenarios by which planetary matter may ultimately end up in white-dwarf atmospheres. First, whereas entire planets rarely come close to a white dwarf, computer experiments suggest that it may indeed sometimes happen. Second, the computer experiments also show that asteroids orbiting in extended belts can change orbits and may either be ejected from the planetary system, or fall into the central star, or be captured by other planets either as new moons or impacting their atmosphere or surface. A third option for white-dwarf contamination with heavier elements is that during the giant phases when most mass is lost, such as the last thermal pulses of intermediate-mass stars or even supernovas of heavy ones, not all of the mass may be thrown entirely free of the planetary

system. Instead, some of that mass may gradually circle back, possibly forming a new disk around what is now a white dwarf or neutron star, in which new solid bodies may condense at a range of distances while volatiles such as hydrogen and helium are blown out. This kind of second-generation planetary system—speculative for now, but plausible—would also be a reservoir for white-dwarf atmospheric contamination.

The rings of white dwarfs

Ultimately, contamination can only happen if the matter actually falls into the white-dwarf atmosphere (Figure 8.2). This is not likely to occur for large bodies: whenever a fairly large body comes near to a white dwarf (within about 50 white-dwarf diameters), the star's very strong gravitational tides will literally pull that approaching body apart into small pieces. These pieces gradually migrate along their orbits, with mutual collisions further grinding them down, thus forming a dusty disk of particles that can then spiral or scatter inward to crash into the white dwarf. In about 4 percent of white dwarfs, these disks have enough mass to be observable. For all white dwarfs with an observed disk, there is also observed atmospheric enrichment with heavy elements, as expected when this is a consequence of matter falling into the atmosphere after fragmentation in such a disk setting. These disks are thought of as rather similar analogs of the ring systems of the giant planets in the Solar System (Figure 8.3): they are comparable in sizes, argued to be similarly thin perpendicular to the disk plane, created by disintegration of larger bodies initially from elsewhere in the planetary system, and possibly displaying a comparable diversity.

Nowadays, the phase of tidal disruption of an exoplanetary system body en route to the white dwarf is actually being observed. One well-studied example is the white dwarf WD 1145+017. Multiple objects transit the white dwarf, blocking part of its light with periods between 4.5 and 5 hours, but changing exactly how

Figure 8.2 Artist's impression of the final phases of a planetary system. In the center of the disk lies a white dwarf, the burned-out ember of what was once a star. Around it gyrates a disk, similar to Saturn's rings (Figure 8.3), made up of debris from destroyed exoplanets and exo-asteroids. Some of that debris may be large fragments of planets, but the fragment on the right appears large only because it is shown close to the imaginary observer. After a phase as a red giant star, dozens of times larger than our Sun, the star shrunk down to a size like the Earth's, a hundred times smaller than our Sun. Fragments of exoplanets that were destroyed during the giant phase, or any remaining asteroids, may be scattered about in their orbits as they sense each other's gravity. Any that come too close to the white dwarf are torn apart by its tidal forces or fragment in collisions with other remnants also there. Eventually, the dust in the debris disk spirals down into the atmosphere of the white dwarf, depositing chemical elements that would not normally be there.

Source: NASA, ESA, and G. Bacon (STScI).

much light is taken away gradually every few months. The light is not obscured measurably directly by the objects, which are too small, but by the clouds of dust around them as they disintegrate, like dust clouds around comets…which is exactly what they are around this white dwarf. The asteroid bodies are likely solid dense matter because gas atmospheres would be pulled off in the gravitational tides, whereas less-dense materials (for example anything

Figure 8.3 Image of the rings of Saturn, shown in a composite of many individual photographs taken by the Cassini spacecraft when it was in the shadow of the giant planet. This particular back-lighting reveals a multitude of diverse rings, all made up of rocks and dust; the debris swirls around Saturn in a system almost 300,000 kilometers (some 180,000 miles) across but typically 0.1 kilometers (300 feet) thick. Debris disks somewhat like this are thought to gyrate around many deceased stars, the white dwarfs; the disk sizes may be like those of the rings of Saturn, but white dwarfs in their center are relatively small, only about one tenth of Saturn's size. Note the small dot just above and to the right of the center of the image, just outside the set of bright rings: that is our home planet Earth viewed from over a billion kilometers (almost a billion miles) away.

Source: The Cassini-Huygens mission, a cooperative project of NASA, ESA, and the Italian Space Agency.

with much water ice) would have been pulled apart already. So the orbiting minor planets or dwarf-planet fragments are most likely composed of what a planet like Earth is predominantly made of. This matches the fact that the white dwarf itself has signatures of at least magnesium, aluminum, silicon, calcium, iron, and nickel in its helium-dominated atmosphere.

Observing evidence of disks around white dwarfs is very challenging simply because the entire disk system is the size a giant planet in the Solar System but viewed across the vast distances on which even stars more than ten times larger can be seen only as a point. In fact, only one such disk was detected prior to 2005, not imaged but through its spectroscopic signatures. Then the Spitzer Space Telescope was launched, which was designed to observe faint sources glowing in the infrared, which is invisible to the human eye, and sensed by us only as warmth. Spitzer found spectral signatures of disks about many white dwarfs with heavy-element contamination. One of the signatures that the Spitzer telescope found was the prevalence of silicate dust, and in particular micron-sized dust from glassy, amorphous rocks. Another finding confirmed the relative paucity of carbon-containing chemicals, as expected from the rarity in the white-dwarf contamination. Plus, for disks with hot components, it appears that dust with volatile elements is rare; those would have evaporated into gas. All in all, these observations of debris material poised to contaminate white-dwarf atmospheres are what you would expect if it had formed out of disintegrated planets and asteroids.

The rate at which heavy elements contaminate white-dwarf atmospheres ranges from 30 million tonnes per year to 100,000 times as much. For bodies with an Earth-like composition, that means a volume equivalent to a single sphere with a diameter ranging from 300 meters (or yards) to 13 kilometers (or 8 miles) per year. It would need to sustain that for dozens of millions of years. If it did so for 30 million years, for example, the total volume of parent asteroids would add up to the equivalent of spheres of about 100 to 4,000 kilometers (60 to 2,500 miles) in diameter. By astronomical standards, that is not much: a single body like Earth's Moon could meet the heavy-element requirements of the most intense dumping onto a white dwarf for tens of millions of years. For an average contaminated white dwarf, the total

mass in the asteroid belt of the Solar System—less than $1/1,000$ of an Earth mass—would be adequate. At some point, available asteroids will run out. This is likely why for white dwarfs older than several hundred million years far fewer show atmospheric contamination.

9

The Worlds of Exoplanets

Many alien worlds have been visited in science fiction writing, television series, and movies. When these fantasy worlds are visited by imaginary humans they are often pictured as quite Earth-like. Not surprisingly, gravity and temperature are set to be tolerable for the human travelers. That is a defensible choice because if these were not compatible with the human physique, or at least compatible with the support of technological aids like space suits, then such a visit would simply be impossible. Commonly, we are presented with fictional planets that in addition to acceptable gravity and temperature have an atmosphere that is breathable and that can sustain the human visitors without cumbersome space suits or even mere breathing apparatus. Expecting this to be true for exoplanets in general would be vastly incongruous with reality, however. One can anticipate that mismatch just by looking at the diverse worlds of the Solar System where we encounter these hospitable conditions only on our own planet. And even on Earth, these conditions have existed for less than half of the planet's lifetime thus far.

The environment on exoplanets

What do we know of conditions on exoplanets? Let's start with surface gravity. What we need in order to determine that property are just the planet's mass and its size underneath its atmosphere. So, this can be determined with reasonable certainty for some Earth-like planets and for what are called super-Earths, which are planets not too much larger than Earth that have thin or

One of Ten Billion Earths. Karel Schrijver, Oxford University Press (2018). © Karel Schrijver.
DOI: 10.1093/oso/9780198799894.001.0001

at least transparent atmospheres. What is required is that they transit their star to inform us about their size and also that they are sufficiently heavy or close in their orbit to pull their star about somewhat so that the planet's mass can be established from the Doppler effect. For non-transiting planets, or for small planets or planets distant from their star, this does not work. Moreover, even if fairly heavy exoplanets were transiting, we would still face a problem if they had thick cloud layers or otherwise opaque atmospheres. In such cases, which are likely common for exoplanets discovered thus far, establishing where the solid or liquid surface exists below the atmosphere is currently very difficult. There are large uncertainties about where the transition from a gaseous atmosphere to a liquid or solid surface might occur and what the chemicals are that would provide ocean-like or continent-like surfaces. In fact, as we shall see in what follows, this is even challenging for the gas and ice giants in our Solar System despite spacecraft having visited these planets to study them in great detail. Another complication is that there may be multiple transitions below the clouds that might be perceived as the planet's surface: liquid oceans (made of water or some other substance) may exist underneath frozen outer layers while they may at the same time lie above yet another ice layer or otherwise solid core.

It is currently also difficult to estimate the surface temperature for exoplanets, because their light cannot as yet be directly observed in almost all cases. Assuming the absence of any atmosphere, an equilibrium temperature may be computed based on the balance of incoming starlight and outgoing visible and infrared radiation. If there is a considerable atmosphere, however, the determination of surface temperatures becomes much more problematic and uncertain, even if we knew where the surface was. There are two main reasons for that. It is in part because an atmosphere reflects some of the incoming starlight, thereby reducing the energy input into the atmosphere. It is also in part because an atmosphere works like a blanket around the planet in which incoming radiation can be converted into

heat which then may be trapped if the atmosphere is not or is only partially transparent for infrared light. On Earth, we refer to this trapping—mostly by water vapor, carbon dioxide, and methane—as the greenhouse effect.

Different gases have different efficiencies in trapping heat, so that the resulting atmospheric temperature distribution depends sensitively on both the mixture of the gases and on their total quantity. One Earth-like planet in which this effect is known to be particularly pronounced is Venus. If its atmosphere did not work as a greenhouse then its equilibrium temperature at the surface would lie somewhere around the freezing point of water, depending somewhat on the value of the atmosphere's reflective properties. In reality, as Venus's atmosphere is mostly the greenhouse gas carbon dioxide mixed with a few percent of nitrogen, the temperature at the planet's surface is a very inhospitable 460 degrees centigrade (or 860 degrees Fahrenheit).

For rocky exoplanets, those that more or less by definition have shallow or transparent atmospheres, we simply do not know the composition and thickness of their atmospheres. Therefore, at present we cannot make reliable estimates of their surface temperatures. For gas giants we can only guess where the surface might be or what it would be composed of. To compound the problem, many of the exoplanets found are so close to their parent star that their rotation has likely been largely, if not entirely, synchronized with their orbit through tidal forces, so that they would present the same face to their star all the time. That would set up asymmetries in their upper atmospheres that promote high-speed winds circulating between the permanently irradiated, hot daytime and the dark, cold night-time hemispheres. For tidally locked exoplanets that have deep atmospheres, efficient circulation and mixing might still result in mild differences between day and night sides at depth in the planet's atmospheres. For those with relatively shallow atmospheres, however, the temperatures on the day and night sides may differ by hundreds of degrees. If they are synchronized rocky exoplanets without a

Figure 9.1 A comparison of landscapes where robots or humans have landed in the Solar System: asteroid Itokawa, Earth's Moon, and Venus, all of which are waterless deserts. More in Figure 9.2 (See Plate 22 for a color version).

Source: Asteroid Itokawa (visited by the Hayabusa probe): ISAS, JAXA, Gordan Ugarkovic; Moon (in this case photographed by the crew of the Apollo 17 mission): NASA; Venus (imaged from the Venera 14 lander): IKI, Don Mitchell, Ted Stryk, Mike Malaska. Composition by Mike Malaska.

substantial atmosphere, the difference between their hemispheres could amount to thousands of degrees.

Visualizing the landscape

Then, there is the landscape to think about. We have learned from exploring our Solar System that the terrains of rocky planets,

Figure 9.2 A comparison of landscapes where robots have landed and where humans have been in the Solar System: the desert planet Mars, the frozen moon Titan orbiting Saturn, and an image somewhere on Earth's oceans that cover 71 percent of the planet's surface area. Continued from Figure 9.1 (See Plate 24 for a color version).

Source: Mars (photographed by the Mars Exploration Rover Spirit): NASA, JPL, Cornell, Mike Malaska; Titan (photographed by ESA's Huygens lander that flew with the Cassini probe): ESA, NASA, JPL, University of Arizona; Earth's seascape Mike Malaska. Composition by Mike Malaska.

asteroids, and moons differ tremendously from each other (see the examples in Figures 9.1 and 9.2). In fact, terrains exhibit a wide variety on any one given planet; just think of Earth's landscapes and oceans. Yet often we find that press releases from research organizations are accompanied by artists' conceptions of the

surface of such planets, depicting clouds, moons, and starrises or starsets in the colored skies. In reality, we have no information upon which such landscapes could be based, other than that we know the appearance of the star or stars above the atmospheres of the planets that orbit them. Such imagined landscapes serve merely as "eye candy" to attract attention to the item on television, the Internet, or in printed media.

Sometimes, such as in the media releases on the occasion of the discovery of the TRAPPIST-1 system (described near the end of this chapter), we are provided with posters such as in Figure 1.5[A] by imaginary travel agencies. One example is NASA's "Exoplanet Travel Bureau" that names a trip to TRAPPIST-1e "Best 'hab zone' vacation within 12 parsecs of Earth." That very attribute highlights the impossibility of its own message: the TRAPPIST-1 system lies at close to 40 light years from Earth, so that if we had the technological capability for long and fast interstellar travel, the universal law that limits travel to below the speed of light means that the trip would bring you back to Earth no sooner than 80 years from the time of departure. We can only dream about such excursions, because there is no evidence that we could find a way to circumvent that law of nature that links distance to time needed to get there and back. If we were to build a spaceship that could reach nearly the speed of light to get there and back, Einstein's theory of special relativity does admit the travelers to age substantially less on their trip than expected by our clocks on Earth, but upon returning home they would find their adult friends all long dead and their grandchildren quite old themselves. In short, such a trip is currently impossible, and likely forever undesirable.

In the announcements of new exoplanets we often see terms that highlight their unique nature: the most Earth-like planet, the most compact system, the most peculiar star or binary around which the planets orbit, and so on. These statements are

---

[A] See https://exoplanets.nasa.gov/resources/2159/.

commonly true, at least at the time of the first public coverage, but they highlight the unusual to the neglect of the common. In this chapter, I attempt a combination of the extreme and the typical, and describe what we know about some of the planets in more detail. What are the common properties of planetary systems, what are some of the properties of their constituent planets, and what is revealed by the unusual and uncommon that enhances our knowledge beyond the typical planetary system and planet?

The assortment of worlds

After the surprise that exoplanet systems are common in the Galaxy, and after the surprise of the great diversity of configurations of these systems, another surprise emerged: a large fraction of the exoplanets observed have sizes between those of the rocky, terrestrial Earth and the volatile-dominated ice and gas giant Neptune, so between one and four times the size of Earth, and because of that size they are often referred to as super-Earths. True, smaller planets are harder to detect, which introduces a bias in the population of observed exoplanets relative to the intrinsic population. Yet the simple fact that such intermediate planets are found at all is quite interesting because there is a substantial difference between the terrestrial planets and the gas and ice giants in our Solar System (Figure 5.9), with nothing in between them.

That gap is not simply one of radius alone. Earth is predominantly made of iron, nickel, and a variety of silicates rich in oxygen, silicon, and various metals. In contrast, much of the mass of Neptune is contained in water, ammonia, and methane. These volatiles, mostly liquid with some solid ices in Neptune's interior, make a thick envelope that appears to outweigh the iron–silicate core underneath it by roughly ten to one, although there is a substantial uncertainty in our knowledge of this number.

Astronomers working to learn about the cores, mantles, and atmospheres of exoplanets are very aware of the biases in the observed sample introduced by where what type of planet is most

readily observed. For example, heavy planets are more readily observed than less massive ones using the Doppler effect. Larger ones are more readily observed than smaller ones when using transit measurements. And planets distant from their star are generally harder to confirm, including those that are transiting their star, because their orbit takes a long time, thus requiring years to centuries of observations to confirm them as planets by seeing them repeat their motion, which is readily longer than the lifetime of a space-based observatory if not of the scientists hunting for such planets.

Consequently, attempts are made to correct for such biases, or at least to acknowledge awareness of biases in any conclusions about exoplanet internal structures. With that in mind, some trends emerge that do seem likely to hold true despite the uncertainties introduced by the available samples. Among them are the following.

Planets that are smaller than about 1.6 times the Earth typically have a high density that suggests that they are largely rocky, with relatively little water or gas. Many of such exoplanets that have been spotted are, however, really close to their parent stars, so that they may have lost much of their volatile components in the heat of their star or by the stripping effects of its dense stellar winds, or both. Or, they may have been formed with but a low percentage of volatiles, as we think happened for the terrestrial planets in the Solar System.

Planets between 1.6 and three times the size of Earth show a trend such that their average density decreases with average size. This must mean that heavier planets tend to have more water or a gaseous atmosphere to envelop their silicate–iron components. Just a few percent of the total planet's mass contained in a hydrogen–helium atmosphere can cause the apparent radius of the exoplanet to be considerably larger than the solid layers underneath that atmosphere, thereby lowering the average density that we compute even though there may be relatively little mass in the atmosphere.

Planets between three and seven times the size of Earth have a density below that of water. To reach such a low average density they must have extended gaseous atmospheres, probably with a lot of hydrogen and helium. The larger ones in this range, and those larger still, are predominantly found to orbit stars that have enhanced amounts of heavy elements: large planets with a lot of hydrogen and helium in their atmospheres apparently form preferentially around stars with a lot of heavy elements.

When comparing the population of planets in the Solar System to that of well-known exoplanets we find that these two populations have in common that heavier planets are typically larger. But what the exoplanets show is that size alone does not suffice to derive the mass. For a planet of given size, the spread in mass can be large. For exoplanets up to the size of Jupiter, or ten times wider than Earth, the masses can differ by a factor of three given a size. For larger exoplanets, masses of the bulk of the exoplanet population range over a factor of ten, while the extreme cases lie a factor of a hundred apart. This brings to light that there is a great variety in the composition and in the thickness of planetary atmospheres and interiors. Comparison with masses expected for hypothetical planets made of iron, rock, or water reveal that whereas exoplanets up to the size of Neptune (about four times the size of Earth) must be made mostly of these heavy components, larger planets must have extended gaseous atmospheres in order to reach the size for their mass. If there is a surface below these atmospheres, and if anyone is there with an equivalent of eyes, they would rarely, if ever, see anything of the Universe surrounding them.

Constraining exoplanet interiors

To learn something about the interiors of planets and exo-planets, scientists need to run a complicated experiment. They may only know the total mass and size of a planet, and with only that they need to envision layered mixes of solids, fluids, and—in

many cases—gases that, when put together, recreate those two overall properties. Their analysis is complicated by the fact that the profile of temperature with height is unknown, and yet that unknown temperature is a key factor in setting the pressure, and pressure in turn combines with gravity to determine how compressed or extended the atmospheric gases are. Moreover, at the high pressures found deep inside planets, the way in which gases compress can be very different from how we know them on Earth; this happens because under great pressure molecular structures can be lost or electrons can become free to move around in unfamiliar quantum fluids that can be superconducting or that move without displaying viscosity. In such circumstances liquids and solids can behave in ways that are either uncertain or surprising. The scientific term for the quantitative description of this behavior that planetary astronomers need to put into their computer codes is the "equation of state." It describes, among other things, the compressibility and the limits of compressibility of substances of a given chemical mix, and also how substances solidify or melt as pressure and temperature are changed. In laboratories, only so much pressure can be put on substances while temperatures can be raised only so high, so that there continue to be new findings on equations of state of substances or mixes of substances as laboratory technologies evolve, while discoveries are also made by work that is done purely in the virtual world of computers.

Having the apparent freedom—simply because of absence of information—to arrange cores, mantles, and atmospheres using a variety of different chemical mixes in their computer experiments leaves astronomers with substantial ambiguities in the solutions that they come up with. This is even true for the internal structure of the relatively nearby giant planets in the Solar System. This ambiguity is amplified for exoplanets where conditions are often markedly different and less well known.

The primary problem is that nothing like terrestrial seismology is available for exoplanets or the Solar System's giant planets.

Even the study of the internal structure of the terrestrial planets in the Solar System other than Earth is challenging because seismology as used on Earth is not yet available on the other terrestrial planets: their seismic activity is expected to be vastly weaker than Earth's so that seismometers are generally not included in lander payloads, while the few that were deployed on Mars by the early Viking landers did not successfully operate; the only properly working seismometers installed beyond Earth to date were placed on the Moon. Other types of seismological signals have been picked up in oscillations of Jupiter's gaseous envelope, and perhaps are seen in density waves propagating through Saturn's rings, but the information that has been extracted from these measurements is limited. Thus planetary conditions that are substantially different from Earth's—which means for just about all known exoplanets—are presently poorly known.

Even the internal structures of Jupiter, Saturn, Uranus, and Neptune are annoyingly uncertain, despite our ability to observe these planets quite well and notwithstanding that space probes have orbited them all. But at least for these Solar-System bodies we can measure the rotation rates and then work out how their spin causes them to deviate from perfect spheres by the effects of centrifugal forces. Space probes can add information to that by measuring the distortion in the gravitational field associated with the deformation of various layers in the interior of planets caused by the centrifugal force of their spinning or by asymmetrically distributed solids. We can also compare the known energy input from the Sun to the measured energy output from the planets. This helps us determine whether there are any internal energy sources such as energy left over from the planet's formation, or warming caused by radionuclide decay, or heating caused by electrical currents driven by space weather around the planets, or energy extracted as chemicals separate (such as by the hypothesized helium that rains out of the hydrogen deep inside Jupiter) or the inverse when weather and ocean currents (re-)mix chemicals. Still, even being able to measure all that,

we lack sufficient information about what exactly goes on in the unexplored, and generally inexplorable, depths of the worlds that surround us.

Just how uncertain things are becomes clear by comparing models made by different research groups or that are found by assuming different chemical mixes and different equations of state. One such comparative evaluation was compiled by Jonathan Fortney and Nadine Nettelmann in 2010. One thing they realized from these comparisons is that interiors of giant planets are not like the environments we know on Earth. For one thing, they note that "Gas giant planets such as Jupiter and Saturn do not consist of gas and icy giant planets such as Uranus and Neptune not of ice." They make clear what they mean by that by pointing out that it seems that what we know as the "gas giant" Jupiter may have a real gas only in its outermost layers containing a mere 0.01 percent of its total mass. For Saturn this may be the case only for the outermost 0.1 percent of its mass. The rest of the interior behaves quite differently from what we know of gases on everyday Earth. Similarly, for the "ice giants" Uranus and Neptune the ice and water states as we know them on Earth are valid for only a small fraction of a percent once below the gaseous atmosphere. So we should really read the descriptors differently: "gas giant" means that the material that pulled together to make the planet was initially mostly in a gaseous state, while "ice giant" means that initially mostly frozen water, methane, and ammonia dominated the material that came together to start the growth of these planets. The present-day interiors of the gas and ice giants are not simply gas, or liquid, or solid ice, but "a warm, dense fluid, characterized by ionization, strong ion coupling and electron degeneracy." That means that it is electrically conducting like a metal is, and that it shows unfamiliar quantum-mechanical effects that strongly affect the behavior of the material. The moderate quantum effects of superfluid helium and superconductor magnets that can be produced by engineers

are highly unfamiliar to us in everyday life, so how such effects play out in the extreme conditions of a planetary core can thus far only be analyzed in computer experiments, there being no direct means to observe these things in the real world.

The various models for planetary interiors suggest that the heavy-element core in Jupiter weighs between zero and 6 Earth masses, with one outlying model suggesting up to 18 Earth masses in the core. Elements other than hydrogen and helium would then make up anywhere from zero to almost 40 Earth masses in the surrounding envelope. The remaining 88 percent to 94 percent of the total of 317.8 Earth masses is almost all made up of hydrogen and helium. That large percentage certainly means that the name "gas giant" is appropriate as interpreted above, but the uncertainties in these numbers are much larger than one would like to see. For Saturn's 95.1 Earth masses similarly uncertain numbers are found for core and heavy-element content, but here, too, "gas giant" is an appropriate description because hydrogen and helium combined make up 70 percent to 86 percent of the total planetary mass.

In contrast to the high percentage of hydrogen and helium in the gas giants, Uranus and Neptune contain a lot of water, either in liquid or solid form, which has earned them the characterization of "ice giants." Models suggest that deep-seated shells made of liquid water, water ice, and unfamiliar high-pressure forms of water around their cores could make up 60–80 percent of Uranus and 70–85 percent of Neptune. These models allow for no more than 20 percent of the total mass of Uranus (with 14.5 Earth masses) in hydrogen and helium, and about 15 percent for Neptune (with 17.1 Earth masses).

Characterizing the interiors of exo-giant planets is, understandably, much more uncertain than for the Solar-System giants. Even so, observations of the large number of thousands of exoplanets is now helping us to develop and validate ideas beyond the eight planets in the Solar System.

Internal structure and history of formation

What we currently know of the gas giants Jupiter and Saturn by measuring their size and mass and by flying satellites around them leaves a large uncertainty on the total mass of the dominant elements within them. An entirely different avenue of finding out what lies deep down uses the formation history of these giants. In order to capture and hold gas gravitationally, a central body must first form that is heavy enough to actually hold onto any gas that falls into its pull. Estimated minimum masses for that initial body range from three to a dozen Earth masses. This provides an additional reason why iron, rock, and frozen or liquid water, methane, and ammonia must make up at least something like 10 Earth masses because otherwise no gas giant would have been able to form.

In this scenario for the formation of giant planets, a dense solid or liquid core has to form first out of initially solid pieces. This core can continue to grow as long as there is more solid, frozen material near its orbit from where it can be pulled in. If there is insufficient initial solid material to form a heavy core, no gas giant can form. Not only should there be ample material initially, but the rate of capture of that material onto the growing protoplanet also needs to be sufficiently high; if the core forms too slowly out of frozen solids, the surrounding gases could be blown out of the proto-planetary disk before a real giant planet can form. It takes a core of at least half a dozen Earth masses or so before it can begin to accumulate and hold on to a gaseous atmosphere. This accumulation accelerates as the planet's mass increases because it pulls in gas faster and from a larger feeding zone as its gravitational reach increases. The planet's growth only ends when there is no more gas to capture. It appears that Uranus and Neptune grew too slowly to reach the phase of rapid atmospheric capture: their growth was curtailed not long after having acquired their solids because by then the gases had been blown out of the nascent

Solar System or captured onto Jupiter and Saturn. The latter two grew faster than the ice giants because they started closer to the Sun where more material was available. Consequently, they had more time remaining after core formation to collect their gaseous atmospheres before the hydrogen and helium surrounding them vanished.

The potentially large fraction of heavy elements in the hydrogen–helium envelopes of the Solar-System gas giants may have its origin in this formation history of the planets. As the planets gathered their gas atmospheres, they also continued to gather solid materials from their surroundings, with clump sizes from dust to full-fledged asteroids. As these bodies fell through the gas atmosphere toward the core, they likely lost some material through friction, they may have been pulled apart, and they or their fragments may have exploded prior to impact into a solid or liquid as meteors falling through Earth's atmosphere frequently do. Much of that material lost from the solids being captured prior to final impact deep inside the growing gas giants could leave the gas atmosphere enriched in heavy elements. How much of that survived subsequent separation by gravitational settling over the ensuing 4.5 billion years is being debated: it depends on the properties of the materials, including the gas motions, which we could call the weather, that could keep materials mixed against the separation by gravitational settling, just as our weather does on Earth.

An alternative scenario for giant planet formation that is still being considered is that somehow dense clouds of solids and gases can form within a pre-planetary disk and then fall together in much the same way that the central star is taking shape. In this scenario gas and dust fall together within their own gravity instead of solids first sticking together in a large core that only later captures gas. This scenario is more speculative as modeling in computers remains too large a challenge to assess its viability. In fact, it may turn out that this should not be viewed as an

alternative, but rather as a complementary scenario that may dominate in phases earlier than the core-accretion scenario or that may reign in heavier disks. This second scenario, referred to as disk instability, may well dominate in stars and disks with little in the way of heavy elements in solid form, and is likely to be involved in the formation of the largest exoplanets, which are just about brown dwarfs, or in the formation of true, small companion stars. This is another avenue of research that, for now, ends with the note that we need to learn more.

The internal structure of exoplanets with masses below about 10 Earth masses (referred to as Earth-like to super-Earth planets depending on their mass) is somewhat more straightforward to figure out than that of giant planets. This includes, for example, a rough determination of the water content of such planets, which enabled the discovery of ocean worlds.

One particularly large challenge in understanding the habitability of rocky exoplanets is that the layers of a terrestrial planet that would interest us most are essentially out of reach of present-day observations. For the Earth, for example, the atmosphere is thin, at only about 1 part in 1,000 of the full size of the planet. The water layer is thinner still and even the crust—crucial in shaping continents, oceans, and the very atmosphere—is small compared to the overall size of the Earth. All together, atmosphere, oceans, and crust span something like 15 kilometers (10 miles) out of the 12,740-kilometer (7,920-mile) diameter, or about 0.23 percent. The mass in these layers, which is mostly in the crust, is 0.4 percent of the Earth's total mass. These tiny layers are just not accessible for exoplanets with present-day instrumentation. But as I said before, astronomers and engineers are determined people: more powerful telescopes are being built, and the study of the visible tops of atmospheres does hold clues about what lies below, so that what now present formidable challenges may be addressed by innovation in techniques and imagination.

The diversity of worlds, revisited

The era of learning about what lies below the surface (or below the cloud-tops) of exoplanets is only in its infancy. Nonetheless, some information has been derived on exoplanet histories and futures, and in some cases we have gleaned information about surface and atmospheric conditions. With what we have learned about planetary systems in general, how they form and evolve in their orbits, and what may happen to them in the future, we can revisit and expand the narratives of the fourteen specimens that we looked at only cursorily in Chapter 4. I am also adding a brief "appraisal" of the planet as a possible home for life in a form somewhat similar to life on Earth, that is, life based on water and organic chemistry. Of the list of fourteen exoplanetary systems discussed earlier, eight are still settling in as mature planets, or are (or have been) subjected to extreme conditions; these are all described in the Notes at the end of this chapter. The six remaining ones—numbered here as before—are more or less comparable to our own Solar System.[1]

3. Kepler-186f, the first known rocky Earth-sized planet in the habitable zone My appraisal as a home: may have liquid water, but if it does then under a greenhouse-gas atmosphere like the early Earth had; acceptable gravity; overall, suitable for further investigation. This planet is listed as one of the exoplanets found by the Kepler satellite that are definitely worth reviewing for any signatures of life (see Figure 9.3).

You can think of Kepler-186 as a dwarf system: Kepler-186f orbits a star half as massive and half the size of our Sun, at a distance of 0.4 times the Sun–Earth distance and with an orbital year of 130 Earth days. This star is considerably fainter and much more red in appearance than our Sun, with a brightness—in terms of total warming power—in Kepler-186f's sky of about 30 percent of that of our Sun.

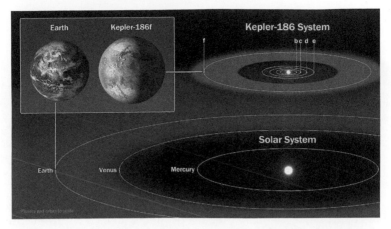

Figure 9.3 A visualization of the Kepler-186 exosystem, alongside one of the Solar System on the same scale. Note the closeness of exoplanets Kepler-186 b...e to their star compared to the terrestrial planets in the Solar System. The gray bands are estimated liquid-water habitability zones around Kepler-186 and the Sun. The image shown for the exoplanet Kepler-186f next to Earth is on the same scale but otherwise purely envisioned by an artist.

Source: NASA/JPL.

There are at least four planets closer in to its star than Kepler-186f, all with sizes only slightly larger than Earth's. All of these inner planets are much closer to their star, the outermost of these being more than three times closer in than Kepler-186f. They orbit so close to their host star that their rotation periods have likely synchronized with their orbits, so that the same sides of these planets always face their star. Kepler-186f, however, is just far enough from its star that this may not (yet) have happened, although it depends on the age of the system, which remains unknown. If the gravitational tides have synchronized orbit and spin, then Kepler-186f would have no day–night cycle, and very different, rather extreme conditions on the starlit and dark sides.

Kepler-186f is probably about the size of the Earth. It looks like the distance to its star puts it within the central star's habitable

zone with a surface temperature for the planet that is neither too high to evaporate all water nor too low to freeze it all. Whether it has liquid water depends on how the planet formed and what its atmosphere is. By 2016 it was one of only some twenty known small planets in the habitable zone of their star, and of those it was the one most likely to have a size very close to that of Earth.

Habitability is a fuzzily defined concept. In a widely accepted definition it means that liquid water could in principle exist on the surface. If there is indeed water on the surface, then atmospheric models can be computed for different mixtures of gases in which water vapor matches the prevailing temperature and pressure, assuming there is always some ocean left below the atmosphere to exchange water with. We do not currently know the atmosphere of Kepler-186f, but it is tempting to explore "what if" scenarios: taking a mixture dominated by nitrogen and carbon dioxide, as likely was the case on the pre-life Earth, astronomers find that liquid water could exist on Kepler-186f provided that the atmosphere is at least three to ten times Earth's atmospheric pressure, in which carbon dioxide would contribute from almost 100 percent at the low-pressure end of that range to only 5 percent at the high-pressure end. Higher pressures would create stronger greenhouse effects and therefore higher surface temperatures. The amount of carbon dioxide needed to maintain possible liquid water is at the low end of what is thought to have existed on the early Earth (more on that in Chapter 10), so that if there is (or was) indeed sufficient water, this planet could indeed be (or have been) habitable by this definition.

4. Kepler-11, the first known compact multi-planet system
My appraisal as a home: definitely iffy. Of the six (or more) planets in the system, it appears that Kepler-11b has a thick atmosphere, maybe dominated by hot water vapor, possibly cooking anything in it, but possibly enveloping an ocean world.

The Kepler-11 system (see Figure 4.1) has at least six planets orbiting a star very much like the Sun: within a few percent,

the central star, Kepler-11, has the same mass, size, and surface temperature as the Sun. Initially, its age was estimated to be almost twice that of the Sun, which would put it only a billion years or so from going into its giant phase, at which time all of these planets would be enveloped within the star's outer layers and subsequently evaporated out of existence. But detailed follow-up observations and analysis suggest that Kepler-11 is much younger—around three billion years with an uncertainty of a billion years. That would make it what is referred to as a solar twin: closely resembling the Sun in mass, radius, temperature, and internal structure, with a spin rate only some 10 percent higher than that of the Sun, and in a similar evolutionary stage.

Five of the six exoplanets in this system orbit their star closer than the closest planet, Mercury, does our Sun, with orbital periods (planetary years) between 10 Earth days and 47 Earth days. Only Kepler-11g orbits a bit further out, with an orbital period of 118 days.

For the innermost five planets, masses were determined between most likely two and nine Earth masses; too little is known about the outermost planet to say much about its mass. Their sizes range from 1.8 to 4.3 times Earth's. The combination of sizes and weights suggests that at least five of these planets have low densities, well below the typical density of rock. In fact, all but Kepler-11b have densities close to that of liquid water. Therefore, if these planets have anything like rocky cores, they would need to have large, gaseous atmospheres to compensate for the cores' density in order to reach their average value. This might make them more like lighter versions of the ice giant Uranus than like the terrestrial planets. And that is what makes them interesting: are they large Earth-like planets that happen to have thick atmospheres, or are they lighter versions of the so-called ice planets?

If they are small versions of ice giants, then current under-standing of planet formation says they would not have been able

to form where they are, because all the known planets in the Kepler-11 system orbit well inside the so-called snow line (the distance from the star beyond which gas-rich planets are thought to readily form starting with solid cores). This has led to the concept that all of these planets were formed much further out, and later migrated inward.

Being relatively close to their star, subject to its rather intense warming, its ultraviolet and X-ray light, and its stellar winds, the planets would have suffered considerable atmospheric loss. This would have reduced their total masses and, in the case of the innermost one, Kepler-11b, may have removed most of its hydrogen and helium, leaving a water-dominated steam atmosphere under present conditions. That is compatible with studies that suggest that all but the innermost planet have envelopes composed predominantly of hydrogen and helium, while Kepler-11b may have an atmosphere that may be largely hot water vapor and a thick ocean envelope weighing, very roughly, as much as half of the total planet mass.

These present-day atmospheres are estimated to be very much heavier than Earth's current atmosphere: Earth's weighs one millionth of the total planetary mass, while those of Kepler-11 are, in the same relative measure, 50,000 to 160,000 times more massive.

5. Kepler-444, the oldest known compact multi-planet system My appraisal as a home: each of the exoplanets in this system has nearby neighbors, but all are way too hot for us. Interesting system though, because these exoplanets have a pair of stars in their skies. Would appreciate a closer look to view this spectacle in the sky, but not to hang around for long.

Kepler-444 is a triple star system: Kepler-444A is a star with a mass and diameter of about three quarters that of the Sun, which is orbited by a binary of two much lighter, cooler, red dwarf stars designated as a pair as Kepler-444BC. The orbit of the BC binary around the A component is highly elliptical. At its furthest (at apastron), the BC pair is almost 70 Sun–Earth distances away

from A, but at its closest (at periastron) it is only five Sun–Earth distances away.

Based on asteroseismic measurements of the A component and detailed modeling of the star's internal structure, the Kepler-444 system is estimated to be approximately 11 billion years old. That is almost two and a half times older than the Solar System, and over 80 percent of the age of the Universe. That planets formed so early in the life of the Universe is interesting because of how heavy chemicals came to be formed after the Big Bang (discussed earlier in Chapter 5). Despite its age, though, the iron abundance, for example, of the stars in the system is only just below a third of that of the Sun; other heavier elements likely have a similar percentage, so there certainly was material to build rocky planets out of, as the existence of the exoplanets in the system confirms beyond doubt.

Scientists working with the Kepler satellite have found five exoplanets, all with sizes between those of Mercury and Venus. All five orbit well within 8 percent of a single Sun–Earth distance, so well inside the periastron of the BC binary from their host star, the A component. The orbital plane of the planets and that of the BC binary are most likely aligned, suggesting that they all formed from a large, all-encompassing disk that fragmented into stars and planets. There may be other planets there further from this star, though, that have escaped detection thus far, but not too much further because the BC binary pair in its elliptical orbit would have truncated the pre-planetary disk to no more than about one or two Sun–Earth distances.

Seen from any of the exoplanets in this system, a binary of two red dwarf stars would swing by once every 200 years, at closest approach appearing reddish at a brightness of well below a percent of how we perceive our Sun. Depending on the phase of the stellar orbit and the position of the planets, the BC binary might be faintly visible in the daytime along with the much brighter primary star, or A might brighten the daysides as the BC binary shines down upon the nightsides of the exoplanets, showing up

between some of the other exoplanets that would appear much brighter than any of those in our Solar System because of their proximity to both the primary star and each other.

The five found planets all orbit their star in close-in orbits with periods between 3.6 Earth days and 9.7 Earth days, and although Kepler-444 is a star that is smaller than our Sun (at 0.75 times its size) and consequently less luminous (at 0.34 times the solar luminosity), it looks like that difference is not so large that these close-in exoplanets can escape extreme heat.

6. Kepler-16b, the first known planet to orbit two stars My appraisal as a home: perhaps like Saturn, but with two stars in its skies over an atmosphere too opaque to see them, with no idea of oceans or landmasses, but not likely Earth-like anywhere. It is freezing cold on this planet, so I shall pass on any invitations to visit.

Kepler-16b (Figure 4.2), or actually Kepler-16 (AB)b, orbits a binary system whose stars are about two thirds and one fifth of the Sun's mass, respectively. The binary components orbit each other once every 41 Earth days. The entire system is remarkably flat, in the sense that the orbital plane of the binary and the orbital plane of the planet are aligned to within a fraction of a degree. The spin axes of the stars in the binary are aligned with these planes to better than a few degrees. All of that is as expected when a planetary system forms through a flat disk phase, barring destabilizations.

Kepler-16b has a mass of about one hundred Earths, so is comparable to Saturn's mass. It has an average density that suggests that it is about half gas and half rock and ice. It orbits its host stars in a year that is 229 Earth days long at 70 percent of the Sun–Earth distance from its stars. Because both stars are much fainter than the Sun (at some 14 percent and 0.5 percent, respectively for A and B), this distance makes for cold weather on Kepler-16 (AB)b.

Although it was initially assumed that planets would have a hard time forming while in orbit around a stellar binary,

observations suggest it happens at least in 1 percent to 10 percent of all binaries. Planets around binaries have to be fairly far from their host stars to avoid having their orbits gravitationally upset by the pair of stars swinging about themselves. This large distance from their stars makes such planets difficult to detect, so the actual percentage may well be higher.

One remarkable feature of this exoplanetary system that has attracted attention in addition to it having a binary star at its center is this: the orbit of the planet is essentially a perfect circle, but the binary stars orbit each other with a substantial eccentricity (0.16) in elliptical orbits. Computer modeling suggests this could happen if the planet initially formed considerably further out from the binary, and then migrated inward as it was interacting with a relatively massive gas-loaded disk. The interaction with that disk would have either circularized the orbit, or kept it circular if it already was.

9. Proxima Centauri b, the nearest exoplanet known (by 2018) My appraisal as a home: this planet has a slowly fading deep-red star that appears large in its sky, may have Earth-like gravity, and has a short year with just a few starrises and starsets in it, if any. It may be orbiting in the star's habitable zone, and may still have some water, although perhaps only on its permanent dayside in a region where its star is almost directly overhead. Overall, definitely worth a look, in part because of the New Year's Eve celebrations that can be expected every 12 Earth days... well, if it had a population interested in that.

Proxima Centauri was discovered in 1915 to be a very nearby star, currently estimated to be located a mere 4.224 light years from Earth. Well, that is "mere" by astronomical standards; it is still 267,000 times further away than the Sun is from Earth, or 8,900 times further than the most distant planet orbiting the Sun, Neptune.

Unfortunately, not very much is known about its one dis-covered planet, Proxima Centauri b, which was hit on a century

after the determination of the star's distance, in 2016, because it is not a transiting planet: its mass exceeds 1.3 Earth masses and is probably less than 3 Earth masses, but how heavy it really is remains undetermined.

We do know its star very well: Proxima Centauri is a cooler, less massive star than our Sun, with a radius of only one seventh that of the Sun and a mass of one eighth of a solar mass. Because it is small and cool, it is not very bright: it has only 0.15 percent of the Sun's luminosity, and much of its light is emitted as invisible infrared light (which we feel as heat), and in the visible deep red.

The planet Proxima Centauri b is, however, much closer to its star than Earth is to the Sun, at about 1/20 of a Sun–Earth distance. Consequently, its stellar energy input is about two thirds of that of Earth, so that its day is fairly warm albeit dim to our eyes. Although this may put it in the habitable zone of planetary orbits, whether it has any liquid water is unknown and would depend on the atmosphere, if any, of the planet. And, of course, it depends on whether there was water on its surface when it formed and how much of that it may have lost over the past billions of years of its history.

The history of the planet's star is quite different from that of our Sun because of its lower mass. The Sun's brightness has been relatively stable, increasing very slowly by a few tens of percent, ever since the Earth became a mature planet, a few tens of millions of years after the formation of the Solar System began. But a star as light as Proxima Centauri evolves far more gradually, although the total change in its luminosity over time is far larger and in the other direction. If we start the clock at the same time as when Earth would have been completed, let us say 30 million years into the life of the star, Proxima Centauri has become dimmer since then by a factor of five. Consequently, its habitable zone moved inward over a large distance, contrary to the Sun's which has slowly shifted outward by a little since that time. Only after a full billion years would Proxima Centauri's brightness have leveled out, maintaining a rather steady level over the subsequent few

billion years. Proxima Centauri b may have come into the liquid-water habitable zone only after the first 100 million years of its existence. But because it was inside that before then, it may well have suffered what is known as a runaway greenhouse effect, which would have evaporated all available water, with much of whatever was there in that phase in effect boiling off the planet. Whether any substantial amount of water was delivered onto the planet to supply an ocean by asteroids impacting it from beyond the snow line, as is thought to have happened on Earth subject to the migration of the giant planets, remains unknown, so Proxima Centauri b may be in the habitable zone but not have any water anyway.

If it did have water, the irradiation by the extreme ultraviolet light emitted due to the star's magnetic activity may have decomposed water into hydrogen—that would have evaporated off the planet—and oxygen. If any oxygen remains after oxidizing the surface, then this may be one way of creating an oxygen-rich atmosphere in the absence of life. Having realized this scenario, scientists no longer consider the existence of atmospheric oxygen by itself an unambiguous signature of life on a planet.

Should the exoplanet in fact have surface water, its real habitability will also depend on whether it is partially or fully synchronized in its rotation with its orbit of 12 Earth days: is it always facing its parent star with the same hemisphere because of the tides that did the same for the Moon facing Earth, or is it still spinning more than that so that the stellar heat can be more efficiently distributed? Computer experiments suggest the tides were strong enough to have induced full synchronization, so that there are no days and nights anywhere on the planet. If so, the nightside might be cold, and the dayside might be in a state where there is open water—provided there is a sufficiently large reservoir of water on the planet overall—only where its star is high in its sky: the "tropics" would not wrap around the planet as they do on Earth, but be limited to a region around the point where the star

is directly overhead (called the sub-stellar region), as in the inset in Figure 1.4. Alternatively, the planet may have been captured in a tidal spin–orbit resonance so that Proxima Centauri b continues to experience starsets and starrises but in which a year on Proxima Centauri b is only a few of its days long, and therefore might also be in a more evenly distributed climatic state.

14. TRAPPIST-1, a packed miniature system My appraisal as a home for TRAPPIST-1e: Earth-like day-side brightness but with a big red sun, Earth-like gravity, possible liquid water on the surface, but likely with a permanent hot day on one side and a permanent cold night on the other (the place most hospitable to potential life might be on the poles), and a year the length of an Earth work week; maybe worth a look (see Figure 4.4). The planets appear to be older than Earth, perhaps by some three billion years, so if there is life it may have evolved well beyond our current stage.

The central star, also known as 2MASS J23062928-0502285, is merely one eighth the size of the Sun, only slightly larger than Jupiter at 0.12 times the Sun's size, and has a brightness that is almost 2,000 times less than that of the Sun.

The star is orbited by seven known planets. Michaël Gillon, the principal investigator of the Transiting Planets and Planetesimals Small Telescope (TRAPPIST) after which the system was named, said about this system "I felt super-excited... amazed by the existence... the very existence of the system... was kind of shock."

The planets are remarkably like Earth in size: the diameters range from about 25 percent smaller to 10 percent larger than Earth's. Their masses have a larger range, with best estimates ranging from 60 percent lower to 40 percent higher than Earth's. The mass estimates are too uncertain to say much about the amounts of water and gas on the planets, except for the fifth one from the star (TRAPPIST-1f) which appears to have a considerable amount of volatiles on it.

All of the TRAPPIST-1 planets are extremely close to their star: the innermost is at 1 percent of the Sun–Earth distance and the outermost at 6 percent of that distance; this places the outermost one almost seven times closer to its star than the innermost planet in the Solar System, Mercury, is to the Sun. The proximity of these planets to their star compensates for its faintness: they receive from 4.3 to 0.13 times as much energy onto their surfaces from innermost to outermost planet as Earth receives from its more distant but much brighter star. The result is that at least three of these seven planets are formally estimated to orbit within the liquid-water habitable zone. However, that does not mean their atmospheres are particularly like Earth's. First of all, the proximity to their stars in all likelihood means that they are tidally locked in their rotation, so that they would always have the same side facing their star, making for two fundamentally different hemispheres. Orbiting while tidally locked means that the exoplanets are rotating once per orbit so that the stars in the skies of their nightsides would show the progression of the constellations similar to what we see throughout Earth's year, but with a planetary "year" of only 1.5 to 20 Earth days going from the innermost to the outermost exoplanet in TRAPPIST-1.

The proximity to their star brings another influence. The star appears to spin at a rate that is about 7.5 times faster than the Sun, which puts its magnetic activity at an elevated level compared to that of our star; it frequently flares with explosive energies as large as the largest that occur only occasionally on the Sun. Not only is the flare frequency higher, but because the planets are much closer to their star, the radiation that they intercept is 300 to 10,000 times enhanced. The stellar wind and coronal mass ejections associated with the increased activity will also be correspondingly enhanced, so that space weather and atmospheric impacts are vastly amplified compared to those on Earth.

Simple climate models show how different these planets could be. If they all had an Earth-like atmosphere, the innermost three

would be expected to have runaway greenhouse atmospheres, in which temperatures would be high and water vapor extremely dense. The next three from the star (planets e, f, and g) could in principle have liquid oceans, particularly if they had very dense atmospheres loaded with carbon dioxide. The outermost one would likely be too cool for any liquids to exist on the surface for any kind of atmosphere if starlight is the dominant atmospheric heat source.

Planet e is of high interest as a habitable planet. Climate models suggest that its atmospheric energy balance might settle on the dayside at a surface temperature at which liquid water could exist, even in the absence of a strong carbon dioxide component in the atmosphere. However, the nightside might be fully frozen, and even much of the dayside might be quite cold, except for the region where the star is high in its sky. This kind of climatic state on a tidally locked exoplanet has been called the "eyeball regime" because looking down at the planet from the direction of its star, it would be frozen and white at its edges, with a warm and dark patch only where in the eye the iris would be (as in the inset of Figure 1.4). These regions could reach out to 45 to 60 degrees from the sub-stellar point.

Now imagine this: seen from TRAPPIST-1e its two nearest neighbors would, on closest approach, appear about as large in its sky as the Moon does to us on Earth. Picture being able to see surface features as well as we can on the Moon, particularly with binoculars or a small telescope. Would we see continents and oceans and clouds? Visualization scientist Robert Hurt, who works at the California Institute of Technology, wondered about this on an evening walk with the assignment to conceptualize TRAPPIST-1 freshly on his to-do list: "I just stopped dead in my tracks, and I just stared at [the Moon]. I was imagining that could be, not our Moon, but the next planet over—what it would be like to be in a system where you could look up and see continental features on the next planet." He and his colleague Tim Pyle envi-

Figure 9.4 Artists's impression of exoplanet TRAPPIST-1f. Its planetary system is so compact that two of its nearest neighboring exoplanets on closest approach in their orbits would appear as large as the Moon seen from Earth. The surface as shown is completely imaginary; the neighboring exoplanets could, at the right time in their orbits, appear as large in the sky as imagined here.

Source: NASA/JPL-Caltech.

sioned just that, and came up with the view depicted in Figure 9.4. Pyle said: "When we're doing these artist's concepts, we're never saying, 'This is what these planets actually look like.' We're doing plausible illustrations of what they could look like, based on what we know so far." Plausible, but still imaginary, at least for now...

In summary, these exoplanets, interesting though they may be, are nothing close to hospitable to terrestrial life, with one possible exception overall, and maybe two more if we focused on what is likely a permanent transition zone between frozen and thawed somewhere between the nightside and the point of high noon. But then, these are the somewhat unusual ones that illustrate what can happen to planets over time as they and their stars evolve. Moreover, they are selected from among the exoplanets that are most easily discovered and therefore are more likely to be heavy, close to their stars, or orbiting low-mass stars.

As our instrumentation and techniques improve over time, many other exoplanets will doubtless be found to be rather more benign to life as we know it, and future instrumentation will tell us not only about their most general characteristics such as orbits and temperatures, but will reveal details about atmospheres, oceans, and—probably not too long from now—about land masses and their characteristic colors. In the meantime, there are relatively close-by planets and moons from which we can learn more about atmospheric evolution and habitability. Specifically, Venus, Mars, and some of the moons of the giant planets in the Solar System provide valuable insights into habitability; these bodies may not be habitable at present for earthlings, but astronomers are finding that Venus and Mars may have been much more hospitable in the distant past, while some of the moons of the giant planets may become so is the distant future. All that is the focus of the next chapter.

Notes

1 Revisiting exoplanet systems: Of the list of fourteen exoplanetary systems discussed in Chapter 4, eight are still settling in as mature planets, or are (or have been subjected) to extreme conditions; these are all described here, numbered as in Chapter 4. Six others, which are more like our own Solar System, are described in the main text.

1. 51 Pegasi b, the first known exoplanet around a Sun-like star My appraisal as a home: way too hot, permanent day and night sides, and permanent disastrous storms.

The International Astronomical Union decided that names other than the dull-sounding catalog designations would be nice, and asked the public for input. With over a hundred billion planetary systems expected to exist within the Galaxy, such a naming competition can obviously only be considered for some special cases, for which this exoplanet clearly qualifies. The IAU selected Dimidium (Latin for half, referring to its mass of half that of Jupiter) as the winning name for 51 Pegasi b from the submitted suggestions. The star 51 Pegasi itself was assigned the name of Helvetios (a tribute to its Swiss discoverers) in addition to its forty other distinct catalog designations!

The planet was found by observing the periodic Doppler swings that revealed that the parent star was orbiting a common center of mass with the nearby heavy planet. More recent observations with a new generation of observatories have revealed signatures of the planet's radiation in the observed joint light. This contains starlight reflected by the planet but with signatures of its own chemistry imprinted on its color spectrum that can be unraveled and interpreted using a spectroscope. Because the planet has its own, larger, and oppositely directed Doppler shifts to the parent star, the two light sources can be told apart.

Spectroscopic determinations of the orbit put the mass of 51 Pegasi b at 151 Earth masses, give or take ten, which is 0.48 times the mass of the gas giant Jupiter. It orbits its star in a planetary "year" of only 4.23 Earth days, with an orbital plane that is rather well aligned with the rotational equator of the host star that is itself only about 10% heavier than the Sun.

The planet appears to have nothing like a "day": a planetary "day" is defined as the time it takes to spin around so that its star is seen again in the same direction in the sky, but the tidal forces from the nearby star of 51 Pegasi b are so strong that its rotation has very likely been synchronized with its orbit around the star (although that has yet to be confirmed observationally). In that sense, it resembles the behavior of our Moon relative to Earth: seen from the near side of our Moon the Earth is always in the sky. If indeed orbit and spin are synchronized, the planet always has the same side facing its sun, so that seen from the dayside of 51 Pegasi b its star is always in the sky and other stars can always be seen from the permanently dark side of the planet.

51 Pegasi b orbits its star at only one twentieth of the distance from the Sun to the Earth, which is just over five times the star's diameter. As the host star has essentially the same radius as the Sun, this means that 51 Pegasi shines brightly with a size twenty times larger than our Sun in Earth's sky. From where the planet orbits, its sun is some 400 times more intense in its sky than we perceive our Sun. Estimates for the temperature in the top layers of the atmosphere of the planet are consequently high, somewhere around 1,000 degrees centigrade (1,800 degrees Fahrenheit). The temperature difference between dayside and nightside may be as large as 500 degrees centigrade (900 degrees Fahrenheit) and winds that circulate between the hot and cooler sides may have sustained speeds of 7,000 kilometers

(or 4,500 miles) per hour (roughly ten times faster than the typical jet aircraft). Deeper inside the atmosphere, however, the transport of heat is adequately efficient to make the temperature much more uniform at any given altitude, but it is too hot for the complex chemistry of life everywhere.

Unraveling the exoplanet's light has revealed the fingerprints of water vapor and carbon monoxide in the somewhat transparent top layers of the atmosphere. No signatures of carbon dioxide or methane have been positively identified. The estimated radius and measured mass of the planet does not enable a constraint on the size of any liquid or solid interior, but do require a heavy, thick gas atmosphere.

The history of 51 Pegasi b remains speculative, but it may have formed much further out from its star than where it currently orbits, to subsequently migrate inward through interaction with the early disk out of which star and planet formed. It possibly ceased its inward spiral more or less where it now orbits because the star's magnetic field swept the region inward of that clean of any material that would enforce an inward spiral. The planet may well survive until the star runs out of fusible hydrogen in its core, but thereafter it will be enveloped and destroyed early on in the star's giant phase, just as Mercury and Venus will be when our Sun runs out of fuel.

2. CoRoT-7b, the first known rocky lava-ocean planet My appraisal as a home: impossibly hot, not much of an atmosphere, vast lava oceans with maybe rocky polar areas.

This is a very strange world: CoRot-7b is so close to its star that it has a year (an orbital period) that is only 20.5 Earth hours long. And because it is so close to its star it pulls on it so much that the star wobbles measurably back and forth as the planet goes around it, which enables astronomers to determine the planet's mass. The planet is 1.6 times larger than Earth and 4.4 times heavier.

Although perhaps similar in internal structure to Earth, researchers of CoRot-7b face a real challenge in figuring out the internal structure, probing a possible ocean layer, and estimating the depth of the atmosphere because there are too many unknowns— they are aiming to determine: the size of the core; the thickness of the mantle and its silicon, magnesium, and iron ingredients; the volume of a possible water layer and of the atmosphere; and even the atmosphere's energy budget and composition. Despite the many unknowns, it often turns out that a trade between various parameters can be made to still learn something. For example, at an

increased iron content in the mantle, the core size may be reduced, or the depth of the oceans may be raised. Increasing the silicon or magnesium content of the mantle at the expense of iron could result in a larger core or may make the possible ocean layer vanish. These are trade-offs that are possible for CoRot-7b and all other Earth-like and super-Earth planets until we have additional information beyond the present-day observations.

And so, despite these challenges, modelers have tried. One thing they can do is to narrow the field of initial choices from which to start the study. They can do this, for example, by assuming that the chemical abundances of the metals would be the same as for the parent star, and that the internal structure should not differ too much from that of Earth. If these assumptions are correct, then CoRot-7b likely has a core that is at most comparable to that of the Earth but may be much smaller, and a mantle that ranges somewhere between 70% of its radius to the full radius. The remaining 30% or less of the planet's size may be a combination of oceans and a dense, non-transparent atmosphere. The ocean could take up to 30% of the radius and 20% of the planet's mass, or it could be negligible or absent altogether; we simply cannot tell with present observations, and that does not leave us in a good position to know anything about the surface layers of CoRot-7b other than that it is pretty hot. And that means that it may turn out not to have much of an atmosphere or ocean layer at all.

The planet is so close to its sun, which is only a bit smaller and redder than our Sun, that that appears several thousand times brighter in its sky than our Sun in ours. Surface temperatures are estimated to lie in the range of some 2,000 to 2,500 degrees centigrade (or 3,500 to 4,500 degrees Fahrenheit), so that the surface is likely largely molten silicate rock. In all likelihood, it consequently has a deep lava ocean. That may exist only on the planet's dayside, though: as in the case of 51 Pegasi b, CoRoT-7b is so close to its star that it is likely to be always showing the same face to its star.

7. HD 80606 b, a planet with a very eccentric orbit My appraisal as a home: like a huge Jupiter, with extreme weather variations high in the atmosphere every four months, and a deep-atmospheric temperature well above where our proteins would denature—we would quickly be cooked. Not appealing.

HD 80606 b is a large, heavy planet, weighing almost 1,300 Earth masses, or four times more than Jupiter. Based on the present sample

of known exoplanets, it has the distinction of having the most non-circular orbit of all known exoplanets: at closest approach in its 111-Earth-day orbit it is only 3 percent as far from its star as the Earth from the Sun, while the largest distance in its elliptical orbit brings it to 88 percent of the Sun–Earth distance, or just about thirty times as far away.

The highly eccentric orbit means that the energy coming in from its star changes from Earth-like at apastron to 800 times that at periastron. The orbital motion near periastron would be swift, but it looks like the dayside temperature could nonetheless rise and then fall again by over 500 degrees centigrade (or 1,000 Fahrenheit) in a matter of hours when the planet moves through its closest approach to its star. The average internal temperature in the planet's high atmosphere may be some 400 degrees centigrade (750 degrees Fahrenheit) as the combined result of stellar irradiation and dissipation of tidal forces during the close approaches in the orbit.

Not only is the orbit a very stretched ellipse, but the orbital plane is tilted away from the rotation axis of the host star by somewhere between 30 and 90 degrees. Both of these peculiarities may well be related: the presence of one or more other bodies can lead to orbital changes in which both inclination and eccentricity are affected. This could hint at the presence of other planets orbiting HD 80606, or maybe reveal the influence of the distant binary companion, HD 80607, which has an average distance to HD 80606 of some 1,200 times the distance from Sun to Earth.

8. V391 Pegasi b, the first known post-apocalyptic planet My appraisal as a home: a heavy Jupiter, likely sterilized long ago by the immense heat during a giant phase of its star.

V391 Pegasi b weighs over 1,000 Earth masses (over three Jupiter masses). It orbits an old, hot star with an orbital year of 1,170 Earth days. That star, V391 Pegasi, has perhaps half a solar mass and a quarter the solar diameter. Its surface is five times hotter than that of the Sun and consequently it glows bright blue with a luminosity fifteen times higher than that of the Sun. This star is much older than our Sun; perhaps more than twice as old. It has gone through a phase that awaits our Sun in some five billion years, namely the giant star phase after it has converted all of its hydrogen fuel into helium.

The maximum size of V391 Pegasi was likely something like 0.7 Sun–Earth distances, and V391 Pegasi b was then at about one Sun–Earth distance. Its star would have been many hundreds of times brighter in its sky, spanning almost 90 degrees.

Some time around its largest phase, V391 Pegasi lost probably over a third of its mass in eruptions, during which the planet would have experienced reducing stellar gravity so that its orbit widened correspondingly. This mass loss involved dense and probably variable outflows that would have buffeted the planet. This may occur in different phases by different processes, and for this type of star by ways that remain particularly poorly understood. After that phase, the remainder of the star shrunk again into the stage that V391 Pegasi is in now, during which it is fueled by helium fusion which should last for another few billion years.

V391 Pegasi lost so much mass as a large red giant that it blew off almost all of its residual hydrogen envelope, leaving less than one percent of the original mass of that envelope. But the state that it is in now is very helpful to find planets: the star pulsates very regularly at a series of frequencies clustered around five minutes. Because of this, it is classified in its current state as a pulsating, extreme horizontal branch sdB dwarf. These pulses are timing markers that reveal the pull back and forth from the orbiting planet by making the distance from the star to the Earth periodically decrease and increase again as the planet orbits the star, thereby changing the arrival time of the pulses in the star's light by up to 10 seconds. This Doppler effect has been used to uncover one planet that orbits it, and to determine a minimum value for its mass.

V391 Pegasi b is one example that can help us understand how a planet can survive this giant phase of a star with its massive mass loss. But that will require more work, including both observations of other such star–planet systems and more theoretical computer-based studies.

This particular star ejected so much of its envelope that it will continue to evolve as a small object, in the end becoming a white dwarf. If it had not, as is to be expected for the Sun and for the vast majority of similar stars, the next phase would again be a giant phase in which it would eject, in a series of pulses amid an irregular sustained wind, another tens of thousands of Earth masses of its outer atmosphere. That would be a still more apocalyptic event for anything near such a star.

10. HR 8799, the first directly imaged multi-planet system My appraisal as a home: too young to expect life, unfriendly atmosphere with ammonia and acetylene, although also with water vapor,

but these are giant planets with thick atmospheres and are, as yet, too hot.

The star HR 8799 is probably some 30 million years old, give or take ten, or more on the longer side. It is about 1.5 times heavier and five times brighter than the Sun. It is surrounded by a dust disk that extends out to 1,000 Sun–Earth distances within which four giant exoplanets were directly seen as point sources with the Keck and Gemini telescopes in Hawaii. The planets glow in the infrared, still radiating away the heat of their formation this early in their lives.

These four planets have similar masses ranging from roughly 1,500 to 5,000 times that of Earth, orbiting at 15, 24, 38, and 68 Sun–Earth distances, with orbital years of 50, 100, 190, and 460 Earth years. Their estimated masses suggest that they are all true planets, being a little too light to be brown dwarfs, although there is a probability that one or more of them should be referred to as brown dwarfs if heavy-hydrogen (deuterium) fusion is taken as the differentiating factor.

Like the giant planets in our Solar System, all four planets of HR 8799 orbit beyond the distance where water could freeze out— that is, beyond the snow line—which lies at about six Sun–Earth distances for HR 8799 and at about three for the Sun.

The formation history of the system is unclear: the debris disk appears to be unusually heavy, so the planets could have migrated substantially from where they formed to where they are now by coupling to the debris in the disk.

The atmospheres of the four planets have signatures of clouds, and look to vary substantially, with different mixes of ingredients dominated by ammonia, acetylene, methane, and carbon dioxide, and with signatures of water and carbon monoxide in one of them.

11. HD 13189 b orbits an evolved, cool giant star My appraisal as a home: huge red giant star in the sky over a really heavy form of Jupiter, maybe fusing deuterium in its deep interior. Nothing like home.

The giant star is estimated to be very roughly four times heavier than the Sun, but it remains quite uncertain, being possibly just as heavy as the Sun or up to six times heavier. The mass of HD 13189 b lies between 2,500 and 6,000 times that of Earth, with a most likely value of 4,500 times as much. The companion is consequently substellar, but it is unclear if it should technically be called a brown dwarf (if heavier than about 4,000 Earth masses) or a true planet (if less massive).

The measurements of the color spectrum of the star look like those of what is formally called a "bright giant" (luminosity class II). The brightness of the star, based on a parallax-based distance measurement, is a few thousand times that of the present-day Sun. The star's radius is some fifty times that of the Sun, give or take a few. Its metal abundance is only a quarter of that of the Sun whereas stars with giant planets typically have a metallicity higher than that of the Sun.

The companion giant planet or small brown dwarf has a "year" of about 470 Earth days, orbiting the star at somewhere between 1.5 and 2.2 Sun–Earth distances, which is only about three to five stellar diameters above the star's surface.

12. PSR 1257+12, the first detected, confirmed exoplanetary system My appraisal as a home for exoplanets c and d: a faint light source in the sky is all that is left of the star(s), which means it is incredibly cold on the planets, with strong planetary gravity and stellar radiation, and the planets are possibly filled with gigantic diamonds and frozen carbon dioxide.

The central object, PSR B1257+12 itself, is a neutron star that spins at an absolutely dizzying rate of 160 times per second. The pulsar is a heavy stellar remnant with about 1.4 solar masses compressed into something with a radius of about 10 kilometers (or 7 miles).

There are three confirmed exoplanets, all within a distance from the pulsar of about the orbit of Mercury around the Sun. The three exoplanets have masses of 0.02, 4.3, and 3.9 Earth masses, and orbital "years" of 25, 67, and 98 Earth days, respectively. The lightest of these three is not quite twice the mass of Earth's Moon. The outer, heavier two are locked into a 3/2 period resonance. The planets orbit at 0.19, 0.36, and 0.46 Sun–Earth distances from the tiny neutron star.

The IAU named the central pulsar Lich (after an undead creature with magical control over others), and the three planets Draugr (an undead creature in Norse mythology), Poltergeist (a supernatural being pestering humans), and Phobetor (after the Greek deity of nightmares).

Each bit of the surface of the neutron star glows brightly owing to its high temperature of almost 30,000 degrees centigrade (55,000 degrees Fahrenheit), which is five times hotter than the solar surface. But these stellar remains are very, very small relative to a typical star. The result is that these planets see a point-like star in their skies with a luminosity fraction of only 10 billionths of our Sun, so it would be

just a faint dot in the sky, as if we were viewing the Sun from a distance of some 11,000 times further than we actually do. It would stand out among the stars in the planet's sky: seen from the outermost, larger two planets, the neutron star would appear about as bright in their skies as Mars would appear in our skies at closest approach, although it would radiate in the ultraviolet and blue rather than look reddish as Mars does to us.

Because the energy in the incoming starlight is so low owing to the small size of the neutron star, only residual internal heat would be available to warm the planet surfaces. Although the neutron star may be faint in terms of photons, there is a lot of dangerous radiation emanating from it. All this together makes the planets most inhospitable places.

The only reason why these exoplanets, the first to be detected, are not generally thought of as post-apocalyptic is because they most likely are not survivors of the formation process of the neutron star. Instead, they are thought to have formed as a delayed consequence of the creation of the neutron star, which itself is the end product of stellar evolution, either directly from a single heavy star that ran out of nuclear fuel or from the merger of two stellar remnants, each of which was the end product of less massive stars that both ran out of fuel. Astronomers consider the latter scenario the more likely for this particular object.

If indeed the pulsar was formed by the tidal disruption of a white dwarf that was pulled onto a compact companion, then the material that would have formed a debris disk around this catastrophe would have been very rich in carbon (as the end product of nuclear fusion in the progenitor stars) and poor in, among others, silicon and iron. Any planets forming in such a setting would be quite different from the terrestrial planets, and markedly different from exoplanets forming out of a molecular cloud. Instead of a rocky mantle and an iron-dominated core they would be composed mostly of carbon and oxygen, and their interiors in fact might just see the carbon compressed into gigantic diamonds.

13. K2-22b, a disintegrating rocky planet My appraisal as a home: no way, because it is evaporating under a huge, bright star in its sky, with the bulk of the planet already evaporated ages ago.

K2-22b (also known as EPIC 201637175 b) orbits a low-mass M-type dwarf star with a size of 0.6 times the Sun's. The planet has an orbital "year" of only 9.15 Earth hours. It orbits its star at 0.9% of the Sun–

Earth distance, which is just over one stellar diameter above the surface of the star. The proximity to the host star causes its surface, likely a few thousand degrees hot, to disintegrate into a vapor that forms a dust cloud once off the surface, which subsequently escapes to form what is a very large comet orbiting very near to the star (see Figure 4.3).

The escape is very different from that of a comet in the Solar System, however: whereas normally the tail trails behind a comet, gradually being blown outward by the solar wind and photon pressures, in this case it has a tail that leads the planet in its orbit. This appears to be the result of it leaving the planet's gravitational field in such a way that it begins its orbiting of the star after leaving the planet's field a little closer to the star than the planet is. Thus, its initial orbit is already faster than that of the planet, so that it leads it as it begins to spiral into the star.

The planet cannot be very heavy because then the dust could not escape its gravity. The mass is lost at an estimated rate of 200,000 tons per second, equivalent to the mass of two of the world's largest aircraft carriers, or the largest oil rig ever built, lost to the exoplanet every second. If the planet's residual mass is somewhere between that of the Moon and of Mercury, then the present mass loss into its comet-like dust environment suggests it could not survive for more than several tens of millions of years. How the planet came to be where it is remains unclear, but it will not be there much longer, measured on planetary timescales.

10

Habitability of Planets and Moons

How do we define "habitable"? A dictionary definition reads "suitable or good enough to live in." Whether something is "suitable" obviously depends on the species. In astrophysics, and as used in preceding chapters, "suitable" was interpreted in the most general form: the "habitable zone" around a star is that where the energy input from that star should result in surface temperatures roughly between the freezing and boiling points of water. We already saw that a surface temperature—if indeed there is a surface underneath a gaseous atmosphere—is set not only by the star's brightness, but also by the properties of the atmosphere, in particular its chemical composition and its total mass, as well as by the presence or absence of large water reservoirs in oceans and seas that can provide water vapor to the atmosphere or receive it from the atmosphere. The atmospheres and oceans of planets change with time, however, so that also habitability, if defined by the actual presence rather than the potential presence of liquid water, will evolve. Moreover, life may make use of other liquids that could then make a planet or moon habitable to that form of life. In this chapter, I explore bodies in our own Solar System that may have been, or that may become, habitable in the sense of having large reservoirs of liquids on a surface in which life in some form or other may find a habitat that is "suitable or good enough."

First stop: Titan

Imagine a landscape with rivers and lakes, mountains and dunes, maybe an isolated volcano or a rare decayed impact crater on the

One of Ten Billion Earths. Karel Schrijver, Oxford University Press (2018). © Karel Schrijver.
DOI: 10.1093/oso/9780198799894.001.0001

horizon, with smog-like clouds overhead that filter sunlight into an orange glow, and a faint, small Sun vaguely visible high in the sky. Not far from the equator, a parachute appears through the haze, descending above a dry lakebed that is sculpted, not out of rock, but predominantly out of rock-hard water ice underneath a layer of grainy, sand-like organic compounds. A beam of light shines down from the small metal structure suspended underneath the parachute. After a while, the craft strikes the ground, making a dent of some 12 cm (5 inches) in the frost-covered top layer of the ground. It slides forward a little, rocking back and forth a few times as it does so, until it settles down, surrounded by pebbles made of water ice. The parachute gently comes to rest next to it some seconds later, its shadow briefly crossing the field of view of the on-board cameras. Dust loosened by the landing craft falls back to the ground. After an hour and a half the radio signals from the craft end. The probe has stopped functioning. An unearthly quiet has returned to the area.

This took place on Titan, the largest moon of Saturn, on January 14, 2005, as the European Huygens space probe descended on the final leg of its seven-year mission from Earth to the gas giant's moon, riding along until near the very end on the American Cassini spacecraft. The landscape (see Figure 9.2) was largely unseen by humans until that time, because a haze, which is now known to be made up of carbon-based chemicals, envelopes Titan. This haze was mostly unpenetrated by telescopes, except for a glimpse of structure seen by the Hubble Space Telescope (Figure 10.1).

The moon now called Titan was chanced upon in 1655 by Christiaan Huygens. It was the first such body discovered after Earth's Moon and after the four largest moons of Jupiter that were spotted by Galileo Galilei in 1610. Huygens found it using a telescope with fairly large lenses that he and his older brother Constantijn had learned to shape, set into an unwieldy 3.5 meter (12 foot) long telescope tube. The lens survives in a museum in Utrecht, the Netherlands. Christiaan proudly engraved part of

an anagram around its edge. He had mailed the entire text to a few close colleagues: "Admovere oculis distantia sidera nostris vvvvvvvcccrrhnbqx." The anagram, starting with a segment of Ovidius's writings, reads: "They brought the distant stars closer to our eyes," followed by leftover letters. The anagram was intended to claim the discovery of Saturn's moon, of which he was clearly proud, but he hid it in this unreadable form in order to have time to further investigate what he saw. Unscrambled, Huygens's anagram reads, in the Latin commonly used in those days: "Saturno luna sua circunducitur diebus sexdecim horis quatuor," or, translated into English: "The moon orbits Saturn every sixteen days and four hours" (where he overestimated the orbital period by a little over four hours). By the way, using the same telescope Huygens also figured out that what had looked like ears on the planet in poorer telescopes (such as Galileo's four decades earlier) were, actually, the thin equatorial rings for which the planet is so famous.

It would not be until a few decades later, between 1671 and 1684, that Giovanni Cassini found the next four moons of Saturn, all of which were less than a third the size of Titan: Dione, Iapetus, Rhea, and Tethys (Figure 3.4). The discoveries by these early scientists caused mission planners in the twentieth century to name their spacecraft and its lander after them, known together as the Cassini–Huygens mission, launched in 1997.

Titan is the second-largest moon in the Solar System, coming only after Jupiter's Ganymede, and it is the only moon with a substantial atmosphere. Its diameter of 5,150 kilometers (3,200 miles) makes it almost 1.5 times the size of Earth's Moon, larger than the planet Mercury, larger than the known dwarf planets, and three quarters the size of Mars. It orbits Saturn once every 15.95 days, always having the same side facing the planet, so its rotation is synchronized with its orbit just as is the case for Earth's Moon. From the planet-facing hemisphere, if it were not for the obscuring deep orange haze, Saturn would appear ten times larger in apparent width than the Sun and the Moon do as seen

from Earth. From high in Titan's atmosphere, the Sun looks just about a tenth the width of what it looks like from Earth; lower down, it would be a blurry albeit slightly brighter patch in the ubiquitous haze.

Gravitationally bound to follow Saturn, Titan orbits the Sun once every 30 years. The distance to the Sun, ranging from nine to ten times the Sun–Earth distance through the orbit, means that sunlight is about a hundred times fainter than at Earth. Consequently, it is freezing cold on Titan: the average temperature lies just under –180 degrees centigrade or –290 degrees Fahrenheit. Surface gravity is 19 percent weaker than on the Moon, so astronauts, should they ever make it there, would move as the Apollo astronauts did, giving the impression of an odd slow, bouncing motion. Titan's atmosphere at its surface is 4.5 times denser than Earth's, and composed predominantly of nitrogen—as is Earth's—with over one percent of methane, a small fraction of a percent of hydrogen, and traces of organic molecules. But there is no molecular oxygen, and not a hint of water vapor.

Whereas Earth's clouds are made of water vapor and water droplets, those on Titan come in three distinctly different types: methane (on Earth the main ingredient of natural gas and its liquefied form, LNG), ethane (a minor component of natural gas on Earth), and a mixture of the extremely poisonous hydrogen cyanide (also known as Prussic acid) and other chemicals. Apparently intense but infrequent methane "rain" and wind storms are responsible for the creation of erosion channels. Occasional strong winds build up fields of dunes up to almost 100 meters (300 feet) high that are not made of silicate-rich sand as on Earth, but rather of hydrocarbon-coated ice particles. And then there are seas and oceans that appear to contain mostly liquid ethane, particularly near the north pole, with depths measured by the radar system on NASA's Cassini satellite of up to 160 meters (500 feet).

This atmosphere may not look like that of present-day Earth, but it does have some resemblance to that of early Earth, apart

from the chill. This was recognized already years ago, with a strong boost by the passage of Voyager 1 in 1980, coming as close as 4,000 kilometers, less than the moon's diameter. That spacecraft was followed by Voyager 2 only nine months later but at a much larger distance. In part because of these pioneering close-ups, Tobias Owen wrote in 1982 that "It seems highly likely that the chemical reactions taking place in the atmosphere of Titan today must resemble in some important respects the chemistry that occurred on the primitive Earth prior to the formation of life. The major difference is of course the absence of liquid water." After arguing for a mission to map Titan and to land on it, he proposed a bold experiment: "Ultimately we will want even more sophisticated equipment that would allow us to perform a variety of chemical experiments on Titan's surface. It may well be that this environment offers our best opportunity for actually dupli-cating conditions on the primitive Earth, once we can devise ways of supplying energy to large parts of the surface and examining the resulting reaction products."

Well, indeed there was a mission to Saturn and its moons, and the Huygens probe landed on Titan. In the meantime, sci-entists have realized that Titan's atmosphere, including some of the trace organic chemicals such as hydrogen cyanide, acetylene (also called ethyne), and cyanoacetylene (known under several unenlightening names), may be even more like that on early Earth than was thought in the 1980s. This would have been similar to the atmosphere that existed when organisms called archaea started developing on Earth billions of years ago, although that environment would have been much warmer, and there would have been abundant liquid water. We earthlings currently con-sider liquid water essential to life because it is for all life we presently know about, but it may not be a critically required ubiquitous substance in general: perhaps another liquid can take on the role of solvent and transportation agent in other worlds, or perhaps locally present water suffices, as it appears to exist on Titan.

Ancient life! Future life?

The new biological "kingdom" of archaea in biology was announced in a revolutionary study published in 1977 by Carl Woese (1928–2012) and George Fox (1945–). In their discovery paper they wrote that "The biologist has customarily structured his world in terms of certain basic dichotomies. Classically, what was not plant was animal. The discovery that bacteria, which initially had been considered plants, resembled both plants and animals less than plants and animals resembled one another led to a reformulation of the issue in terms of a yet more basic dichotomy, that of eukaryote [which has organelles wrapped into separate membranes, including a cell nucleus] versus prokaryote [which has no such structures]." They then argued that a third kingdom was needed based on a type of anaerobic organism (that is, one that does not need or even tolerate oxygen) that was found to "possess a unique metabolism based on the reduction of carbon dioxide to methane." They introduced the name of "archaebacteria," nowadays known as "archaea" for short, from the Greek base for ancient, primitive, or from the beginning. They chose that name because it looked like an ancient, simple life form that "seems well suited to the type of environment presumed to exist on earth 3–4 billion years ago."

Archaea were known from their very discovery to generate methane by processing carbon dioxide. Ronald Oremland, in 1989, noted that archaea grow when they have access to acetylene, so that because worlds like Titan "have relatively high levels of acetylene within their atmospheres, the possibility exists that an anaerobic microbial food chain exists therein which exploits this available carbon and energy source." But it could not be like what we see on Earth because of the lack of liquid water. Then again, the existence of phenomena called cryovolcanoes may indicate that liquid water exists locally on Titan.

What are "cryovolcanoes"? The Cassini spacecraft and the Huygens probe have found evidence that there is a liquid ocean

below Titan's surface, which itself is comprised primarily of water ice. The subsurface regions could have a temperature that is much more hospitable—in terms of terrestrial life—than the frozen outer shell: radioactive decay deep inside Titan could warm subsurface layers to somewhere around –20 degrees centigrade, possibly up to +30 degrees centigrade in the warm spots (about 0 to 90 degrees Fahrenheit). Above such warm spots there may be volcano-like mountains of ice in which liquid water may play a role similar to what is silicate-rich magma in Earth's volcanoes, leading to eruptions and associated temporary surface flows that build up the volcanoes. The existence of such cryovolcanoes is still being researched, but substantial evidence has been provided by the combination of spectrometry and stereoscopic radar mapping that was possible from the Cassini spacecraft.

No signs of life have (yet) been found on Titan, but then very little of this moon has been explored. That said, it is not formally in the habitable zone as currently widely defined because surface water exists only in a rock-hard frozen form, except in transient flows from cryovolcanoes. With the knowledge now available for Titan, however, the introduction of a "methane habitable zone" is being considered. Investigating Titan in detail is clearly of interest and feasible: Carl Sagan already wrote in 1996 that he expected a time when "we should not only have multiply-capable roving vehicles—sleek planetary tanks and perhaps submarines—wandering through the Titanian wilderness. Likely, we will have returned sample missions so that the organic chemistry and radiometric dating of Titan can be done with the fullest sophistication of which our species is capable." But he envisioned that this might take a century to achieve.

The current cold state of Titan will change in the distant future, during the Sun's phase as a bright, giant star. Whether that phase will last long enough and be sufficiently stable in terms of solar warming of Titan to allow surface life to develop with then-available liquid water is, and will remain, a point of academic debate. Proof is likely to remain elusive until well after our Earth

Figure 10.1 This infrared image of Saturn's moon Titan, taken by the Cassini spacecraft, shows sunlight glinting off seas near the north pole. These bitterly cold seas consist of a mixture of primarily liquid methane and ethane. Toward the upper left is the bright reflection off Kraken Mare; around this sea a bright peripheral ring shows up, likely formed by reflection off salt deposits left behind by evaporating methane and ethane. Clouds are visible; these are thought to contain mostly droplets of methane, which may lead to "rainfall" of what we know on Earth as liquified natural gas. The image is a composite of narrow color bands in the infrared in which Titan's hazy atmosphere is somewhat transparent; visible light would only show an orange haze (See Plate 26 for a color version).

Source: NASA/JPL-Caltech/University of Arizona/University of Idaho.

has been sterilized by the Sun's blazing heat and after possibly being engulfed by the expanding Sun in its terminal gasps for fuel.

Second stop: Venus

Almost 30 years before the Huygens probe landed on Titan, another spacecraft descended into an opaque atmosphere. The lander eventually came down dangling below three parachutes, gliding through some 40 kilometers (25 miles) of clouds of sulfuric acid (also known as oil of vitriol), and a cloud base some 30 kilometers above the ground, moving through an atmosphere with mostly carbon dioxide, a few percent nitrogen, and traces

of hydrochloric acid, hydrofluoric acid, bromine, and iodine. It touched down on the rocky surface under an atmospheric pressure over ninety times higher than on Earth. Its camera transmitted the first images from another planetary surface, from a desert planet with a surface temperature of very nearly 500 degrees centigrade (or 900 degrees Fahrenheit). This was the lander from the Russian Venera 9 spacecraft, touching down on October 22, 1975. Three days later, the Venera 10 lander touched down almost 2,200 kilometers (1,400 miles) away. Other landers came before (Venera 7, which crashed in 1970) and after (Venera 11 through 14, launched between 1979 and 1981; a Venera 14 image is shown in Figure 9.1); no landers have come since 1982, and none of the landers survived for more than two hours in Venus's hostile environment.

Clearly, Venus today is not a particularly habitable planet, but this may have been different in the distant past. Computer experiments can be performed for the global circulation of the atmosphere, making assumptions about the chemical composition, the depth of hypothetical oceans and seas, and estimates of the energy input from the Sun billions of years ago. This work suggests that it is possible that Venus was once a moderately pleasant planet: a surface air temperature like in Hawaii, and liquid water over much of the planet. Interestingly, this shows that the slow spin of Venus would be critical in this state: the slow rotation allows air masses to circulate over the entire planet, rather than being directed into latitudinal bands as is the case on Earth because of the forces associated with our planet's relatively rapid spin. So, no tropical, mid-latitude and polar bands on Venus, but a single globe-encompassing weather pattern all over the planet—there would only be limited differences in the weather forecasts, no matter where one was located.

If these assumptions and computations are correct, the weather on that ancient Venus would have been relatively similar over much of its surface, with permanent cloud cover on the day side dampening the rise in temperature that would have

occurred otherwise on this very slowly rotating planet: it has a "solar day" of 116 Earth days—long compared to its orbital "year" of 225 Earth days, and therefore within a factor of two from full tidal synchronization that exists in the Earth–Moon and Saturn– Titan systems. Clearer skies on the night side could have allowed effective cooling of the planet, with the day–night temperature difference being reduced by strong large-scale winds. This state of Earth-like habitability may have lasted until perhaps 700 million years ago, not too long before the so-called Cambrian Explosion some 540 to 520 million years ago during which the diversity of life greatly expanded on Earth, moving from mostly single-celled organisms sometimes organized in colonies to a diversity of multi-celled species including the first animals.[1]

Whether the real Venus was ever in this mild state of the simulated Venus remains to be seen. There is evidence supporting it, not only in the evolution of the Sun's brightness, but also in a special property of Venus's atmosphere. The latter helps us understand why there is hardly any water on present-day Venus: there is nothing in liquid form, and but a little in gaseous state in the atmosphere. The overall water mass of Venus on the surface and in the atmosphere is estimated to be some 200,000 times lower than that of Earth. There are reasons to think that Venus had a considerable amount of water once, though. The main one is the high fraction of heavy hydrogen (deuterium) on Venus: it is roughly 150 times higher than on Earth. Why is that a hint to what happened to Venus's water in the past? Water molecules contain two hydrogen atoms and one oxygen atom each. The hydrogen comes in two forms: each has only a single proton in its core, but one also has a neutron, which makes it twice as heavy. In any water molecule, the hydrogen can be the lighter form (which is simply called hydrogen) or the heavy form (deuterium). When water vapor high in a planet's atmosphere is subjected to the Sun's ultraviolet light, the chemical bonds can be broken. The separated atoms will generally eventually bind to

something else, but in the meantime the lighter form of hydrogen has a somewhat larger chance of evaporating off the planet than the heavier form. The water thus lost from the atmosphere is replaced by water evaporated from the ocean reservoir below it as long as that lasts. The difference in the probability of heavy and light hydrogen from broken-up water molecules leaving the planet means that, over time, the ratio of heavy to light forms of any remaining hydrogen gradually increases as ocean water is lost to space.

To reach a deuterium-to-hydrogen ratio as high as seen for Venus, oceans' worth of water must have been lost. Exactly when that would have taken place on Venus remains uncertain. It could hypothetically have happened over a very long period, with lost water being replaced by new water supplied to the planet by a long-lasting sequence of occasional impacts by comets or asteroids. If, instead, it happened by the loss of a large, ancient water reservoir without further water coming in, then Venus may have been a habitable world well before Earth became one. That is the scenario expected from the Nice model in which giant-planet migration would have caused large water deposits through impacts by minor planets in the early phases of the Solar System, but then nothing more in significant amounts.

Regardless of the scenario for its loss of water, we used to consider Venus as lying outside the habitable zone when only simple estimates of its surface temperature were made. With present-day three-dimensional climate models for the atmosphere and its coupling to oceans, however, we now realize that its thick atmosphere and slow rotation could have led to a pleasant temperature on ancient Venus. Whether that ever led to the development of life in the luke-warm oceans and uncommonly long days of Venus, in whatever form it may have taken, remains to be determined. Could it be that we missed life on Venus by what seems a long time for humans but what may have been less than a percent of the overall lifetime of the Solar System?

Third stop: Mars

We can look at other spacecraft coming down on yet another
body in the Solar System: landers have successfully touched down
on seven sites on Mars, and rovers have explored four of them
(Figure 10.2 shows the Sojourner rover, the first of four Martian
rovers thus far, seen from the Mars Pathfinder lander). These landers, plus detailed studies from orbiting spacecraft, have shown
that Mars used to have a much thicker atmosphere, and significant liquid water on its surface. Perhaps Mars offered an environment within which life could have developed some four billion
years ago, and perhaps that lasted for a few hundred million years,
which may suffice for life to gain a foothold into evolution. But
then things went downhill rather quickly: Mars was never heavy
enough to hold onto an atmosphere for very long, nor to sustain
a magnetic shield against the detrimental exposure to cosmic rays
of any life it may have supported for a brief time—astronomically
speaking—on its surface.

Even today, Mars is formally still within, or very near to, the
outer edge of the habitable zone of the Solar System. However,
most of the Martian atmosphere, including the carbon dioxide
that could have kept the planet warm and its water liquid, was
long ago blown off the planet. This happened by a combination
of the Sun's ultraviolet light that breaks up chemicals and the
stripping effect of the solar wind that touches on the topmost
layers of the atmosphere that is, in contrast to Earth's, unshielded
by a substantial planetary magnetic field. This atmospheric loss
continues to this day.

After billions of years of having its atmosphere gradually dis-
mantled, the present-day Martian surface lies below a trifling
envelope of gases with an average surface pressure of 0.6 percent
of Earth's. It is rather unprotected from cosmic rays under that
thin layer of gas and the planet's weak, patchy magnetic field;
when exposed to cosmic rays, organic matter is easily damaged,
and this can be lethal to life forms. Moreover, Mars has at best only

Figure 10.2 The rover Sojourner photographed by the Mars Pathfinder on the surface of Mars in 1997. NASA/JPL wrote about this image: "Based on the first direct measurements ever obtained of Martian rocks and terrain, scientists on NASA's Mars Pathfinder mission report [...] that the red planet may have once been much more like Earth, with liquid water streaming through channels and nourishing a much thicker atmosphere. [...This] panoramic view of Pathfinder's Ares Vallis landing site [...] reveals traces of this warmer, wetter past, showing a floodplain covered with a variety of rock types, boulders, rounded and semi-rounded cobbles and pebbles. These rocks and pebbles are thought to have been swept down and deposited by floods which occurred early in Mars' evolution in the Ares and Tiu regions near the Pathfinder landing site" (See Plate 27 for a color version).

Source: NASA/JPL.

a sporadic local presence of liquid water. So, it may be formally in the habitable zone, but it is by no means a welcoming place. By the way, the winds in the Martian dust storms that are occasionally seen to envelop the planet are too insubstantial to be forceful

enough to blow over a heavy rocket, as was suggested to be possible in Ridley Scott's movie *The Martian*.

I have discussed these examples of Titan, Venus, and Mars to make it clear that we need be mindful of what we mean when we define the "habitability zone." Titan may be (or may become) habitable to life forms that we do not have here on Earth; the metabolism of archaea suggests that this could in principle work in Titan's atmosphere. Plus, that moon may well be "habitable" in its oceans beneath the icy surface. Venus and Mars have no liquid water now, but they likely had them in the distant past. These examples show that habitability, even if defined as simply as by the possibility of having water in liquid form on the surface, depends on time as much as on place and on the atmospheric and planetary conditions, including the chemical composition, and even on the planet's rate of spin.

Magnetism and lost oceans

The loss of water from a planetary atmosphere, such as happened to Venus, is a slow process that is the result of the irradiation of the planet's atmosphere by the Sun's ultraviolet light and the stripping by the solar wind. Both of these effects are related to the Sun's magnetism, and both affect all of the planets to some degree, depending on their distance from the Sun and on their own magnetism and gravity. Similar processes, driven by solar-like magnetism in most other stars, alter the atmospheres and the habitability of exoplanets, but at present what we know about these builds on the lessons from the bodies in our own Solar System.

The magnetism of the Sun and of stars like the Sun has been the main focus of my work as an astrophysicist. This includes trying to understand how the Sun's magnetic dynamics have changed over time by looking at stars with different ages, and also how the Sun's magnetism has affected Earth over time, including the present when this influence is commonly described under the

terms space weather and, for longer-term effects, space climate. One of the activities that I helped organize for six years was a NASA summer school program for beginning researchers in which they could learn about stars, planets, and their connections. That science, known as heliophysics, goes far beyond what more senior researchers have been taught as students: instead of focusing on a particular specialty area, heliophysics is about the big picture; the connections that tie together astrophysics, planetary physics, climate physics, and geophysics, for the Solar System and for exosystems. Without that summer school, I would not have been aware of many of the things described in this book, including many of the effects of stellar magnetism on planetary atmospheres. Through that activity, I was able to learn, along with the students, from an international team of experts in a variety of fields, including new developments about exoplanets, Earth's long-term climate changes, and planetary habitability.

Stellar magnetism has its roots deep inside a star, where movements in the hot gas stretch and turn and twist budding magnetism, just as happens by the hot liquid nickel–iron mixture deep inside Earth. In a spinning star, the chaotic gas motions have a slight preference for order because they are affected by the rotation. On Earth, too, gas flows—that we call winds—respond to a force associated with the planet's rotation; this causes, for example, the characteristic swirls of large atmospheric depressions, which circulate in opposite directions on the northern and southern hemispheres because of Earth's spin. In the thick outer envelope of the Sun where gas motions occur, these effects create order in, and give a pattern to, the magnetic field. That order shows up on the largest scales by giving the Sun a global field that reaches out into interplanetary space.

Wherever the Sun's internally generated magnetic field buoys to the surface, sunspots can form (Figure 10.3). Over the largest of these, gigantic magnetic storms and explosions can occur (Figure 10.4), the largest of which involve energies equivalent to 100 billion simultaneously ignited hydrogen bombs. On the

Figure 10.3 The Sun in visible light (displayed in enhanced yellow) observed by the HMI instrument on-board NASA's Solar Dynamics Observatory (SDO). This picture was taken on June 7, 2011. Sunspots— comparable in size to the Earth—are visible in several places. These sites of strong clusters of magnetic field are surrounded by weaker patches that light up slightly brighter than the solar surface around them, particularly when viewed near the Sun's edge. A comparison image of the atmosphere above the Sun's surface taken in the extreme ultraviolet is shown in Figure 10.4. When the Sun was young and spinning much more rapidly, sunspots and their surrounding magnetic patches were much larger, more frequent, and often showing up towards the Sun's polar regions (compare Figure 5.5) which never happens any more on the present-day Sun (See Plate 28 for a color version).

present-day Sun, this activity comes and goes with the fairly regular beat of the 11-year sunspot cycle. Along with the number of sunspots, the effects of this magnetic see-saw are also changing the Sun's outer atmosphere: the magnetic field that threads the outer atmosphere causes it to heat up and to expand, more or less in step with the rhythm of the sunspot number. The hot gases

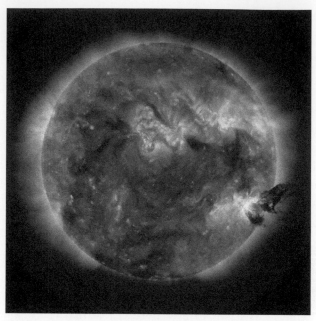

Figure 10.4 The Sun in extreme ultraviolet (EUV) light observed by the AIA instrument on-board NASA's Solar Dynamics Observatory (SDO). This image is a composite of three pictures, each picking out a range in temperatures, and each translating EUV light into a visible color: blue is coolest at around a million degrees, green is predominantly a million and a half, and red is at least two million degrees. These images were taken on June 7, 2011, around 6:36 UT. The Sun's surface (shown in Figure 10.3) is too cool to shine in EUV light, and so the Sun itself shows up as a dark sphere underneath its bright atmosphere. Wherever the Sun's surface magnetic field is strongest, the EUV is brightest and the atmosphere warmest. At the time of this image, the largest eruption observed by SDO was in progress over a region in the lower right. The cool, dark material rose and then mostly fell back, but a coronal mass ejection moved out into interplanetary space. An associated explosion, called a solar flare, occurred, which made the EUV Sun so bright at its location that it saturated the cameras, as can be seen in the green artifact just to the left of the eruption (See Plate 29 for a color version).

reach temperatures that can be a thousand times higher than that of the surface: the outermost solar atmosphere is several millions of degrees while the surface is "only" a few thousand. The glow of that atmosphere comes out as ultraviolet (UV) light, extreme ultraviolet (EUV) light, and as X-rays, whereas the surface glows mostly in what we call the visible range, thus named simply because we can see it with the unaided eye.

The heat of the high atmosphere also causes a wind to blow off the Sun. This solar wind carries strands of stellar magnetism within it, moving outward to beyond the furthest planets at speeds of something like a million to three million kilometers per hour (or half a million to two million miles per hour).

Both the Sun's brightness in UV, EUV, and X-ray light and the speed of the solar wind are highly variable, particularly when large explosions, called "solar flares" and "coronal mass ejections," occur in the Sun's magnetized atmosphere. Then the solar wind can exhibit gusts two to five times faster than its background breeze, while the whole Sun can brighten by up to tens to thousands of times in EUV and X-rays. The Sun's brightness in visible light does not noticeably change during such flares, because even the largest flare is small compared to the overall output of solar energy. In (E)UV and X-rays this is not the case: there is so little of that light in quiescence that during an explosion the flare readily outshines the whole outer atmosphere.

The level of activity on a star depends primarily on its age and on its mass. All very young stars spin rapidly, and all of these stars are magnetically very active. The enhanced brightness in the UV and X-rays of young stars adds to the forces that evaporate the gases from the protoplanetary disk that surrounds each of them early on. Moreover, the magnetic field of a rapidly rotating young star can whip the gases in such disks out of their immediate environments, thereby creating what looks like a hole in the center of the disk. It is at the edge of such a hole that any migrating young planets would likely halt their spiraling inward toward the star to settle instead into a long-lasting orbit.

While young, the Sun probably spun about its axis faster than once a day rather than once every 25 days as it does now. The magnetic activity would have caused large numbers of enormous sunspots to dot the surface; on the youngest stars, we see that half of the stellar surface may be covered by spots, often clustering around the stellar rotational poles. The present-day Sun has never shown such large, high-latitude spots; it often looks as it appears in Figure 10.3, with just a few spots at mid-latitudes. The extensive coverage with clusters of large starspots on young stars would be a very strong influence on planetary weather: imagine the entire Sun brightening and dimming markedly every few days as gigantic sunspots rotate onto and off the hemisphere shining down onto the young Earth and its sibling planets.

Magnetic explosions on young Sun-like stars are frequent and large: the largest flares on the young Sun could have been up to a thousand times more energetic than those that occur on the present-day Sun, and they would have occurred every few days rather than once per decade as they do for the current middle-aged Sun. The solar wind blowing off the young Sun probably carried on average something like a hundred times more mass. Gusts in the wind of young, active stars are faster, denser, and lug stronger magnetism. The most energetic events, when enveloping a planet, would cause bright auroral glows around most of the planet. In contrast, present-day space storms cause auroras only at high geographic latitudes; an example is shown in Figure 10.5.

A stellar wind carries with it not only magnetic field, but also rotational momentum. This causes the Sun, and any star like it, to slow its rotation with time. At first, this magnetic brake would apply quite strongly, but with ebbing magnetic activity, the braking action also lets up. That brake still works on the present-day Sun, but its action now shows up only over timescales of billions of years rather than millions of years as it did early on.

The UV and X-ray light from a magnetically active star shines onto the atmosphere of exoplanets, just as the Sun's light does onto the planets in the Solar System. It typically cannot penetrate

Figure 10.5 A picture of the southern lights (aurora australis) taken by astronauts on the International Space Station. Such auroras are powered by the coupling between the magnetized solar wind and the Earth's magnetic field. They are most spectacular when a coronal mass ejection temporarily envelopes the Earth's magnetic field, particularly when the gusts in the solar wind are strong and the interplanetary magnetic field blowing past the Earth is opposite to the Earth's magnetic field direction. The Sun's X-ray and ultraviolet light combines with the cause of the auroras—the magnetized solar wind—to gradually erode a planet's atmosphere in the course of billions of years—that is how the lightweight Mars lost most of its atmosphere long ago, while the heavier Earth continues to lose it much more slowly (See Plate 23 for a color version).

Source: NASA/ISS Expedition 29 crew.

into an atmosphere very deeply before it is absorbed. As this happens, the light has sufficient energy to break chemical bonds and to kick some of the knocked-free atoms up to enhanced velocities. Some may go so fast that they can escape the planet's gravity, or reach high enough to be picked up by the stellar wind that flows past the planet. Given hundreds of millions of years, or even billions of years, this can strip a planet of much of its atmosphere. These processes can affect some chemicals in

particular, like when they broke up Venus's water molecules and took away its lighter hydrogen fragments, eventually robbing it of all its oceans. For other planets, either in the densest winds very close to their star or with inadequate gravity to hold onto their gases, stellar activity and wind can work to gradually remove almost the entire atmosphere well within a planet's lifetime, as happened to Mars.

If the planet has its own magnetic field, this tends to keep the solar wind off the planet's atmosphere. This is what happens even today on Earth: the solar wind typically stays away from the Earth's atmosphere by a few Earth diameters. But now something else happens: the magnetic fields of Earth and solar wind interact. As is the case for magnetic fields everywhere, the coupling depends on the relative directions of the fields. In one orientation, they can preclude coupling to each other so that the solar wind simply glides by Earth's field like water around the bow of a ship. In another orientation, they can couple, triggering magnetic storms and electrical currents around the Earth. This can lead to pretty northern and southern lights or to annoying (and sometimes damaging) effects on the electrical power grids that we all rely on—these influences are part of space weather. These changing fields and currents also heat up the outermost layers of Earth's atmosphere and push the gases about, so that more or less of the atmosphere is stripped off, depending on space weather conditions. How much the gustiness of the solar wind affects atmospheric stripping and how effective a planetary magnetic field is in protecting the planet are still being investigated: we know planets lose their atmospheres subject to the magnetic activity of their star, but we do not have enough diversity in the nearby planets to know exactly what accelerates and what slows that loss. How different the loss of atmospheres of Solar-System planets was subject to solar activity in the distant past, how it depends on a planet's properties, and how it works in exoworlds...well, that is all an exciting, growing branch of exoplanet science, bringing together experts from terrestrial

space weather, stellar magnetism, and atmospheric sciences. All of the interaction processes of stars and planets need to be understood in order to analyze Earth's paleoclimate and to appraise the climates of exoplanets.

Interestingly, note that the trends over time for (E)UV and X-ray light and for visible light are opposite. The Sun's surface is gradually brightening at visible colors by about 10 percent for every billion years as the fuel supply in its core is consumed. The (E)UV and X-rays from the Sun's outer atmosphere, in contrast, are fading as its magnetism dies out with its decreasing spin rate: the X-ray brightness was once more than a thousand times higher and the UV brightness close to a hundred times. Consequently, the planetary effects are shifting in balance between high and low in the terrestrial atmosphere over the course of billions of years. Similar changes affect all planets and exoplanets.

The final thing to mention here, only briefly, is that even the giant planets and the Moon have their role to play in Earth's atmosphere. Their combined gravitational pull causes slight changes in Earth's orbit and causes the spin axis of the Earth to slowly migrate and wobble, just as spinning tops do. These changes affect climate, contributing to the coming and going of global ice ages. Such cyclic changes over thousands of years were first worked out by Mulitin Milanković (1879–1958) in the 1920s, and are therefore known as Milankovitch cycles. These same processes also cause slow cyclic changes in the climates of, among others, Mars and Titan.

Fourth stop: GJ581d, unconfirmed

When we look at exoplanets we find conditions that do not occur in the Solar System, including tidal synchronization of planetary spin and orbit around its star, a different mix of colors from the star, or sustained intense stellar magnetism. The lessons that scientists have learned by observing and modeling past and present climates for the terrestrial planets in the Solar System are

now being applied to exoplanets with such unfamiliar properties. To show how this can be done, let me introduce you to Gliese 581d (or GJ581d). GJ581 is a cool dwarf star just over 20 light years away, with a mass of about one third and a luminosity that is only 1.3 percent of that of the Sun. Among its multiple planets is GJ581d. The status of GJ581d is formally "unconfirmed": it may turn out to be an artifact of some other effect that was interpreted as an exoplanet. But if it is real, then it should be a planet that is at least about six times heavier than the Earth and which orbits its faint star at one quarter of the Sun–Earth distance.

The reason why the signal attributed to an exoplanet may be an artifact lies hidden in the method used to study GJ581 and in stellar magnetism. Exoplanets pull on their parent star as they orbit each other, causing a wobble in the star's position that is— for a non-transiting system such as GJ581—detectable by the alternating blue–red–blue shift in the stellar light through the Doppler effect. However, starspots can introduce a similar signal. When a starspot shows up from the back side of the rotating star, its dark area results in a little less light coming from the side of the star moving toward the observer than from the side moving away. Once the star's spin has moved the starspot to the other side, the opposite effect occurs. The net effect, therefore, is also a swing in the Doppler signal, one that is quasi-periodic for at least as long as a starspot survives. Clusters of starspots complicate the signal, so that it may look as if one or more "exoplanets" exist where these are in reality artifacts of stellar activity.

This "unconfirmed," possibly Neptune-sized exoplanet has been subjected to several studies related to habitability, two of which I highlight here because they are great illustrations of what is being put to the test. The first is work by Robin Wordsworth and colleagues. Should GJ581d be confirmed to exist, it would be a special case because then it would be "the first confirmed super-Earth (exoplanet of 2–10 Earth masses) in the habitable zone." They applied advanced fully three-dimensional global climate

models to this presumed exoplanet. If indeed it exists, GJ581d would be so close to its parent star that the gravitational tides will have modified the planet's rotation, either locking it to its orbit around the star such as is the case for the Moon as it goes around the Earth, or making it such that not one but rather a low number of full planetary rotations occur within an orbit of the planet around its star. In other words, one should expect this planet to either always have the same side facing its star or it might spin differently but then far more slowly than Earth, in fact somewhat like Venus. This slow rotation would mean that the nightside of GJ581d could become very cold, so cold that water and even carbon dioxide might freeze out on that cold side such that any flowing from the warm to the cold side would then become trapped there as ice. Their computer model shows that, provided a sufficient reservoir of carbon dioxide exists in the atmosphere, GJ581d could have temperatures above freezing, regardless of whether it were mostly land covered or largely ocean dominated. This load of carbon dioxide would act as a strong greenhouse component, along with water vapor, to slow the heat loss from the atmosphere. Such an atmosphere would also be sufficiently heavy to effectively circulate dayside heat around to the night side to keep that from freezing under its perhaps permanently dark sky. It would require a pressure of carbon dioxide of more than ten times Earth's atmospheric pressure, so quite a considerable bulk, but not a disproportionate amount compared to what once existed on Earth and what now exists on Venus.

Another study looked into the role of chemical weathering that removes carbon dioxide from an atmosphere by dissolving it into a weakly acidic rain that can wear down rocks. Through a sequence of chemical reactions this eventually results in the transformation of the original carbon dioxide into deposits of calcium carbonate in oceans and seas. If GJ581d has been a wet world for a good part of its history, then this chemical weathering might by now have removed the carbon dioxide needed to keep

it a warm, wet world. This has been assessed by Werner von Bloh and colleagues. They looked not only at the liquid-water habitable zone but at what is called the photosynthetic habitable zone (or pHZ), which encompasses the range of distances from the star that enables a planet to have sufficient carbon dioxide in its atmosphere to keep terrestrial plant life going. Their computations include approximations for chemical weathering of carbon dioxide out of the atmosphere. But they also look into the reintroduction of carbon dioxide: if a planet has tectonic activity, the calcium carbonate deposits will eventually be subducted into the planet's interior, and after this material has resided there for some time it may resurface processed in part into carbon dioxide through volcanic activity—more on that in Chapter 12.

Therefore, von Bloh and his colleagues also investigate effects of tectonic plate movements, which may be "lubricated" to some degree by water that is contained within rocks, as they review tectonic motions driven by heat supplied from within GJ581d and lost from its surface, the effects of the exoplanet's spin rates by tidal forces, and consequences of different relative sizes of oceans and continents that are important to the chemical weathering. They even include an estimate of the influence of biological activity on the rates of chemical weathering. Their atmospheric climate model is not a full three-dimensional model, though, but something much simpler and thus considerably more approximate. However, as did their colleagues described above, they too concluded that, for a carbon dioxide pressure of some ten times Earth's total atmospheric pressure, GJ581d could sustain liquid water on its surface, and that this would last longest for a planet with a relatively small total area in its continents. The dependence they find on continental area is quite strong: the total time that GJ581d might be in the pHZ could change from 0.3 billion years if it were 90 percent land mass, so with a lot of area for chemical weathering to remove carbon dioxide, up to nine billion years if it were 90 percent ocean initially, with much slower chemical weathering on the smaller land mass.

The faint, lightweight star GJ581 is classified by astronomers as an M3V star. The "V" in the classification means that it is a mature adult star in a stable phase of its life on the main sequence in a Herzsprung–Russel diagram (see Figure 5.8). Such stars are commonly called "dwarf" stars but that is not because they are small, but really only because they are not "giant" stars. The "3" in the classification is a subclass within "M," while "M" is the astronomers' way of saying that it is much more red than the Sun (which is a G2V star) and that its surface is quite cool for a star. M-type stars are, however, far more common than warmer classes of stars, so the insights just discussed for GJ581d, whether the candidate exoplanet turns out to be real or not, also hold for many real exoplanets with similar properties.

There are other things to consider for the numerous red M-type stars. One is precisely that their light comes out mostly as red and infrared light (see the imagined world in Figure 1.4), rather than the white-yellow of the Sun. Photons of red light have less energy, and the photosynthesis that powers most plants on Earth would not work in such light. Some life forms on Earth do use infrared light, including some types of green and purple bacteria that digest hydrogen sulfide. This type of metabolism does not generate free oxygen, so it is possible that life on exoplanets around M-type red dwarf stars exists without creating the atmospheric oxygen that researchers are looking for as a biosignature.

Another thing to consider is that magnetic activity of M-type stars persists much longer than on Sun-like stars. For the Sun, the activity of its flares and its background ultraviolet and X-ray emission came down quite fast owing to efficient magnetic braking of its spin, which diminished considerably within the first billion years. For M-type stars, this drop in activity can take several times as long. This leaves the average M-type star in the Galaxy quite active compared to the Sun. For example, for a star such as GJ581 that weighs one third as much as the Sun, the X-ray brightness powered by its magnetic activity is somewhere between one part

in 10,000 to one part in 100 of the total light coming from the star. For the present-day Sun, it is less than one part in a million.

The relatively large stellar magnetic activity combined with the possible absence of atmospheric oxygen even if there is life on an M-star exoplanet may leave such life much less well protected from the star's dangerous ultraviolet and X-ray light: in the absence of free oxygen, there might be no, or much less, shielding by an ozone layer—which is a type of oxygen. Is that a problem for life? It would be for life on Earth, but that has evolved to the conditions that it is in, which include a shielding ozone layer. Other life forms elsewhere, under different skies, may well evolve their own shielding or repair mechanisms, or may stay out of harm's way under water. We just do not know.

The challenge of exoplanet climatology

If all of this makes it sound like climate modeling for exoplanets over periods of billions of years is hard, then the message has come across. In a review of atmospheres of Earth-like planets, Adam Showman and colleagues wrote about the daunting diversity of conditions on exoworlds: "Given the diversity emerging among more massive exoplanets, it is almost certain that terrestrial exoplanets will occupy an incredible range of orbital and physical parameters, including orbital semi-major axis [the maximum distance from the star], orbital eccentricity, incident stellar flux [total brightness of the star in the planet's sky], incident stellar spectrum [the distribution of the colors in the starlight], planetary rotation rate, obliquity [the tilt of the planet's rotation axis relative to its orbit around the star], atmospheric mass, composition, volatile inventory (including existence and mass of oceans), and evolutionary history. Since all of these parameters affect the atmospheric dynamics, it is likewise probable that terrestrial exoplanets will exhibit incredible diversity in the specific details of their atmospheric circulation patterns and climate." What happens when all these options are explored? Patterns become clear that will

hint at what is important, and what is not, to exoplanet climates and habitability, and where things are most likely to end up most frequently within all the possible conditions.

In a way, for now, having all these possibilities lie open in front of us is OK, because it means that researchers are exploring what worlds might look like with a lot of "what if" experiments. Although nowadays often supported by computer experiments, these are essentially what are known as "thought experiments." It is what Albert Einstein (calling them "Gedankenexperimente" in his native German) did to think about what would happen if one traveled along with a ray of light, or what Erwin Schrödinger (1887–1961) did when thinking about quantum mechanics with the parallel of a simultaneously possibly dead and possibly alive cat in a box, at least as long as the box remained closed to observation. Such experiments may not reveal the true characteristics of any particular exoplanet, but they help scientists realize where the boundaries of their knowledge lie, and where they really need observations to lift them out of logical predicaments. And that, in turn, helps them devise future instruments to come to their aid. After all, there is too much to observe in the Universe and too little time and money to do it all, but at the same time observations are, in the end, the only way to uncover the physical workings of the many worlds that exist around the Galaxy. Ultimately, one would like to understand all of the possible worlds out there, but for now these thought experiments are truly valuable in that they guide us to where and how to look for habitable worlds in our quest to find extraterrestrial life.

Notes

1 A study by N. Wickramasinghe and J. Wickramasinghe published in 2008 (*Astrophysics and Space Science*, volume 317, pages 133–7) considered the possibility that microorganisms could actually be blown away from high in Venus's atmosphere, caught in the solar wind, and transferred to Earth whenever that was in the downwind position.

This remains speculation, subject to the hazards of space travel and the very low fraction of the solar wind that actually connects to the Earth's atmosphere. Such interplanetary seeding of life had been proposed earlier also: microbes could have hitchhiked on rocks ejected by meteorite impacts on Mars and found their way to Earth. Such Martian rocks have been found on Earth. Whether life was aboard and, if it was, survived impact is pure speculation (see an article in *Discover* magazine, August 2001). Although intriguing, we should look further into this only once any of these speculative hypotheses are provided a solid evidence-based foundation.

11

The Long View of Planetary Systems

The Universe is three times older than the Solar System, and it will long outlast the Sun's nuclear fuel. So, there must be many planetary systems that are far older than ours, and there will be many that are yet to be created. Where does our own Solar System fit in? How many planetary systems formed before ours did, and what were these like? How many will yet follow, and what will the Universe look like to any life that may develop on habitable planets within these future systems?

Astrophysicists have learned so much in recent years that they can now formulate a scenario for the past and the future of planetary systems. In general terms, the life of the population of planetary systems within our Milky Way Galaxy looks like this. At first, within the Galaxy that itself was still forming, there would have been insufficient elements out of which planets could be made at all. In this phase, stars might have lightweight companions orbiting them, but these would be brown dwarfs composed in essence of only hydrogen and helium. It would take a few generations of heavy stars ending their lives in supernova explosions or series of less catastrophic mass ejections before substantial amounts of elements including carbon, oxygen, silicon, and iron would be available for the formation of what we think of as planets. The formation of rocky planets would then have started and later, with more of the heavy elements becoming available, would have ramped up dramatically. The frequency at which exoplanets formed would then follow the formation frequency of stars themselves, which appears to have reached a peak some four billion years after the Big Bang that now lies 13.8 billion years in

One of Ten Billion Earths. Karel Schrijver, Oxford University Press (2018). © Karel Schrijver.
DOI: 10.1093/oso/9780198799894.001.0001

our past. Giant planets need substantial amounts of ices and other solids to initiate their rapid growth so that they can then capture nearby gas in their gravity. These ices contain heavy elements that would be increasingly available after yet more generations of stars exploded their processed nuclear fuel back into the interstellar gas, enriching it with heavier elements. Consequently, giant planets started forming in larger numbers only after a delay of a few billion years compared to smaller planets: the formation rate of giant planets appears to have reached a peak only when the Universe was seven to nine billion years old. Our Solar System with its four giant planets formed after the formation rate of planets of any size peaked throughout our Galaxy; we live in a planetary system that took shape relatively late.

Then there are the free-floating nomad planets. Earlier we saw that observations and computer experiments reveal that exoplanets can be ejected from their systems. This can be the result either of planet–planet interactions in an early evolutionary phase of the system, of star–star interactions particularly in the densely populated birth setting of star clusters, or of stellar evolution late in the life of planetary systems. Such ejected planetary bodies are likely to live forever as migrating solitary nomads because the probability of either collision with or of capture into another planetary system is very small. The Galactic population of nomad bodies would therefore have been building up to ever larger numbers over time, and will continue to do so as long as stars continue to form and die.

In the future, the Galaxy will see its rate of star formation dwindle as it runs out of gas in molecular clouds. The bright blue and yellow stars will disappear first as their nuclear fuel runs out, ending their lives in giant explosions or as compact white dwarfs after a phase as giants. These end-of-life changes of the stars will destroy all the planets in compact orbits around them, while more distant planets—as far out as Jupiter and beyond— will increasingly find themselves orbiting white dwarfs, and perhaps neutron stars if they survive the supernova of their host

stars. So, there will be residual planetary systems around white dwarfs, but with any inner planets that might have existed earlier cleared out by the intervening giant-star phase. The exo-asteroid reservoirs in such systems are, over time, ground down into rocky fragments, dust, and gas, roughly half of which will spiral into the white dwarf that they happen to be orbiting while the other half is ejected into interstellar space. Compact planetary systems around faint, orange-red stars will increasingly dominate the overall population of planetary systems, until they will be the only such compact systems that exist in the Galaxy. The stars in the night sky will eventually be mostly fading white dwarfs, generally too faint to readily spot from one planetary system to the next, nestled among a multitude of comparably feeble and dimming deep-red stars.

All of the other galaxies surrounding ours will evolve similarly. Eventually, the entire Universe will likely fade out. It may have another end, perhaps a spectacular one, but the story of the end of the Universe belongs in other books. Whatever the end may be like, it is unlikely to be viewed by humans, although Douglas Adams did speculate about that in his slogan for "one of the most extraordinary ventures in the entire history of catering": "If you have done six impossible things this morning, why not round it off with breakfast at Milliways, the Restaurant at the End of the Universe?"

Populations of exosystems

Among the questions that we can ask about planetary systems are these: "how many systems are there?" "how old are exoplanets typically?" and "how do exoplanet systems change with time?" If we wanted to know how many planetary systems have formed and been destroyed over the course of the history of the Universe, we might at first recoil at the very thought of tackling this problem. Just about everything described up to this point in the book would have to be included, plus all the studies upon

which that work is based: How do stars form in a galaxy? How does a star evolve over time? How much of the matter captured in a given star is recycled into material available for the next generation of stars by stellar winds, ejected shells, and supernovas? How is the heavy-element content of the interstellar medium changing with time, and how does that influence the formation of planetary systems? How can we deal with well over a hundred billion stars in the present-day Galaxy, and combine that with all the stars that came before? And looking forward, can we know how many more planetary systems will form? Can we combine all that information to say something about the typical age range of planetary systems in today's Galaxy so that we may know how many planetary systems have had time to potentially develop life as happened on Earth?

Given the considerable knowledge of the stars that astronomers have gathered over the past century, it turns out that all these questions can be answered with fair accuracy. In order to do so, however, we need to shift from looking at the evolution of individual stars to the ensemble of all stars in entire galaxies. In other words, the focus needs to be on population statistics. In effect, astrophysicists would have to work analogous to, say, statisticians in insurance companies or in transportation companies such as airlines: they work with knowledge of large numbers of people that range from the ages of their death in studies related to life insurance to the anticipated numbers of people traveling on weekdays, weekends, summer vacations, and winter holidays for the tourism and business sectors. These statisticians do not analyze each traveler in any detail, but look at overall patterns.

Astrophysicists take a similar approach in looking at how many stars form over time. The first step, and the first equation, to deal with is the one that describes how the total mass evolves for matter that remains trapped in stars that have not yet ejected some or all of their mass back into their galaxy, as well as that which will forever remain trapped in their remnants, namely the white

dwarfs, neutron stars, and black holes; that rate is proportional to the rate of star formation, but corrected for the fraction of matter that is returned in stellar winds, pulsation-driven ejections, and supernova explosions. The second equation deals with the rate of star formation; this is proportional to the total mass available in interstellar gas, which decreases over time by the same amount as what is left trapped in the longest-lived stars and stellar remnants. In order to know more about the planetary systems that can form out of the interstellar gas clouds, a third equation quantifies the increase in the heavier elements made inside stars, which is proportional to the amount of gas thrown back from all stars over time multiplied by the fraction of that gas made of elements heavier than helium.

Each of the components of these equations is known, albeit with differing degrees of certainty, so that star formation rates and heavy-element fractions over time can be estimated for galaxies as a whole. Provided, that is, that star formation in galaxies does not differ too much from that in our own or its nearest neighbors. This is because in order to quantify the total amount of mass going into star formation, you have to see the full range from the brightest, heaviest ones to those that are substantially less massive than our Sun. Most of the mass ends up in the frequent lower-mass stars, but such stars are too faint and too numerous to be resolved anywhere but in our own Galactic neighborhood. Even exploring all of our own Galaxy is a challenge, because of all the gas and dust in it that precludes us from seeing it in its entirety. Hence, astronomers must combine information from our neighborhood within our Galaxy with observations of galaxies like our own to fine-tune and validate their computations. And then they need to look further and further away to see ever deeper into the history of the Universe by using the light that has traveled up to many billions of years to reach our telescopes. One of the problems that pops up here is not only that the more distant galaxies show little detail, but that the light from them is attenuated depending on

how much and where dust resides in a galaxy, because a weaker galaxy could mean fewer stars or more dust, or some mixture of these.

For the average galaxy, it appears that star formation starts out at a rate that increases with time going forward from the Big Bang. About half a billion years after the beginning of the Universe, it was likely comparable to what it is at present, but then rising rather than falling as it is now. The peak formation rate was reached when the Universe was some 3.5 billion years old, or 10 billion years in the past. Star formation in the Universe likely peaked at a rate that was some nine times higher than it is today. About 25 percent of all the mass currently contained in stars and stellar remnants was already a star before the first 3.5 billion years of the Universe had passed. Another 25 percent was captured into stars in the latter half of the Universe's age, that is in the most recent seven billion years. The rest was formed in between these ages of 3.5 billion and 6.5 billion years. Overall, researchers found that well over three quarters of the stars and stellar remnants that ever existed and will exist in the Universe were formed before the Solar System formed; we live in a relative latecomer among stars and their planetary systems.

The questions can be focused more by not merely asking about the rate of planet formation overall, which roughly follows the rate of star formation, but by differentiating between Earth-like planets and gas or ice giants. Their formation rates over time behave differently because whereas the formation of Earth-like planets is hardly dependent on the abundance of heavy elements, provided there is enough to begin with, giant planets in contrast form much more readily when the abundance of heavy elements is high. So, given a gas cloud of the same initial mass, size, and other properties in the young Universe—with little heavy elements—giant planets would form far less readily than in the present Universe. In the future, in contrast, giant planets will form more readily because then even more heavy elements will have been thrown into the interstellar medium by evolved stars

which will provide more solid matter to initiate increasingly rapid growth of the initial cores that can then attract gases and grow into giant planets.

The result of differentiating between these Earth-like and giant planets shows that the peak formation rates of rocky and giant planets in, say, our own Galaxy differ by about 2.5 billion years. At their peaks, the giant planet formation rate in the Galaxy appears to have exceeded that of Earth-like rocky planets by a factor of three, and currently it exceeds it by a factor of ten. Despite the fact that the formation of giant planets initially lagged behind that of Earth-like planets, the total number of giant planets that ever formed is estimated to be several times larger than that of Earth-like planets around Sun-like stars (of spectral types F, G, and K).

The quantitative details of planet formation are as yet quite uncertain, though, so that using this information to estimate how many planets might exist in the Galaxy overall is difficult to do with certainty. This is compounded by the fact that even the total number of stars in the Galaxy remains subject to uncertainty. Nonetheless, astronomers have tried to estimate the numbers: their models suggest that there may be from one to over ten billion Earth-like planets orbiting Sun-like stars in the Galaxy, and a few times more giant planets. The number of terrestrial planets around the lighter, fainter stars (of spectral type M) that dominate the population of stars in the Galaxy is several times larger still than either of the above.

Our planet is a fairly late arrival in the overall ensemble of planets in the Galaxy: most planetary systems that ever existed to date formed well before Earth formed: Earth formed after 80 percent of Earth-like planets formed that ever existed in the Galaxy to date. Many of the planetary systems that formed earlier will, however, have been destroyed either entirely or at least partly as their host stars evolved at the ends of their lives. We can get an impression of how old planetary systems are by looking at the average age of stars according to statistical studies. Some 75 percent of all mass in existing stars and their remnants was

captured into forming stars along with their planetary systems within the past 10 billion years, and those older than 11.5 billion years contribute only 8 percent of the total mass in stars. The average age of stars and stellar remnants in the Galaxy is just about 8 billion years, and roughly the same is true for planetary systems. Our Galaxy was a little slower than most others, so that the typical Galactic planetary system is about three billion years older than our Sun and its planets. That is worth remembering: if life formed in exoplanetary systems as readily as in ours, then most life in the Galaxy has had on average three billion years longer to evolve than we have. I, for one, cannot imagine what that would mean for the evolution of species and their technologies.

The life cycle of a planetary system

What happens to individual planetary systems over eons? First, let us look at the period after the disk has been cleared of gas and most asteroids have disappeared from the planetary system up to the point where the host star swells up into a giant star into a phase where it loses a substantial fraction of its mass. For our Solar System, it seems highly likely that all eight planets will survive this long-lived phase of some 10 billion years without major changes in their orbits. For exoplanetary systems, the computer experiments needed to see what might happen are on one hand simpler and on the other less reliable. They are simpler because all exoplanet systems have fewer than eight confirmed exoplanets, although a few are now known with five, six, or seven bodies, while Kepler-90 has eight confirmed exoplanets, putting it on a par with our own solar system. HD 10180 has six confirmed exoplanets, but possibly three more; if the unconfirmed exoplanets were to prove real, then this would make it a record holder with one more planet than our own Sun.

Most exoplanetary systems have only two or three confirmed planets. This does not mean, of course, that there necessarily are only a few planets in most exoplanet systems; it is more likely

that the low number of exoplanets in exosystems is an artifact reflecting the difficulty of detection of lower-mass or smaller planets and for planets that orbit quite far from their host star. But other than establishing that many configurations are not stable and might result in planetary collisions, descent into their star, or ejection from their system, we do not yet have enough information to establish general rules.

Once stars begin to run out of fusible fuel, they increase in size. In their final phases as giant stars they also throw out a large fraction of their mass. Close-in exoplanets will be captured into the stellar atmosphere and most likely destroyed. More distant exoplanets may survive the stellar giant phase, but the mass loss may cause orbital changes that lead to instabilities. The "contamination" of white dwarf atmospheres with exoplanet debris shows two things. First, that some exoplanetary bodies—maybe entire planets but possibly only exoplanet fragments, exo-asteroids, or exo-comets—survive the giant phase, but second that their orbits are changing: perturbations in the orbits of bodies in such systems must occur to enable the "contaminating" bodies to be torn apart by gravitational tides and then crash onto the white dwarf.

We can gain insight into what happens when white dwarfs are contaminated by looking at our own Solar System. There, too, there are debris disks: the rings of the gas giants, the asteroid belt, and the Kuiper Belt. This particular story started when Daniel Kirkwood (1814–1895), a mathematician, looked in detail at the eighty-seven asteroids of which the orbits were known by the middle of the nineteenth century. He first presented his results at a meeting of the American Association for the Advancement of Science in 1866, after their meetings had been postposed since the last such planned meeting in Nashville, Tennessee, in 1861 because by then the American Civil War had broken out. By the time he published his results in the *Monthly Notices of the Royal Astronomical Society*, he had worked with 100 orbits; in contrast, the Minor Planet Center[A] currently lists close

---

[A] IAU Minor Planet Center: http://www.minorplanetcenter.net.

to 730,000 orbits—a result of better telescopes and recording instruments, and automated searches and orbit comparisons. Kirkwood noted gaps in the distribution of distances from the Sun for these asteroids. He correctly concluded that these gaps occurred wherever the orbital periods formed a ratio of low integer numbers relative to that of the largest planet, Jupiter, like 2:1, 3:1, 4:1, and also 5:2 and 7:3.

Kirkwood explained the existence of gaps, or "chasms" as he called them, at specific period ratios with Jupiter by arguing that for any particle on such orbits, say one in which it completes three orbits for every one of Jupiter, such an asteroid would come "always into conjunction [closest approach] with that planet in the same parts of its path. Consequently [because of the pull of the massive planet always in the same phase] its orbit would become more and more eccentric until the particle would unite [that is, collide] with others, [on orbits that were] either interior or exterior, thus forming the nucleus of an asteroid. Even should the disturbed body not come in contact with other matter, the action of Jupiter would ultimately change its mean distance, and thus destroy the commensurability of the periodic times. In either case the primitive [original] orbit of the particle would be left destitute of matter. The same reasoning is, of course, applicable to other intervals." Once the orbit is disturbed sufficiently, the body may cross orbits with other asteroids or—if the orbit is sufficiently eccentric—with another planet, upon which it will be scattered into a different orbit, possibly on its way out of the planetary system, or toward the central star or its relict, the white dwarf.

A white dwarf is very small, so the probability of hitting it directly is likewise very small. But when an asteroid comes close enough, the gravitational tides will pull it apart: if the difference between stellar gravity at the points on an asteroid nearest to the star and furthest from the star exceed the force by which the asteroid holds itself together, it will be pulled apart. Thereafter, the fragments may collide with others and—because they now have even less gravity to hold them together—shatter into more

pieces, eventually building up a debris disk with a lot of micron-sized dust. Fragments that suffer collisions or that are scattered about by gravitational forces in near misses may eventually end up in the white dwarf's atmosphere, thereby leading to the observed contamination.

This plausible scenario requires two things to work. One, there should be a population of bodies that can be scattered into close proximity with the white dwarf; the general term for these bodies is "minor planets," which formally includes everything except (exo-)planets, their moons, and the population of comets, although in this particular case comets may also be included. Two, there must be at least one substantial planet left in orbit about the white dwarf to enable the scattering to occur efficiently enough that a measurable contamination level of the white-dwarf atmosphere can be built up. An asteroid belt such as in the Solar System is one such reservoir, and the more distant Kuiper Belt another. In both, orbits can be relatively stable for billions of years until the central star loses a substantial fraction of its mass in the giant phase, upon which orbital perturbations of the minor planets can occur. If the mass loss were to occur asymmetrically around the star, then the resulting rebound kick might even cause the star to drift away from its initial position, thus shifting the system's center of mass so that all orbits would be affected, with ample possibilities of major perturbations away from near-circular orbits of exosystem equivalents of the Kuiper Belt objects at the far fringes of the Solar System.

However, white dwarf contamination looks to be of rock-dominated rather than ice-dominated origin, suggesting that the population of bodies causing the heavy-element enrichment was formed closer to the star, within the ice line, and thus more likely to originate from a planetesimal disk like our asteroid belt. But to have sufficient material to explain the observed contamination levels, there should be hundreds to thousands of times more matter in an exo-asteroid belt to be consistent with white-dwarf observations. This presents a problem that is yet to be resolved:

the composition of the contaminating bodies favors an origin nearer to the star but astronomers do not understand how there could be enough mass there, while there is sufficient mass in an exo-Kuiper Belt but then the composition seems to be wrong being biased too much toward volatiles rather than the observed refractories.

One thing that is clear is that for any object from afar to fall close enough to the white dwarf to be disrupted, it has to be on a highly eccentric orbit. When such a body is broken up, it appears likely that roughly comparable amounts of material eventually fall onto the white dwarf as are thrown free of the planetary system. Contaminated white dwarf systems thus effectively work as shredders of minor planets from which roughly half the mass is thrown into interstellar space as clumps of rocky matter of a wide range of sizes. Not only are fairly large bodies created by gravitational tides, but subsequent collisions between bodies grind them down to dust, while sublimation under the heat glow from the white dwarf of material off the boulders and dust also creates a gas component.

How will it all end?

Currently still 2.5 million light years away, the Andromeda galaxy (also known as M31) is racing toward the Milky Way—well, actually, the two galaxies are falling toward each other due to their mutual gravitational pull—at over 100 kilometers per second (70 miles per second). It is expected to collide with our Galaxy about four billion years from now. The Andromeda galaxy is estimated to contain about twice the number of stars of our own Galaxy. Because of the vast distances between stars in both galaxies, such a galactic-scale collision is likely to eventually result in a merger of the two galaxies—which may take two billion years to complete—with relatively few actual collisions between the hundreds of billions of stars in the two. With so many stars in play, however, it is entirely possible that the passage of one or more

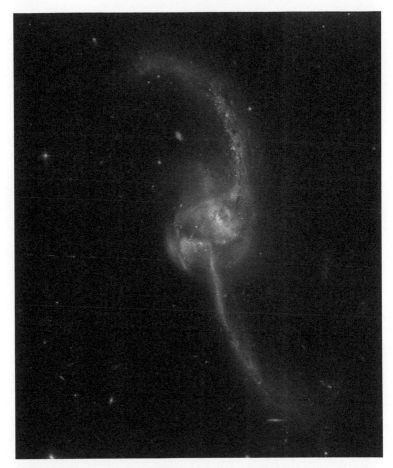

Figure 11.1 The object cataloged as NGC 2623 is in reality a pair of galaxies, 300 million light years away from us, that collided and are now merging into a single large galaxy about the size of our own Milky Way Galaxy. The collision has initiated star formation both in the interior and in the drawn-out arms, revealing itself in multitudes of young, heavy stars glowing blue. Around the compound galaxy many other more distant galaxies show up in the background, along with some foreground stars within our Galaxy. Some four billion years in the future, the Andromeda galaxy and our own Galaxy are expected to similarly collide and merge (See Plate 30 for a color version).

Source: Hubble Legacy Archive, ESA, NASA.

stars will upset at least the outer parts of the Solar System by their gravitational pull.

In the far distant future, well past the time that the Sun will have ended up as a white dwarf, the population of stars in our Galaxy, and indeed in all galaxies, will be dominated by the longest-lived ones: the majority will be red dwarf stars, as is already the case, but all giant blue stars and the average yellow stars will have vanished in supernova explosions with neutron stars or black holes as residuals, or will have turned into fading white dwarfs.

Long, long past the time that the Andromeda galaxy will have merged with our Milky Way, any astronomer peering into the skies will see fewer and fewer galaxies as time passes. Exactly when and how this will happen remains a matter of speculation to some extent, because we have yet to learn the nature of space, time, and matter on the largest, longest scales of things. But it appears entirely possible that this will happen: for one thing, everything but our direct neighbors in the Local Group of galaxies will have moved far, far away, and from our perspective will have become very faint. But there is more: if the expansion of the Universe continues as it now appears, accelerating with time, then visible light can only reach our Milky Way/Andromeda galaxy from ever smaller distances. This is because the expansion of the Universe will continue to accelerate, and light from some distance away is Doppler-shifted progressively toward the red and even the invisible infrared. So, the light from the galaxies will fade because of both distance and discoloration. By the time the Universe reaches something like its 100 billionth birthday, only very few galaxies will be left visible to study, namely those that are now members of the gravitationally bound Local Group. Most of these are dwarf galaxies, which are satellites to the largest ones, dominated by the Andromeda and Milky Way galaxies that will then be one unit, with shared satellite galaxies should these survive without themselves merging with the large merged Galaxy. By then, a Universe with 2 trillion visible galaxies, as we now see it, will be unheard

of, unless it is enshrined in acquired knowledge from the distant past. If it is not, however, then those far-future astronomers will not be able to learn by observations about the Big Bang and the associated phenomenon of an expanding Universe but will have to resort to theoretical arguments only—and perhaps to aspects of physics that have as yet not become part of our thinking.

12

Living on a Pale Blue Dot

In 1990, two years before the first exoplanets were discovered, the spacecraft Voyager 1 was commanded to point its cameras back toward the Solar System. At the time, the vehicle was 40 Sun–Earth distances away, had moved beyond the orbit of the outermost planet, Neptune, and was about to have its cameras turned off so that other instruments could make use of the on-board power and memory. From afar, the Voyager 1 took sixty pictures that were later compiled into a mosaic "family portrait" of the Solar System (Figure 12.1) showing Neptune, Uranus, Saturn, Jupiter, and—appearing faintly, close to the distant Sun—Venus and Earth. Carl Sagan, a member of the Voyager science team, looked at the pale blue dot that was Earth and later noted: "That's here. That's home. That's us. On it everyone you love,

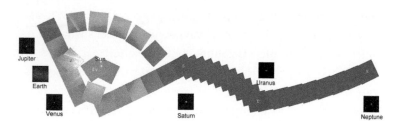

Figure 12.1 Mosaic of images taken by the Voyager 1 spacecraft in 1990 as it was at 40 Sun–Earth distances, well beyond the orbit of the most distant planet, and just before its cameras were permanently turned off. The planets, their positions labeled in the images by a letter, are too small to see in this print; the insets show them at the best resolution of the camera.

Source: Modified after an image by NASA/JPL. https://voyager.jpl.nasa.gov/galleries/images-voyager-took/solar-system-portrait/

One of Ten Billion Earths. Karel Schrijver, Oxford University Press (2018). © Karel Schrijver.
DOI: 10.1093/oso/9780198799894.001.0001

everyone you know, everyone you ever heard of, every human being who ever was, lived out their lives." Even this mosaic does not convey the immensity of distances and the emptiness of space: when displayed at the full power of the best camera on Voyager 1, Earth looks too large, because the camera's smallest-resolution element is still eight times larger than Earth should appear. Although Voyager 1 was at a vast distance of 6 billion kilometers (4 billion miles), that is less than 1/6,000 of the distance to the nearest star, Proxima Centauri, which might be orbited by a planet from where another civilization would have the best view of our home world from the outside.

Life on Earth endures in a largely self-maintained niche that is wedged between empty space and solid rock. The discovery of multitudes of exoplanets is stimulating the quest to understand how life on Earth could originate, how it is sustained, and where else in the Universe we might encounter conditions favorable to life in any form vaguely akin to that on Earth. That quest requires that scientists who are highly specialized in their field see beyond the perimeters of their disciplines that generally have taken shape over many centuries: for example, solar astrophysicists need to be aware of planetary atmospheric processes, planetary scientists need to exchange information with biologists and climatologists, and experts on stellar evolution need to know how dust coagulates into planets. There are always some scientists whose work touches on multiple disciplines, but the quest for life in the Galaxy requires that many now reach across the divides built into the education system, into university departmental structures, and into government funding processes: understanding (exo-)planetary habitability is fundamentally an interdisciplinary activity.

Interdisciplinary discipline

The hurdles faced by interdisciplinary scientists are not only those of the structure of their cultural environment, however.

A successful specialist is continually challenged to keep up with research even within her or his own discipline: there are thousands of research publications that appear each year within any of the traditional disciplines and it requires much time to read any of these in depth. To keep an additional close eye on multiple other disciplines has become a practical impossibility. Whereas at the start of the Renaissance a few centuries ago one could still be a real polymath approaching the old ideal of the "uomo universalis" with expertise in all things scientific, this is no longer possible simply because our collective knowledge has expanded tremendously. That very expansion, however, is a challenge to scientists in general, including those interested in the new field of exoplanet habitability, because an enormous diversity of knowledge needs to be mustered to come to grips with that issue.

Many choose to address the challenge of the diversity of expertise needed to advance exoplanet science by teaming up with colleagues from different disciplines to pool knowledge and, in due course, to be educated in the process. Interdisciplinary work faces not only the problem of knowing who around the world to most successfully collaborate with, but also that there is a selection bias that works against interdisciplinary proposals for funding and telescope time; this may be because expertise in peer-review panels is not necessarily well matched with such proposals, or because it is harder to argue in a cogent and focused manner for the multifaceted study being proposed. Despite this selection bias, scientists recognize there is much to be gained by trying to obtain funding for working with colleagues with other areas of expertise for the purpose of integrating and thereby furthering general knowledge. Consequently, studies on (exo-)planetary habitability have authors not only from departments of physics and astronomy, as most other publications on astrophysics do, but also from astronautics, atmospheric sciences (including meteorology and climatology), chemistry, computer and information sciences (including supercomputational physics), Earth sciences (also known as geology and geophysics, including oceanography,

and terrestrial magnetism) life sciences (astrobiology and biology, including bacteriology, botany, genetics, paleobiology, and zoology) mathematics, meteoritics, and from departments whose very names already indicate their interdisciplinary nature, such as departments of geomicrobiology, integrative biology, and geophysics and planetary sciences. Whereas many of these studies address exoplanetary habitability, they build on a foundation of knowledge that has its surest footing in what has been learned about Earth's habitability over its multibillion-year history: in the end, the best-known example of a habitable planet is Earth.

Habitability, photosynthesis, and volcanism

The ancient Earth had an atmosphere without free oxygen and with a very much larger amount of carbon dioxide mixed with nitrogen. Over time, this changed. First, most of that carbon dioxide was removed from the atmosphere in the early life of the planet by chemical weathering of rocks due to acidic rain that transformed atmospheric carbon dioxide into sedimentary limestone and its relatives. Then life started to influence the overall makeup of the atmosphere, eventually leading to a large percentage of free oxygen as a byproduct of the metabolism of early life's constituents. These changes form a formidable challenge for researchers who try to understand the climate of Earth in periods that lie hundreds of millions to billions of years in the past: climate depends on the composition of a planet's atmosphere at the time, but the development of these changes is difficult to derive from geological records.

The process that removes carbon dioxide from Earth's atmosphere is one form of chemical weathering. Atmospheric carbon dioxide dissolves in rain drops in the form of carbonic acid. When that falls onto rocks, this relatively weak acid gradually dissolves the rocks. This frees calcium, magnesium, potassium, and sodium that eventually flow into the oceans along with clusters of oxygen and carbon bound into (bi-)carbonate that

were originally contained in atmospheric carbon dioxide. Once in the oceans, calcium and the bicarbonate ions combine into deposits of calcium carbonate, thereby trapping the once atmospheric carbon dioxide. Compressed layers of this form thick sediments of limestone, which can transform into marble in the deep Earth. If this process were the only one affecting the carbon dioxide level in Earth's atmosphere then eventually all of that would vanish, which would make plant life impossible. Over the past century, humans have been injecting a lot of carbon dioxide into the air by burning fossil fuels, but geologically speaking that is a very brief anomaly. That carbon dioxide has not vanished from the atmosphere in the oceans of time prior to the industrial revolution is because of movements in the hot interior of the Earth in what is known as tectonic activity.

Tectonic activity is a process that, in the long run, is critical to keeping life on Earth going through its associated volcanoes: carbon dioxide from within the Earth is released into the atmosphere in eruptions and through vents, porous rocks, volcanic lakes, and hot springs. It may be counterintuitive, but the carbon dioxide that is being released by volcanoes has been paramount in keeping the atmosphere "breathable" to plant life for billions of years. Thereby it indirectly supports all animals, too, because they, in turn, depend on plants, both for their oxygen and as food, be it directly or, for carnivores, indirectly as nourishment of their ultimate food source in the chain. Until the start of the industrial revolution in the middle of the nineteenth century, volcanic activity put back, on average, as much carbon dioxide as was taken out by weathering in what is known as the carbonate–silicate cycle. At present humans are the main contributors to this cycle because of their burning of fossil fuels: this puts about sixty times more carbon dioxide into the atmosphere per year than present-era volcanic activity. The result is that the level of carbon dioxide in the atmosphere is shooting up, having risen by a third over the natural level seen at the beginning of the heavy use of coal that powered the industrial revolution.

As the Sun will gradually brighten in the course of billions of years, the planet will warm up considerably. This will unavoidably play out on a timescale vastly longer than our current predicament of human-induced but avoidable climate change. The warming of the planet in the distant future will lead to increased evaporation of water into the atmosphere, and from there intensified loss of that water from the top of the atmosphere (as happened on Venus, described earlier). Eventually, there will be less water dissolved in the continental plates, which will cause tectonic movements to slow and in the end stop altogether. Because much of the volcanic activity will vanish when tectonic motions cease, the remaining surface water will likely lead to the removal of the residual carbon dioxide through chemical weathering, so that Earth's atmosphere will end up with insufficient carbon dioxide to sustain plant life as we know it. Or maybe Earth will end up without water first, thereby ending life as it presently exists. Either way, it is likely that both flora and fauna will be exterminated on Earth well before the Sun's size and heat become the primary threat to the inhabiting biological kingdoms that are now on the planet. That point may be reached one to two billion years into the future, so is hardly of immediate concern for our species that has existed so far for at most 200,000 years, only a fraction of one part in 10,000 of the time remaining.

The process of weathering has captured enormous amounts of carbon dioxide from the atmosphere. The entire inventory of captured carbon dioxide estimated from deposits of limestone and its chemical relatives is so large that Earth's initial atmosphere must have been completely dominated by it. In fact, there is so much limestone, and so much that has been subducted into the Earth by tectonic motions, that the young Earth likely had an atmosphere over twenty times denser than today. In that early atmosphere, nitrogen would have been but a few percent of the total. This early-Earth chemical mixture—not the total amount— before substantial rain-driven chemical weathering and before the emergence of plant life would have been rather similar to the

present-day atmospheres of Venus—with a surface pressure almost a hundred times higher than Earth's—and Mars—with a surface pressure over a hundred times lower than Earth's.

All of the phenomena related to plate tectonics, including volcanism, are powered by heat from deep beneath the Earth's crust that is transported to, and escapes from, the surface by overturning motions of highly viscous fluid-like substances. When two plates floating on the surface are moved apart by these motions, the opening space between them is filled up with cooling magma from below, frequently forming volcanoes in the process. Other types of volcanoes form where one plate is forced to move underneath another, or at so-called hot spots where plumes of rising magma are particularly pushy and can breach the continental or oceanic plates. The internal heat that powers this on Earth stems from a combination of the heat created by the collisions that formed the Earth in the first place, plus heat from the radioactive decay of unstable isotopes that have their origin in supernovas that preceded the formation of the Solar System.

Both these forms of heat also exist in other bodies in the Solar System. In some of the moons of the giant planets, tidal forces may provide a third energy source adding to their internal heat. Whether these heat reservoirs and sources suffice to drive volcanism, after such a long time since original formation, is a question of how fast energy is being lost from the surface relative to the total heat contained or generated inside. Larger bodies have an advantage because they have more volume in which heat is contained or created relative to their surface area from which it is lost. Bodies with more rapid tectonic motions have their internal energy transported to the surface faster, and therefore cool more quickly. Bodies with lower water content have slower or stalled tectonic movements, so although heat may be there at depth, the surface no longer displays tectonic activity. These qualitative arguments are consistent with Mars having no active volcanoes or plate tectonics: it is much smaller than Earth, and is argued to have run out of internal energy much sooner as a result. The same

is true for Mercury. Venus is similar in size to Earth, but it has lost its water already because it is closer to the Sun. Consequently, not being "greased" by water, plate tectonics stalled on Venus, leaving it with only limited local geological activity without large-scale plate motions.

Lava flows associated with volcanic activity over time erase other features in the landscape by flowing over them or by filling them in with newly solidified basaltic rock. Given an adequate amount of time, all meteorite impact craters are erased, one after another. The surfaces of Venus, Earth, Mars, and also Jupiter's moon Io (Figure 12.2) reveal signs of volcanic activity by the relatively low number of impact craters on their surfaces if indeed

Figure 12.2 A volcanic eruption on Jupiter's third-largest moon, Io, in March 2007, observed by the New Horizons spacecraft as it passed by on the way to Pluto. The volcanic eruption from the Tvashtar volcano reaches a height of 300 kilometers (200 miles). Io's surface (covered in yellow sulfur, a whitish frosting of sulfur dioxide, and exhibiting a variety of other colors from sulfur and other compounds), shows many black areas in the depressed central regions of volcanoes. Overall, the moon has more than 400 active volcanoes (See Plate 25 for a color version).

Source: NASA's Goddard Space Flight Center and NASA/JPL/University of Arizona.

volcanism is not observed directly in action during eruptions. The Moon has vast plains indicative of ancient lava flows primarily into impact basins (see Figure 3.1), and the same is possibly true for Mercury.

True tectonic plate movements do not appear to occur anywhere in the Solar System except on Earth. Only here do we find subduction zones and strings of volcanoes along the ridges where plates separate. However, Venus, Mars, and Io do exhibit evidence of volcanoes; these are of a type that forms not as a result of colliding plates but instead toward the interior domains of plates on Earth, in hot spots, with volcanoes like those of the Hawaiian islands. Venus, for example, has such volcanoes that number almost 1,600 over the planet counting only the major ones.

Bringing it home

Climates and conditions of habitability change over time for all planets in the Solar System as they do for exoplanets. This is abundantly clear even if only looking into Earth's history. Long before human activity, and with only slow changes in the atmosphere, the global climate shifted dramatically across what have become known as tipping points. For instance, there are times in the not-so-distant past—by astronomical standards—when the entire Earth appears to have been frozen over, with ice sheets and glaciers hundreds of meters (or yards) thick all the way from the poles to the equator. This did not occur just once, but multiple times, with some of these deep freezes lasting for millions of years. And yet, we are talking about the same planet that has only a gradually changing atmosphere, and which is warmed by a Sun that is very slowly and quite gently brightening.

The climate system of a planet like Earth is very complicated, and would be so even if it were to be studied in the absence of life: different types of clouds differentially block incoming sunlight, ice on the oceans and snow on land reflect sunlight differently from deserts or forests, greenhouse gases (including water vapor)

can trap heat, and winds and ocean currents transport energy over large distances and a range of heights and depths, deflected by mountain ranges both on land and on the seabed and also by the forces associated with the planet's spin. All these effects need to be taken into account. Until around 2010, computers were really not capable of doing a full three-dimensional model that included both the atmosphere and the oceans, so that major simplifications had to be made: the models would ignore the ocean or at least the changes in ocean currents, they would use low spatial resolution so that complex topography such as mountain ranges could not be dealt with in detail, or they would be terminated in a state that might appear to be where things settled but that could not be verified by continuing for centuries or millennia of model time to see if there might be changes much later on. Or, in the most limiting approximation used early on, the models would use only a so-called one-dimensional approach: a layered atmosphere and a reservoir of water at the base in which light, heat, and gases could be transported and mixed only up or down but never sideways; the entire atmosphere was assumed to be the same everywhere all over the globe, and there could be no winds blowing across the virtual surface.

Starting around 2011, the first fully three-dimensional climate models have become available on supercomputers. Full ocean descriptions remain hard, in part because the oceans change very slowly compared to the changes in the atmosphere which need to be tracked, so that long and computationally expensive runs are needed to see how the full atmosphere–ocean system responds. Also, physical descriptions of cloud behavior and of energy transport through the atmosphere by gas motions and by thermal radiation remain challenging. Nonetheless, meaningful experiments are possible from which lessons can be learned.

Tipping points

Among the experimenters using such state-of-the-art computer codes were Eric Wolf and his colleagues. They wanted to

investigate differences in climates between an Earth-like planet orbiting a star like the Sun and such planets orbiting other stars that would be somewhat more or less massive. They did not look at the least massive stars, because there the habitable zone overlaps with the distance range within which planets are thought to be tidally locked to always have the same face toward their star; the codes that had been developed for application to conditions on Earth simply were not designed or tested for such slow planetary spin and—for fully synchronized planets—utterly unbalanced energy inputs between the two hemispheres.

From these Earth-rooted models one realizes that even with identical planets (just like today's Earth, but without life on it) and identical total heat inputs from the star (which means closer to less massive stars and more distant from more massive stars than the Earth is to the Sun), climate is slightly different for otherwise identical planets, except for the fact that they are orbiting different stars. The reason for this difference is solely that less massive stars shine much more in red and infrared light and less in blue light, while heavier ones have more blue in their glow. This shift in color pattern changes how the light is blocked or reflected by clouds, by ice, and by the greenhouse gases carbon dioxide, methane, and water vapor. With everything equal except for the incoming colors of the starlight, surface temperatures on the modeled planets increase by about 15 to 60 degrees centigrade (30 to 100 degrees Fahrenheit)—depending on which computer approximations are used—when the stellar surface temperature is lowered from nearly 7, 000 degrees centigrade, which shines just leaning to bluish, to just over 5, 000 degrees centigrade, which is more orange than the Sun.

The ultimate goal of their experiment was to see how climate would change subject to the aging of the stars around which these planets orbit. They found, not surprisingly, that as the star aged and became brighter, the model planets all became warmer. But they also found that although in some eras the planetary temperature responded gradually to the gradual changes in the star,

at other times the planet's climate state would almost jump from one state to another. There seemed to be four dominant climate states: it was either a "snowball" planet in which essentially all oceans are frozen over, a "waterbelt" planet in which there is open water around the equator but ice and snow reaching from the polar regions into the subtropics, a "temperate" planet as Earth is now, or a "moist greenhouse" in which much of the oceans would evaporate into the hot atmosphere. In their calculations, the models tended to avoid the temperature range between the waterbelt and temperate worlds (between −25 degrees centigrade or −13 degrees Fahrenheit and just above the freezing temperature of water) and away from the range between a temperate state and a moist greenhouse state (from about +40 to +60 degrees centigrade, or about +100 to +140 degrees Fahrenheit), pretty much regardless of the type of star around which the model planet orbited. They found that rather than changing smoothly, the planetary climate toggles between below and above these avoided temperature ranges because of the properties of sea ice and the greenhouse effect of water vapor.

What complicates things still more is that, for the present-day Earth for example, the climate could settle into either snowball or temperate conditions at the same solar illumination depending on whether that illumination had been increasing or decreasing before. The real Sun's brightness will only go up, so this experiment is rather academic in nature. But there are other ways to modify the energy budget of the climate system. For example, changes in cloud cover, greenhouse gases, or land use (involving, on geological time scales, also the re-arrangement of continental areas) can hamper or ease the loss of heat from the Earth. Another way would be by changing Earth's orbit, which can happen by mild but persistent tugs from the giant planets. Any of these modifications would shift the balance between heat coming in from the Sun and being lost by Earth, which could knock Earth's climate out of its present pleasant temperate climate state into a snowball mode. What

these experimenters and their colleagues really found was that if somehow the Earth's climate were pushed into one of these states, the models suggested that it could stay there, even with the present-day atmosphere and with the present-day sunshine: the Earth's climate might remain in either a temperate or a snowball state. Moreover, this behavior pattern has apparently now existed for hundreds of millions of years, and will remain possible for hundreds of millions of years into the future.

Different research groups come up with somewhat different results, though, because the processes are coded into their computers differently, so that when the array of included processes interfere with each other in complicated ways the outcome is not identical although often the trends are comparable. As scientists now have much better ideas about how such temperate-to-snowball and snowball-to-temperate transitions might happen, the ultimate question is whether these transitions occurred for the real-world Earth. Climatologists already knew that this did actually happen and the consequences for life on Earth may well have been dramatic: whereas single-celled life emerged on Earth possibly 3.9 billion years ago, the modern-day multi-celled plants and animals did not emerge until some time around 500 million years ago, when the long period with occasional snowball events was well in the past.

Snowball Earth

Snowball-Earth events are not in the same league as the moderate cold spells like the most recent "ice ages," the last of which ended some 11,000 years ago. During these ice ages, ice layers of a few kilometers thickness might cover the continents in the northern hemisphere, for example, down to beyond the US–Canadian border in the Americas and to south of the United Kingdom in Western Europe. A true snowball event, would see ice sheets on the continents well into the tropics even at sea level, reaching into the equatorial oceans, while the oceans in general may well

have been frozen over across almost the entire globe. The last two such global events ended around 720 and 635 million years ago, in an era when life was limited to single-celled organisms, mostly in the oceans; imagine a mostly frozen planet on which there is no vegetation, only bacterial mats, anywhere where you could walk. What can be gleaned from ancient geological records is limited, and consequently climate scientists are still trying to understand the details of why the Snowball-Earth periods happened, how they ended, and why they have not manifested over the past hundreds of millions of years.

By the way, if you could travel in time, you might just survive walking around on Earth during the most recent snowball era: oxygen levels in the atmosphere had at the time likely risen above halfway of present-day values, although there is uncertainty about the percentage over time. If the oxygen level was about half the present-day value, your coordination and judgement would be impaired, but if it were only just below halfway, you would not last more than minutes; so best keep the visit brief and monitored! What you could have visited was mostly a cluster of large land-masses, all in close proximity, located mostly around the equator and southward of it. Apart from offering relatively breathable air, though, this Earth would not be supportive of humans. Besides the fact that there would be no food for us other than in bacterial form, it would be very, very cold: the equatorial regions might reach around –30 degrees centigrade (–20 degrees Fahrenheit), but mid-latitude temperatures might be around –60 to –100 degrees centigrade (–80 to –150 degrees Fahrenheit) in winter months. These frigid temperatures touch on the range in which carbon dioxide would condense out of the atmosphere to form dry ice. This is seen on the present-day Martian poles during their winters, adding a few feet of dry ice to the permanent layers of water ice that are up to three kilometers or two miles thick (see Figure 12.3). Ice sheets on the historical Snowball Earth might have ranged from some 900 meters (3,000 feet) on land near the equator to 3 kilometers (10,000 feet) at the polar caps; the thickness of equatorial sea ice remains poorly known.

Figure 12.3 A 2-kilometer (more than a mile) thick ice cap extends over 1,000 kilometers (more than 600 miles) in Mars's north polar region. Radar imaging with the Mars Reconnaissance Orbiter revealed that the ice has thick layers alternating with dust and sand. These are a consequence of climate cycles owing to changes in the tilt of Mars's rotation axis and in orbital eccentricity. In the northern winter months carbon dioxide freezes in the cold atmosphere and adds a layer of dry ice onto the permanently frozen water ice, that sublimates away again in the summer months when the freezing shifts to the south polar region. In the center of the image, a large gap in the ice cover extending toward the lower left is the Chasma Boreale, a 2-kilometer (1.2-mile) deep and 560-kilometer (350-mile) long "canyon" comparable in size to the North American Grand Canyon.

Courtesy: SVS/GSFC/NASA.

How did Earth get out of a snowball state? One option being considered is an increase in the greenhouse gas carbon dioxide as a result of continued volcanism but reduced chemical weathering. The thick ice sheets on the land masses of a Snowball Earth would severely slow, if not essentially stop, the removal of the greenhouse gas carbon dioxide from the atmosphere because its

dissolved form would not rain onto silicate rocks upon which it could be bound into another chemical form; instead, snow would fall onto snow-covered rocks, preventing weathering. In the meantime, however, the volcanoes would still continue to put carbon dioxide into the atmosphere. As a result, the greenhouse effect would increase, which could (and possibly did) terminate the snowball state and push the global climate back into a temperate condition, possibly after an initial warm spell. But scientists are thinking of other mechanisms that might terminate the snowball state, too. Among them is the "mudball Earth" hypothesis: dust from, for example, volcanic eruptions or scooped up by glaciers, is expected to be transported by the global net sea-ice movement toward the equator, where it could gather in gigantic terminal moraines that might collect sufficient warmth from the Sun's light to push climate out of the snowball state. Such a "thaw" might take hundreds of years where the ice cover is thinnest, and likely multiple millennia for the entire globe. It simply takes a long time to melt the ice masses that accumulate over a few to a dozen million years of a snowball state.

Same Earth, different Earth

From the perspective of the present story, I would say that the main lessons are these: (1) planet Earth has gone through several cycles from an extremely cold Snowball Earth to a temperate water-rich world and back with but moderate changes in its atmospheric constituents and in incoming solar energy, (2) while we now realize that there are multiple agents that contribute to such extreme glaciation and deglaciation, none of these need to change things particularly strongly to have major global consequences, and (3) that life did continue to exist through these snowball states, although it was a simple life form: the last Snowball-Earth state appears to have ended before life forms for the first time evolved into multicellular, differentiated plants and animals.

This discussion of Snowball-Earth periods illustrates that global climate can change dramatically, even when not much happens within the Solar System: the Sun did not change too much in brightness from before to after the last two major Snowball-Earth eras that transpired some 600 to 700 million years ago, nor did Earth's orbit. Nonetheless, during such deep-freeze periods, Earth might not, technically be seen as habitable in the sense that liquid water might not exist at the surface: maybe there was some open water around the equatorial regions then, but maybe not.

When looking into planetary climates over billions of years, even the slowest and most gradual effects eventually let their presence be known. These processes can be internal to the planet, such as changes in volcanism, or external, which could be related to the evolution of the star, to the gravitational tug by other planets, or to the passage of the entire planetary system through an interstellar molecular cloud. These processes can affect each other, so that one can either dampen or amplify one or more others. These changes can push a climate over a tipping point with relatively little change in any of the conditions in particular. All of those paleoclimatic studies start out, of course, with the initial state of the atmosphere and oceans for any given planet, which depends on the initial orbit and on planetary spin rate and on the tilt of the spin axis; the latter is the cause of the seasons through the orbital year on Earth and contributes to those on Mars.

The climate of life

Life is a major player in setting the planetary climate on planet Earth. Life in the oceans developed first and, possibly with more microbial life on land, caused oxygen levels to reach a few percent some 2.1 billion years after the formation of the Solar System. More complex, multicellular life was first detected for an age of Earth of 3.7 billion years, perhaps one or a few hundred million

years before oxygen levels reached near-present-day levels. Only when Earth was 4.1 billion years old did so-called vascular plants, such as trees, which transport water and organic compounds in a network of ducts, appear on land, followed a little later by animals. So an Earth that we would recognize as related to the present-day Earth developed only after the planet had lived some 90 percent of its current age of 4.6 billion years.

Life as we now know it is based on photosynthesis; that is, the capture of sunlight into energy-bearing chemicals. In this process, plants "breathe in" atmospheric carbon dioxide and "exhale" oxygen. Animal life also depends on photosynthesis because plants provide both the food they need for energy and the oxygen they need to breathe. At some point in the future, chemical weathering will let the carbon dioxide levels drop below what present-day plants need to survive. This will occur because some time in the distant future plate motions are expected to slow down and eventually vanish, just as happened on Venus long ago: plate motion seems to be "lubricated" by water that is captured into the crust, but as the Sun brightens with time, it will cause the oceans to evaporate (perhaps gently, perhaps catastrophically) and the Sun's ultraviolet light will break the chemical bonds of water, while the solar wind will strip the released hydrogen away. Sometime within the next billion years, perhaps faster than we anticipate today, the combination of chemical weathering, ocean loss, and the end of plate tectonics will end plant life as we know it, and with it animal life will vanish.

The Earth itself may survive for maybe another seven or eight billion years up to the Sun's end-of-life gasps. During all that time from Earth's birth to its demise, the Earth in something like its present-day state, with similar atmospheric and climatic conditions, and life in some way related to what now exists will be around for a billion years, or maybe 1.5 billion years. If alien life forms were to observe (or visit) our planet at a random time in its overall life cycle, they would see something akin to the present-day Earth for at most some 10 percent of the planet's lifetime.

How does our Solar System compare?

With all the information from observations and computer experiments about planetary systems, one question that we can begin to answer is this: How common is a planetary system such as our own? One way to address that question is to look at the combination of a Sun-like star and a fairly distant Jupiter-like planet, because, as Alessandro Morbidelli and Sean Raymond noted in a review of developments, "Jupiter is the only Solar System planet within the reach of current observations [of other planetary systems]. Based on statistically-sound exoplanet observational surveys, the Sun–Jupiter system is special at roughly the level of one in a thousand. First, the Sun is an unusually massive star; the most common type of star are M dwarfs, with masses of 10–50 percent of the Sun's. The Sun is unusual at the [about] 10 percent level, depending on the definition of 'Sun-like' and [just how many stars of what mass formed in the past]. Second, only [about] 10 percent of Sun-like stars have gas giant planets with orbits shorter than a few to 10 [Sun–Earth distances]. Third, only about 10 percent of giant exoplanets have orbits wider than [one Sun–Earth distance that are moreover near-circular]. Taken together, these constraints suggest the Sun–Jupiter system is a [one-in-a-thousand] case."

Then, of course, there is the question of the formation of terrestrial planets, the timing of the formation of Jupiter slightly ahead of that of Saturn such that the migration of the two together would be first inward and later outward, so that Jupiter would keep Uranus and Neptune from getting in the way of their tiny cousins such as Earth, eventually followed by an upset in asteroid orbits that helped put water on the otherwise parched terrestrial planets. All of these are among the many characteristics that determined the shaping of our Solar System in general and of Earth in particular, which means that quantitative comparisons require that we specify exactly what we mean when we ask about planetary systems "such as our own." In their review,

Morbidelli and Raymond pointed out that "Just like any individual person, the Solar System has its own history. The probability of any other planetary system following an identical blueprint is zero. But how typical was the Solar System's evolutionary path? (*Did it go to college, get married and get a normal job?*) Or was the Solar System's path unusual in some way? (*Did it quit high school to teach scuba diving in Belize?*)"

Jupiter-mass exoplanets at hundreds of Earth masses are easier to detect than lightweights like Earth. Nonetheless, astronomers have attempted to estimate how many of the latter there are. The difficulty of their detection means that estimating just how hard they are to find becomes very important: you know how many you found, but you also need to know how many you could not have found even if they existed because of the instruments and the methods that you are using. Estimates are then made for a range in planet masses combined with a range of orbital periods (planetary "years") so the numbers published by researchers differ in part because of variations among what they call "Earth-like" as well as because of different estimates of the corrections that need to be made. Nevertheless, the numbers come out in a fairly limited range. I mean, they could have found the probability of roughly Earth-mass planets with roughly Earth-like years to be one in a million or one in a thousand. But different groups, using different instruments and methods, come up with answers that are somewhere between one in eight and one in two for planets between roughly half and twice as large as Earth with years somewhere between 10 and 100 Earth days. These orbit durations are shorter than an Earth year and astronomers are well aware of that, of course, but the longer the exoplanetary year, the further the planet is from its star, and the harder it is to detect.

By present-day knowledge, it appears that a little over one in five "Sun-like" stars in the Galaxy have something like a terrestrial planet with a size between one and two times Earth's orbiting between one half and two Sun–Earth distance from their star. Such planets would be more or less in their stars' habitable

zones, although that would also depend on their atmospheric and surface conditions. Overall for the Galaxy, with anywhere from 100 to 400 billion stars of which about 20 percent are quite "Sun-like" (that is, of spectral types G and K), this would mean that there could be between four billion and eighteen billion such planets. In the class of "Earth-like planets" we could allow for somewhat smaller planets also, and moreover we could extend the category of "Sun-like stars" to encompass the relatively infrequent somewhat warmer stars or the abundant somewhat cooler stars. Then again, maybe the above range in orbital distances from half to double the Sun–Earth distance is a bit too optimistic for long-term planetary habitability (for comparison, Venus and Mars are on average about 0.7 and 1.5 times as far from the Sun as Earth). All in all, the calculations are at present uncertain easily to three times fewer or three times more than any of the published numbers, but at least it gives us an idea of how common Earth-like planets really are. Using a simple round number in the middle of the range, we can estimate that *there are of order ten billion Earth-sized planets in astronomically defined habitable zones around roughly Sun-like stars within just our own Galaxy.*

Imagine that: ten billion worlds like Earth! This is simply staggering, even with the large uncertainty in this number, and despite the fact that the phrase "like Earth" should allow quite an assortment in what these planets actually look like (which we simply cannot know at present). Ten billion Earth-sized planets in the Galaxy means that there are almost ten thousand for every letter in this book. Every one on Earth could have one and there would still be many to spare. Try to picture that!

Earth: common and exceptional

On the one hand, this large number of Earth-sized planets in moderate proximity to their stars tells us that the formation of such planets is rather common throughout the Galaxy. On the other hand, the lessons about the formation processes of

planetary systems are revealing that many factors influence whether such planets can offer the stable, water-rich settings that we enjoy on Earth. It appears that for Earth's smaller sibling Mars, the likely rather special "Grand Tack" movement of the Jupiter–Saturn pair of giant planets swept nearby asteroids away so that Mars's growth was prematurely aborted for lack of building materials. However, what ended Mars's assembly may at the same time have provided the terrestrial planets with their ultimate complement of water. We have also seen that over the past billions of years the planetary system itself has been quite stable, with little dangerous debris in its inner regions. Whereas the habitability zone extends more or less from the orbits of Venus to Mars, Venus has lost its water because the Sun's ultraviolet light is so intense at Venus's distance that it led to its destruction on a molecular level, while Mars's gravity and likely its weak magnetism have allowed almost its entire atmosphere to evaporate off the planet. In addition to all these steps that shaped Earth's ultimate habitability there is the Earth's internal structure: geological activity is sufficiently moderate that life is not in jeopardy of extinction, but at the same time strong enough that it has continually injected carbon dioxide in a fairly slow and steady stream for billions of years to compensate the loss of that gas by chemical weathering. Moreover, Earth's Moon has likely stabilized Earth's spin axis although whether that is critically important to life we simply do not know. Finally, the Sun is a moderately bright star that enables planets to be in its habitable zone while remaining far enough away beyond the grip of its tidal forces; within the habitable zones of the faint M-type stars these tidal forces would inevitably work to synchronize spin and orbit and, as a consequence, would lead to extreme asymmetries between these planets' dayside and nightside. The evolutions of both Sun and Earth are sufficiently slow to have enabled life for billions of years, and likely will continue to do so for another one to two billion years. But then the Earth's water supply will likely run out, or maybe the atmospheric carbon dioxide levels

will drop below the level required for photosynthesis before that; either way, life as we know it will end there. Provided that the amalgamating "collision" with the Andromeda galaxy does not disrupt the Solar System before then, very roughly four billion years after tectonic motions fade out, Mercury and Venus will be engulfed by the Sun. The Earth may just also be engulfed, but even if not, it will at least be sterilized and likely partially vaporized by the Sun in its final giant stage before contracting into a white dwarf.

The prolonged transience of our yin-yang home

The yin-yang concept of Chinese philosophy embodies the complementary duality of apparent opposites that shape the world in their interdependence. One such duality is that of good versus bad, which is viewed as a perceived rather than an intrinsic contrast. In many ways, our existence on Earth reflects the yin and yang of the forces in the Universe as they are both beneficial and detrimental to life: the Sun's steady glow maintains Earth as a comfortable home in the otherwise hostile settings throughout the Solar System, but ultimately the Sun will sterilize if not swallow Earth; the solar wind keeps harmful cosmic rays at bay, but gradually erodes planetary atmospheres; the migration of the giant planets could have shepherded the terrestrial planets to their demise in the Sun's heat during the early phases of the Solar System, but in going only part of the way, they prompted delivery of water essential to life; volcanism can be locally disastrous but in contrast it releases carbon dioxide that, in the long run, sustains plant life.

After what we have seen in this book about planetary systems in space and time, it may seem that this duality of processes displays a preponderance of ending badly for us. Although that is true in the long run, our inaptitude with big numbers and our fascination with disaster stories plays a trick on us: we are

remarkably safe on this Earth, provided that we ourselves do not inflict irreparable damage to the planet's climate and ecosystem. After all, the last time a major asteroid destroyed a significant fraction of life on the planet appears to have been the Chicxulub impact, which happened some 65 million years ago and wiped out some three quarters of all plant and animal species, including most of the large dinosaurs. The oldest cities on Earth, such as Byblos, Damascus, and Jericho, are some 5,000 years old; the Chicxulub impact occurred 13,000 times longer ago than the foundation of these cities. Another type of hazard is that a relatively nearby star may explode as a supernova in the future; six aged, heavy stars have been identified as candidates within about 1,000 light years. Analysis of past such events suggests that any future ones do not appear to pose a significant danger to life, based on the identification of twenty-three supernovas that have likely happened over the past 300,000 years within that same distance range. The human race in its modern form as Homo sapiens sapiens has existed a mere 100,000 to 200,000 years. If as described above, Earth does indeed lose its water and carbon dioxide in a billion years or so, that leaves us 5,000 to 10,000 times more time with a habitable Earth than our species has existed in to date. In other words, apart from what we humans and our evolutionary successors may do to our planet, we are poised to enjoy our comfortable hermitage.

From the relative safety of Earth, scientists from an array of disciplines are embarking on another phase of galactic exploration: doubtless, many more exoplanetary systems, near and far, will be discovered, but the view is expanding toward understanding the origin and evolution of planetary atmospheres, and along with that the search for life outside Earth is intensifying. We have learned much about our planetary system by exploring it and by comparing it to the multitude of others that have been discovered. We have realized that rocky planets are rather common around the Galaxy, occurring by the billions. And yet, the many particular properties of Earth

and of its setting within the Solar System that are beneficial to, if not critical for, life in the single setting in which we know it exists caution us that the common occurrence in the Galaxy of Earth-mass planets at Sun–Earth-like distances from a star like that around which Earth orbits should not be taken to mean that Earth look-alikes are common. With its abundant life, and within the volume of space that is within reach of human travelers under the laws of physics as we presently know them, Earth, our cozy cocoon in the inhospitable immensity of space, is preciously unique.

Further Reading

Chapter 1: From One to Astronomical

Allegro, J. J.: 2016, "The bottom of the universe: Flat earth science in the Age of Encounter," *History of Science,* doi: 10.1177/0073275316681799.

Ashby, N.: 2002, "Relativity and the Global Positioning System," *Physics Today* May 2002.

Asimov, I.: 1989, "The relativity of wrong," *The Skeptical Inquirer* 14, 35.

Behroozi, P. & Peeples, M. S.: 2015, "On the history and future of cosmic planet formation," *Mon. Not. R. Astron. Soc.* 454, 1811, doi: 10.1093/mnras/stv1817, http://arxiv.org/abs/1508.01202.

Bessel, F. W.: 1838, "On the parallax of 61 Cygni," *Mon. Not. R. Astron. Soc.* 4, 152, doi: 10.1093/mnras/4.17.152.

Bornmann, L. & Mutz, R.: 2015, "Growth rates of modern science: A bibliometric analysis based on the number of publications and cited references," *J. Assn. Inf. Sci. Tec.* 66, 2215, doi: 10.1002/asi.23329, http://arxiv.org/abs/1402.4578.

Conselice, C. J., Wilkinson, A., Duncan, K., & Mortlock, A.: 2016, "The Evolution of Galaxy Number Density at z > 8 and Its Implications," *The Astrophysical Journal* 830, 83, doi: 10.3847/0004-637X/830/2/83, http://arxiv.org/abs/1607.03909.

Danielson, D. R.: 2001, "The great Copernican cliché," *American Journal of Physics* 69, 1029, doi: 10.1119/1.1379734.

Goldstein, B. R.: 1967, "The Arabic Version of Ptolomy's Planetary Hypotheses," *Trans. of the American Phil. Soc.* 57, 3.

Goldstein, B. R.: 2002, "Copernicus and the origin of his heliocentric system," *Journal for the History of Astronomy* 33, 219.

Granada, M. A.: 2008, "Kepler and Bruno on the Infinity of the Universe and of Solar Systems," *Journal for the History of Astronomy* 39, 469.

Guicciardini, N.: 1998, "Did Newton use his calculus in the Principia?," *Centaurus* 40, 303–344, doi: 10.1111/j.1600-0498.1998.tb00536.x.

Hoffleit, D. & Jaschek, C.: 1991, "*The Bright Star Catalogue,*" New Haven, CT: Yale University Observatory, 5th rev. edn.

Huygens, Chr.: 1699, *DE WERELD-BESCHOUWER VAN CHRISTI-AAN HUGENS, Of Gissingen over de Hemelsche Aard-klooten, en derzelver*

cieraad, geschreven aan KONSTANTIJN HUGENS, zijn Broeder. TWEEDE BOEK. (Cosmotheoros van 1698, vertaald door Pieter Rabus), Barend Bos, Rotterdam.

Newton, I.: 1962, *I. Newton MS. Add. 4005, fols. 21âL"22*, in A. R. Hall and M. Boas Hall (eds.), *Unpublished scientific papers of Isaac Newton: A selection from the Portsmouth Collection in the University Library Cambridge*, Cambridge University Press, Cambridge, 374.

Popper, K.: 1959, "The Logic of Scientific Discovery," Hutchinson and Company, London, UK.

Schrijver, C. J., Bagenal, F., & Sojka, J. J. (eds.): 2016, "Heliophysics: Active Stars, their Astrospheres, and Impacts on Planetary Environments," Cambridge University Press, Cambridge.

Tasker, E., Tan, J., Heng, K., Kane, S., Spiegel, D., Brasser, R., Casey, A., Desch, S., Dorn, C., Hernlund, J., Houser, C., Laneuville, M., Lasbleis, M., Libert, A.-S., Noack, L., Unterborn, C., & Wicks, J.: 2017, "The language of exoplanet ranking metrics needs to change," *Nature Astronomy* 1, 0042, doi: 10.1038/s41550-017-0042.

von Hippel, T. & von Hippel, C.: 2015, "To Apply or Not to Apply: A Survey Analysis of Grant Writing Costs and Benefits," *PLoS ONE* 10, e0118494, doi: 10.1371/journal.pone.0118494.

Warner, B.: 2005, "Thomas Henderson and α Centauri," in D. W. Kurtz (ed.), *IAU Colloq. 196: Transits of Venus: New Views of the Solar System and Galaxy*, p. 198.

Wuchty, S., Jones, B. F., & Uzzi, B.: 2007, "The increasing dominance of teams in production of knowledge," *Science* 316, 1036, doi: 10.1126/science.1136099.

Chapter 2: One Step Short of Life

Barnes, R., Meadows, V. S., & Evans, N.: 2015, "Comparative Habitability of Transiting Exoplanets," *The Astrophysical Journal* 814, 91, doi: 10.1088/0004-637X/814/2/91, http://arxiv.org/abs/1509.08922.

Deming, D., Wilkins, A., McCullough, P., Burrows, A., Fortney, J. J., Agol, E., Dobbs-Dixon, I., Madhusudhan, N., Crouzet, N., Desert, J.-M., Gilliland, R. L., Haynes, K., Knutson, H. A., Line, M., Magic, Z., Mandell, A. M., Ranjan, S., Charbonneau, D., Clampin, M., Seager, S., & Showman, A. P.: 2013, "Infrared Transmission Spectroscopy of the Exoplanets HD 209458b and XO-1b Using the Wide Field Camera-3 on the Hubble Space Telescope," *The Astrophysical Journal* 774, 95, doi: 10.1088/0004-637X/774/2/95, http://arxiv.org/abs/1302.1141.

Deming, L. D. & Seager, S.: 2017, "Illusion and reality in the atmospheres of exoplanets," *Journal of Geophysical Research (Planets)* 122, 53, doi: 10.1002/2016JE005155.

Domagal-Goldman, S. D., Wright, K. E., Adamala, K., Arina de la Rubia, L., Bond, J., Dartnell, L. R., Goldman, A. D., Lynch, K., Naud, M.-E., Paulino-Lima, I. G., Singer, K., Walter-Antonio, M., Abrevaya, X. C., Anderson, R., Arney, G., Atri, D., Azúa-Bustos, A., Bowman, J. S., Brazelton, W. J., Brennecka, G. A., Carns, R., Chopra, A., Colangelo-Lillis, J., Crockett, C. J., DeMarines, J., Frank, E. A., Frantz, C., de la Fuente, E., Galante, D., Glass, J., Gleeson, D., Glein, C. R., Goldblatt, C., Horak, R., Horodyskyj, L., Kaçar, B., Kereszturi, A., Knowles, E., Mayeur, P., McGlynn, S., Miguel, Y., Montgomery, M., Neish, C., Noack, L., Rugheimer, S., Stüeken, E. E., Tamez-Hidalgo, P., Walker, S. I., & Wong, T.: 2016, "The Astrobiology Primer v2.0," *Astrobiology* 16, 561, doi: 10.1089/ast.2015.1460.

Drake, F.: 2011, "The search for extra-terrestrial intelligence," *Phil. Trans. R. Soc. A* 369, 633, doi: 10.1098/rsta.2010.0282.

Drake, F. & Sobel, D.: 1992, *Is anyone out there? The scientific search for extraterrestrial intelligence*, New York, Delacorte Press.

Drake, F. & Sobel, D.: 2010, *Astronomy Beat* 46, page 1. Publisher: Astron. Soc. of the Pacific, San Francisco.

Editorial: 2009, "SETI at 50," *Nature* 461, 316, doi: 10.1038/461316a.

Jones, E. M.: 1985, "Where is everybody? an account of Fermi's question," NASA STI/Recon Technical Report N 85.

Kane, S. R., Hill, M. L., Kasting, J. F., Kopparapu, R. K., Quintana, E. V., Barclay, T., Batalha, N. M., Borucki, W. J., Ciardi, D. R., Haghighipour, N., Hinkel, N. R., Kaltenegger, L., Selsis, F., & Torres, G.: 2016, "A Catalog of Kepler Habitable Zone Exoplanet Candidates," *The Astrophysical Journal* 830, 1, doi: 10.3847/0004-637X/830/1/1, http://arxiv.org/abs/1608.00620.

Madhusudhan, N., Agúndez, M., Moses, J. I., & Hu, Y.: 2016, "Exoplanetary Atmospheres—Chemistry, Formation Conditions, and Habitability," *Space Sci. Rev.,* 205, 285, doi: 10.1007/s11214-016-0254-3, http://arxiv.org/abs/1604.06092.

Schrijver, K. & Schrijver, I.: 2015, *Living with the stars: How the Human Body Is Connected to the Life Cycles of the Earth, the Planets, and the Stars,* Oxford University Press, Oxford.

Tirard, S., Morange, M., & Lazcano, A.: 2010, "The Definition of Life: A Brief History of an Elusive Scientific Endeavor," *Astrobiology* 10, 1003, doi: 10.1089/ast.2010.0535.

Ward, P. & Brownlee, D.: 2000, "Rare Earth: Why Complex Life is Uncommon in the Universe," Copernicus, Springer-Verlag, New York.

Whitmire, D. P.: 2017, "Implications of our technological species being first and early," *International Journal of Astrobiology*, doi: 10.1017/S1473550417000271.

Zackrisson, E., Calissendorff, P., Gonzalez, J., Benson, A., Johansen, A., & Janson, M.: 2016, "Terrestrial planets across space and time," *The Astrophysical Journal* 833(2), 214.

Chapter 3: Exploring the Solar System

Aldrin, E. A.: 1969, "Lunar dust smelled just like gunpowder," *Life* issue of August 22.

Armstrong, N. A.: 1969, "The moon had been awaiting us a long time," *Life* issue of August 22.

Buchwald, V. F.: 1992, "On the use of iron by the Eskimos of Greenland," *Materials Characterization* 29, 139, doi: 10.1016/1044-5803(92)90112-U.

Cernan, E. & Davis, D. A.: 2000, "The Last Man on the Moon," St Martin's Griffin, New York (for an interview, see https://www.jsc.nasa.gov/history/oral_histories/CernanEA/CernanEA_12- 11-07.htm).

Cochran, A. L., Levasseur-Regourd, A.-C., Cordiner, M., Hadamcik, E., Lasue, J., Gicquel, A., Schleicher, D. G., Charnley, S. B., Mumma, M. J., Paganini, L., Bockelée-Morvan, D., Biver, N., & Kuan, Y.-J.: 2015, "The Composition of Comets," *Space Sci. Rev.* 197, 9, doi: 10.1007/s11214-015-0183-6, http://arxiv.org/abs/1507.00761.

Connolly, H. C.: 2005, "From Stars to Dust: Looking into a Circumstellar Disk Through Chondritic Meteorites," *Science* 307, 75, doi: 10.1126/science.1108284.

Downs, C., Linker, J. A., Mikić, Z., Riley, P., Schrijver, C. J., & Saint-Hilaire, P.: 2013, "Probing the Solar Magnetic Field with a Sun-Grazing Comet," *Science* 340, 1196, doi: 10.1126/science.1236550.

Heller, R. & Barnes, R.: 2015, "Runaway greenhouse effect on exomoons due to irradiation from hot, young giant planets," *International Journal of Astrobiology* 14, 335, doi: 10.1017/S1473550413000463, http://arxiv.org/abs/1311.0292.

Hughes, D. W.: 2001, "Six stages in the history of the astronomical unit," *Journal of Astronomical History and Heritage* 4, 15.

Hughes, D. W.: 2002, "Measuring the Moon's mass (125th Anniversary Review)," *The Observatory* 122, 61.

Luu, J. X. & Jewitt, D. C.: 2002, "Kuiper Belt Objects: Relics from the Accretion Disk of the Sun," *Ann. Rev. Astron. Astrophys.* 40, 63, doi: 10.1146/annurev.astro.40.060401.093818.

Olmsted, J. W.: 1942, "The scientific expedition of Jean Richer to Cayenne (1672–1673)," *Isis. Journal of the History of Science Society* 34, 117.

Sawyer Hogg, H.: 1962, "Out of Old Books. (The Tunguska Meteoric Event)," *Journal of the Royal Astronomical Society of Canada* 56, 174.

Schrijver, C. J., Bagenal, F., & Sojka, J. J. (eds.): 2016, "Heliophysics: Active Stars, their Astrospheres, and Impacts on Planetary Environments," Cambridge University Press, Cambridge.

Schrijver, C. J., Brown, J. C., Battams, K., Saint-Hilaire, P., Liu, W., Hudson, H., & Pesnell, W. D.: 2012, "Destruction of Sun-Grazing Comet C/2011 N3 (SOHO). Within the Low Solar Corona," *Science* 335, 324, doi: 10.1126/science.1211688.

Schrijver, C. J. & Siscoe, G. L. (eds.): 2011, *Heliophysics: Plasma Physics of the Local Cosmos*, Cambridge University Press, Cambridge.

Schrijver, C. J. & Siscoe, G. L. (eds.): 2012, *Heliophysics: Evolving Solar Activity and the Climates of Space and Earth*, Cambridge University Press, Cambridge.

Schrijver, K. & Schrijver, I.: 2015, *Living with the Stars: How the Human Body Is Connected to the Life Cycles of the Earth, the Planets, and the Stars*, Oxford University Press, Oxford.

Sekanina, Z., Chodas, P. W., & Yeomans, D. K.: 1998, "Secondary fragmentation of comet Shoemaker–Levy 9 and the ramifications for the progenitor's breakup in July 1992," *Planet. Space Sci.* 46, 21, doi: 10.1016/S0032-0633(97)00115-3.

van Dishoeck, E. F., Bergin, E. A., Lis, D. C., & Lunine, J. I.: 2014, "Water: From Clouds to Planets," Protostars and Planets VI 835–858, doi: 10.2458/azu_uapress_9780816531240-ch036, http://arxiv.org/abs/1401.8103.

van Helden, A.: 1985, "Cassini, Flamsteed, and Halley on the Dimensions of the Solar System, in Bulletin of the American Astronomical Society," vol. 17 of *Bull. Amer. Astron. Soc.*, 844.

Walsh, K. J., Morbidelli, A., Raymond, S. N., O'Brien, D. P., & Mandell, A. M.: 2011, "A low mass for Mars from Jupiter's early gas-driven migration," *Nature* 475, 206, doi: 10.1038/nature10201, http://arxiv.org/abs/1201.5177.

Wood, J. A.: 2005, "The Chondrite Types and Their Origins, in A. N. Krot, E. R. D. Scott, and B. Reipurth (eds.), Chondrites and the

Protoplanetary Disk," vol. 341 of *Astronomical Society of the Pacific Conference Series*, 953.

Zellner, N. E. B.: 2017, "Cataclysm No More: New Views on the Timing and Delivery of Lunar Impactors," Origins of Life and Evolution of the Biosphere, doi: 10.1007/s11084-017-9536-3, http://arxiv.org/abs/1704.06694.

Chapter 4: Exoplanet Systems and Their Stars

Baines, E. K., McAlister, H. A., ten Brummelaar, T. A., Turner, N. H., Sturmann, J., Sturmann, L., Goldfinger, P. J., & Ridgway, S. T.: 2008, CHARA "Array Measurements of the Angular Diameters of Exoplanet Host Stars," *The Astrophysical Journal* 680, 728, doi: 10.1086/588009, http://arxiv.org/abs/0803.1411.

Baraffe, I., Chabrier, G., & Barman, T.: 2010, "The physical properties of extra-solar planets," *Reports on Progress in Physics* 73(1), 016901, doi: 10.1088/0034-4885/73/1/016901, http://arxiv.org/abs/1001.3577.

Bedell, M., Bean, J. L., Meléndez, J., Mills, S. M., Fabrycky, D. C., Freitas, F. C., Ramírez, I., Asplund, M., Liu, F., & Yong, D.: 2017, "Kepler-11 is a Solar Twin: Revising the Masses and Radii of Benchmark Planets via Precise Stellar Characterization," *The Astrophysical Journal* 839(2), 94.

Campante, T. L., Barclay, T., Swift, J. J., Huber, D., Adibekyan, V. Z., Cochran, W., Burke, C. J., Isaacson, H., Quintana, E. V., Davies, G. R., Silva Aguirre, V., Ragozzine, D., Riddle, R., Baranec, C., Basu, S., Chaplin, W. J., Christensen-Dalsgaard, J., Metcalfe, T. S., Bedding, T. R., Handberg, R., Stello, D., Brewer, J. M., Hekker, S., Karoff, C., Kolbl, R., Law, N. M., Lundkvist, M., Miglio, A., Rowe, J. F., Santos, N. C., Van Laerhoven, C., Arentoft, T., Elsworth, Y. P., Fischer, D. A., Kawaler, S. D., Kjeldsen, H., Lund, M. N., Marcy, G. W., Sousa, S. G., Sozzetti, A., & White, T. R.: 2015, "An Ancient Extrasolar System with Five Sub-Earth-size Planets," *The Astrophysical Journal* 799, 170, doi: 10.1088/0004-637X/799/2/170, http://arxiv.org/abs/1501.06227.

Campante, T. L., Lund, M. N., Kuszlewicz, J. S., Davies, G. R., Chaplin, W. J., Albrecht, S., Winn, J. N., Bedding, T. R., Benomar, O., Bossini, D., Handberg, R., Santos, A. R. G., Van Eylen, V., Basu, S., Christensen-Dalsgaard, J., Elsworth, Y. P., Hekker, S., Hirano, T., Huber, D., Karoff, C., Kjeldsen, H., Lundkvist, M. S., North, T. S. H., Silva Aguirre, V., Stello, D., & White, T. R.: 2016, "Spin–Orbit Alignment of Exoplanet Systems: Ensemble Analysis Using

Asteroseismology," *The Astrophysical Journal* 819, 85, doi: 10.3847/0004-637X/819/1/85, http://arxiv.org/abs/1601.06052.

Cassan, A., Kubas, D., Beaulieu, J.-P., Dominik, M., Horne, K., Greenhill, J., Wambsganss, J., Menzies, J., Williams, A., Jørgensen, U. G., Udalski, A., Bennett, D. P., Albrow, M. D., Batista, V., Brillant, S., Caldwell, J. A. R., Cole, A., Coutures, C., Cook, K. H., Dieters, S., Prester, D. D., Donatowicz, J., Fouqué, P., Hill, K., Kains, N., Kane, S., Marquette, J.-B., Martin, R., Pollard, K. R., Sahu, K. C., Vinter, C., Warren, D., Watson, B., Zub, M., Sumi, T., Szymański, M. K., Kubiak, M., Poleski, R., Soszynski, I., Ulaczyk, K., Pietrzyński, G., & Wyrzykowski, Ł.: 2012, "One or more bound planets per Milky Way star from microlensing observations," *Nature* 481, 167, doi: 10.1038/nature10684, http://arxiv.org/abs/1202.0903.

Charbonneau, D., Brown, T. M., Latham, D. W., & Mayor, M.: 2000, "Detection of Planetary Transits Across a Sun-like Star," *Astrophysical Journal Letters* 529, L45, doi: 10.1086/312457, http://arxiv.org/abs/astro-ph/9911436.

Coleman, G. A. L. & Nelson, R. P.: 2016, "Giant planet formation in radially structured protoplanetary discs," *Mon. Not. R. Astron. Soc.* 460, 2779, doi: 10.1093/mnras/stw1177, http://arxiv.org/abs/1604.05191.

Doyle, L. R., Carter, J. A., Fabrycky, D. C., Slawson, R. W., Howell, S. B., Winn, J. N., Orosz, J. A., Prsa, A., Welsh, W. F., Quinn, S. N., Latham, D., Torres, G., Buchhave, L. A., Marcy, G. W., Fortney, J. J., Shporer, A., Ford, E. B., Lissauer, J. J., Ragozzine, D., Rucker, M., Batalha, N., Jenkins, J. M., Borucki, W. J., Koch, D., Middour, C. K., Hall, J. R., McCauliff, S., Fanelli, M. N., Quintana, E. V., Holman, M. J., Caldwell, D. A., Still, M., Stefanik, R. P., Brown, W. R., Esquerdo, G. A., Tang, S., Furesz, G., Geary, J. C., Berlind, P., Calkins, M. L., Short, D. R., Steffen, J. H., Sasselov, D., Dunham, E. W., Cochran, W. D., Boss, A., Haas, M. R., Buzasi, D., & Fischer, D.: 2011, "Kepler-16: A Transiting Circumbinary Planet," *Science* 333, 1602, doi: 10.1126/science.1210923, http://arxiv.org/abs/1109.3432.

Gillon, M., Triaud, A. H. M. J., Demory, B., Jehin, E., Agol, E., Deck, K. M., Lederer, S. M., de Wit, J., Burdanov, A., Ingalls, J. G., Bolmont, E, Leconte, J., Raymond, S. N., Selsis, F., Turbet, M., Barkaoui, K., Burgasser, A., Burleigh, M. R., Carey, S. J., Chaushev, A., Copperwheat, C. M., Delrez, L., Fernandes, C. S., Holdsworth, D. L., Kotze, E. J., Van Grootel, V., Almleaky, Y., Benkhaldoun, Z., Magain, P., & Queloz, D.: 2017, "Seven temperate terrestrial planets around the

nearby ultracool dwarf star TRAPPIST-1," *Nature* 542, 456, doi: 10.1038/ nature21360.

Hatzes, A. P., Guenther, E. W., Endl, M., Cochran, W. D., Döllinger, M. P., & Bedalov, A.: 2005, "A giant planet around the massive giant star HD 13189," *A&A* 437, 743, doi: 10.1051/0004-6361:20052850.

Henry, G. W., Marcy, G. W., Butler, R. P., & Vogt, S. S.: 2000, "A Transiting '51 Peg-like' Planet," *Astrophysical Journal Letters* 529, L41, doi: 10.1086/312458.

Kuzuhara, M., Tamura, M., Kudo, T., Janson, M., Kandori, R., Brandt, T. D., Thalmann, C., Spiegel, D., Biller, B., Carson, J., Hori, Y., Suzuki, R., Burrows, A., Henning, T., Turner, E. L., McElwain, M. W., Moro-Martín, A., Suenaga, T., Takahashi, Y. H., Kwon, J., Lucas, P., Abe, L., Brandner, W., Egner, S., Feldt, M., Fujiwara, H., Goto, M., Grady, C. A., Guyon, O., Hashimoto, J., Hayano, Y., Hayashi, M., Hayashi, S. S., Hodapp, K. W., Ishii, M., Iye, M., Knapp, G. R., Matsuo, T., Mayama, S., Miyama, S., Morino, J.-I., Nishikawa, J., Nishimura, T., Kotani, T., Kusakabe, N., Pyo, T.-S., Serabyn, E., Suto, H., Takami, M., Takato, N., Terada, H., Tomono, D., Watanabe, M., Wisniewski, J. P., Yamada, T., Takami, H., & Usuda, T.: 2013, "Direct Imaging of a Cold Jovian Exoplanet in Orbit around the Sun-like Star GJ 504," *The Astrophysical Journal* 774, 11, doi: 10.1088/0004-637X/774/1/11, http://arxiv.org/abs/1307.2886.

Lada, C. J.: 2006, "Stellar Multiplicity and the Initial Mass Function: Most Stars Are Single," *Astrophysical Journal Letters* 640, L63, doi: 10.1086/503158, http://arxiv.org/abs/astro-ph/0601375.

Lammer, H., Blanc, M., Benz, W., Fridlund, M., Foresto, V. C. d., Güdel, M., Rauer, H., Udry, S., Bonnet, R.-M., Falanga, M., Charbonneau, D., Helled, R., Kley, W., Linsky, J., Elkins-Tanton, L. T., Alibert, Y., Chassefière, E., Encrenaz, T., Hatzes, A. P., Lin, D., Liseau, R., Lorenzen, W., & Raymond, S. N.: 2013, "The Science of Exoplanets and Their Systems," *Astrobiology* 13, 793, doi: 10.1089/ast.2013.0997.

Léger, A., Grasset, O., Fegley, B., Codron, F., Albarede, A. F., Barge, P., Barnes, R., Cance, P., Carpy, S., Catalano, F., Cavarroc, C., Demangeon, O., Ferraz-Mello, S., Gabor, P., Grießmeier, J.-M., Leibacher, J., Libourel, G., Maurin, A.-S., Raymond, S. N., Rouan, D., Samuel, B., Schaefer, L., Schneider, J., Schuller, P. A., Selsis, F., & Sotin, C.: 2011, "The extreme physical properties of the CoRoT-7b super-Earth," *Icarus* 213, 1, doi: 10.1016/j.icarus.2011.02.004, http://arxiv.org/abs/1102.1629.

Marois, C., Zuckerman, B., Konopacky, Q. M., Macintosh, B., & Barman, T.: 2010, "Images of a fourth planet orbiting HR 8799," Nature 468, 1080, doi: 10.1038/nature09684, http://arxiv.org/abs/1011.4918.

Neveu-VanMalle, M., Queloz, D., Anderson, D. R., Brown, D. J. A., Collier Cameron, A., Delrez, L., Díaz, R. F., Gillon, M., Hellier, C., Jehin, E., Lister, T., Pepe, F., Rojo, P., Ségransan, D., Triaud, A. H. M. J., Turner, O. D., & Udry, S.: 2016, "Hot Jupiters with relatives: discovery of additional planets in orbit around WASP-41 and WASP-47," A&A 586, A93, doi: 10.1051/0004-6361/201526965, http://arxiv.org/abs/1509.07750.

Sanchis-Ojeda, R., Rappaport, S., Pallè, E., Delrez, L., DeVore, J., Gandolfi, D., Fukui, A., Ribas, I., Stassun, K. G., Albrecht, S., Dai, F., Gaidos, E., Gillon, M., Hirano, T., Holman, M., Howard, A. W., Isaacson, H., Jehin, E., Kuzuhara, M., Mann, A. W., Marcy, G. W., Miles-Páez, P. A., Montañés-Rodríguez, P., Murgas, F., Narita, N., Nowak, G., Onitsuka, M., Paegert, M., Van Eylen, V., Winn, J. N., & Yu, L.: 2015, "The K2-ESPRINT Project I: Discovery of the Disintegrating Rocky Planet K2-22b with a Cometary Head and Leading Tail," The Astrophysical Journal 812, 112, doi: 10.1088/0004-637X/812/2/112, http://arxiv.org/abs/1504.04379.

Schrijver, C. J., Bagenal, F., & Sojka, J. J. (eds.): 2016, Heliophysics: Active Stars, their Astrospheres, and Impacts on Planetary Environments, Cambridge University Press, Cambridge.

Schrijver, C. J. & Siscoe, G. L. (eds.): 2012, Heliophysics: Evolving Solar Activity and the Climates of Space and Earth, Cambridge University Press, Cambridge.

Schuler, S. C., Kim, J. H., Tinker, Jr., M. C., King, J. R., Hatzes, A. P., & Guenther, E. W.: 2005, "High-Resolution Spectroscopy of the Planetary Host HD 13189: Highly Evolved and Metal-poor," Astrophysical Journal Letters 632, L131, doi: 10.1086/497988, http://arxiv.org/abs/astro-ph/0509270.

Silva Aguirre, V., Davies, G. R., Basu, S., Christensen-Dalsgaard, J., Creevey, O., Metcalfe, T. S., Bedding, T. R., Casagrande, L., Handberg, R., Lund, M. N., Nissen, P. E., Chaplin, W. J., Huber, D., Serenelli, A. M., Stello, D., Van Eylen, V., Campante, T. L., Elsworth, Y., Gilliland, R. L., Hekker, S., Karoff, C., Kawaler, S. D., Kjeldsen, H., & Lundkvist, M. S.: 2015, "Ages and fundamental properties of Kepler exoplanet host stars from asteroseismology," Mon. Not. R. Astron. Soc. 452, 2127, doi: 10.1093/mnras/stv1388, http://arxiv.org/abs/1504.07992.

Silvotti, R., Schuh, S., Janulis, R., Solheim, J.-E., Bernabei, S., Østensen, R., Oswalt, T. D., Bruni, I., Gualandi, R., Bonanno, A., Vauclair, G., Reed, M., Chen, C.-W., Leibowitz, E., Paparo, M., Baran, A., Charpinet, S., Dolez, N., Kawaler, S., Kurtz, D., Moskalik, P., Riddle, R., & Zola, S.: 2007, "A giant planet orbiting the 'extreme horizontal branch' star V391 Pegasi," *Nature* 449, 189, doi: 10.1038/nature06143.

Tokovinin, A., Thomas, S., Sterzik, M., & Udry, S.: 2006, "Tertiary companions to close spectroscopic binaries," *A&A* 450, 681, doi: 10.1051/0004-6361:20054427, http://arxiv.org/abs/astro-ph/0601518.

Wells, R., Poppenhaeger, K., Watson, C. A., & Heller, R.: 2018, "Transit visibility zones of the Solar system planets," *Mon. Not. R. Astron. Soc.* 473, 345, doi: 10.1093/mnras/stx2077, http://arxiv.org/abs/1709.02211.

Winn, J. N. & Fabrycky, D. C.: 2015, "The Occurrence and Architecture of Exoplanetary Systems," *Ann. Rev. Astron. Astrophys.* 53, 409, doi: 10.1146/annurev-astro-082214-122246, http://arxiv.org/abs/1410.4199.

Winters, J.: 1996, "The planet at 51 Peg," *Discover Magazine* isse for January.

Wolszczan, A.: 2008, "Fifteen years of the neutron star planet research," *Physica Scripta* Volume T 130(1), 014005, doi: 10.1088/0031-8949/2008/T130/014005.

Wolszczan, A. & Frail, D. A.: 1992, "A planetary system around the millisecond pulsar PSR1257 + 12," *Nature* 355, 145, doi: 10.1038/355145a0.

Chapter 5: The Birth of Stars and Planets

Akiyama, E., Hasegawa, Y., Hayashi, M., & Iguchi, S.: 2016, "Planetary System Formation in the Protoplanetary Disk around HL Tauri," *The Astrophysical Journal* 818, 158, doi: 10.3847/0004-637X/818/2/158, http://arxiv.org/abs/1511.04822.

Anglés-Alcázar, D., Faucher-Giguère, C.-A., Kereš, D., Hopkins, P. F., Quataert, E., & Murray, N.: 2017, "The cosmic baryon cycle and galaxy mass assembly in the FIRE simulations," *Mon. Not. R. Astron. Soc.* 470, 4698, doi: 10.1093/mnras/stx1517, http://arxiv.org/abs/1610.08523.

Arny, T.: 1990, "The star makers: A history of the theories of stellar structure and evolution," *Vistas in Astronomy* 33, 211, doi: 10.1016/0083-6656(90)90021-Y.

Bartusiak, M.: 2004, *Archives of the Universe: A Treasury of astronomy's historic works of discovery*, Pantheon Books, New York.

Basri, G.: 2004, "The discovery of brown dwarfs," *Scientific American* issue of September 1 (updated from the April 2000 issue).

Benz, W., Ida, S., Alibert, Y., Lin, D., & Mordasini, C.: 2014, "Planet Population Synthesis," *Protostars and Planets VI* 691–713, doi: 10.2458/ azu_uapress_9780816531240-ch030, http://arxiv.org/abs/ 1402.7086.

Chabrier, G., Johansen, A., Janson, M., & Rafikov, R.: 2014, "Giant Planet and Brown Dwarf Formation," *Protostars and Planets VI* 619–642, doi: 10.2458/azu_uapress_9780816531240-ch027, http://arxiv.org/ abs/ 1401.7559.

Clerke, A. M.: 1890, "The System of the Stars," Longmans, Green, and Co., London.

Desch, S. J.: 2007, "Mass Distribution and Planet Formation in the Solar Nebula," *The Astrophysical Journal* 671, 878, doi: 10.1086/522825.

Doherty, C. L., Gil-Pons, P., Siess, L., Lattanzio, J. C., & Lau, H. H. B.: 2015, "Super- and massive AGB stars—IV. Final fates—initial-to-final mass relation," *Mon. Not. R. Astron. Soc.* 446, 2599, doi: 10.1093/mnras/stu2180, http://arxiv.org/abs/1410.5431.

Eddington, A.: 1920, "The internal constitution of the stars," *The Observatory* 43, 341.

Haisch, Jr., K. E., Lada, E. A., & Lada, C. J.: 2001, "Disk Frequencies and Lifetimes in Young Clusters," *Astrophysical Journal Letters* 553, L153, doi: 10.1086/320685, http://arxiv.org/abs/astro-ph/0104347.

Helled, R., Bodenheimer, P., Podolak, M., Boley, A., Meru, F., Nayakshin, S., Fortney, J. J., Mayer, L., Alibert, Y., & Boss, A. P.: 2014, "Giant Planet Formation, Evolution, and Internal Structure," *Protostars and Planets VI* 643–665, doi: 10.2458/azu_uapress_9780816531240-ch028, http://arxiv.org/abs/1311.1142.

Joergens, V., Bonnefoy, M., Liu, Y., Bayo, A., Wolf, S., Chauvin, G., & Rojo, P.: 2013, "OTS44: Disk and accretion at the planetary border," *A&A* 558, L7, doi: 10.1051/0004-6361/201322432, http://arxiv.org/ abs/1310.1936.

Kenyon, S. J., Najita, J. R., & Bromley, B. C.: 2016, "Rocky Planet Formation: Quick and Neat," *The Astrophysical Journal* 831, 8, doi: 10.3847/0004-637X/831/1/8.

Lecar, M., Podolak, M., Sasselov, D., & Chiang, E.: 2006, "On the Location of the Snow Line in a Protoplanetary Disk," *The Astrophysical Journal* 640, 1115, doi: 10.1086/500287, http://arxiv.org/abs/astro-ph/0602217.

Li, Z.-Y., Banerjee, R., Pudritz, R. E., Jørgensen, J. K., Shang, H., Krasnopolsky, R., & Maury, A.: 2014, "The Earliest Stages of Star and Planet Formation: Core Collapse, and the Formation of Disks and Outflows," *Protostars and Planets VI* 173–194, doi: 10.2458/azu_ uapress_9780816531240-ch008, http://arxiv.org/abs/1401.2219.

Lin, D. N. C., Bodenheimer, P., & Richardson, D. C.: 1996, "Orbital migration of the planetary companion of 51 Pegasi to its present location," *Nature* 380, 606, doi: 10.1038/380606a0.

Ma, B. & Ge, J.: 2014, "Statistical properties of brown dwarf companions: implications for different formation mechanisms," *Mon. Not. R. Astron. Soc.* 439, 2781, doi: 10.1093/mnras/stu134, http://arxiv.org/abs/1303.6442.

Marboeuf, U., Thiabaud, A., Alibert, Y., Cabral, N., & Benz, W.: 2014, "From stellar nebula to planetesimals," *A&A* 570, A35, doi: 10.1051/0004-6361/201322207, http://arxiv.org/abs/1407.7271.

McKee, C. F. & Ostriker, E. C.: 2007, "Theory of Star Formation," *Ann. Rev. Astron. Astrophys.* 45, 565, doi: 10.1146/annurev.astro.45.051806.110602, http://arxiv.org/abs/0707.3514.

Morbidelli, A., Lunine, J. I., O'Brien, D. P., Raymond, S. N., & Walsh, K. J.: 2012, "Building Terrestrial Planets," *Annual Review of Earth and Planetary Sciences* 40, 251, doi: 10.1146/annurev-earth-042711-105319, http://arxiv.org/abs/1208.4694.

Moulton, F. R.: 1905, "On the Evolution of the Solar System," *The Astrophysical Journal* 22, 165, doi: 10.1086/141260.

Murray, N.: 2011, "Star Formation Efficiencies and Lifetimes of Giant Molecular Clouds in the Milky Way," *The Astrophysical Journal* 729, 133, doi: 10.1088/0004-637X/729/2/133, http://arxiv.org/abs/1007.3270.

Payne, C. H.: 1925, Ph.D. thesis, Radcliffe College (an interview on her experiences can be found at https://www.aip.org/history-programs/niels-bohr-library/oral-histories/4620).

Perryman, M.: 2012, "The origin of the architecture of the solar system," *European Review* 20, 276, http://arxiv.org/abs/1111.1286.

Pfalzner, S., Davies, M. B., Gounelle, M., Johansen, A., Münker, C., Lacerda, P., Portegies Zwart, S., Testi, L., Trieloff, M., & Veras, D.: 2015, "The formation of the solar system," *Physica Scripta* 90(6), 068001, doi: 10.1088/0031-8949/90/6/068001, http://arxiv.org/abs/1501.03101.

Russell, S. S., Gounelle, M., & Hutchison, R.: 2001, "Origin of short-lived radionuclides," *Philosophical Transactions of the Royal Society of London Series A* 359, 1991, doi: 10.1098/rsta.2001.0893.

Sagan, C.: 1980, *Cosmos*, Random House, New York.

Salo, H. & Laurikainen, E.: 1999, "A Multiple Encounter Model of M51," *Astrophysics and Space Science* 269, 663, doi: 10.1023/A:1017002909665.

Schrijver, C. J., Bagenal, F., & Sojka, J. J. (eds.): 2016, *Heliophysics: Active Stars, their Astrospheres, and Impacts on Planetary Environments*, Cambridge University Press, Cambridge.

Schrijver, C. J. & Siscoe, G. L. (eds.): 2012, *Heliophysics: Evolving Solar Activity and the Climates of Space and Earth*, Cambridge University Press, Cambridge.

Schrijver, K. & Schrijver, I.: 2015, *Living with the Stars. How the Human Body Is Connected to the Life Cycles of the Earth, the Planets, and the Stars*, Oxford University Press, Oxford.

Stassun, K. G., Feiden, G. A., & Torres, G.: 2014, "Empirical tests of pre-main-sequence stellar evolution models with eclipsing binaries," *New Astron. Rev.* 60, 1, doi: 10.1016/j.newar.2014.06.001, http://arxiv.org/abs/1406.3788.

Tokovinin, A., Thomas, S., Sterzik, M., & Udry, S.: 2006, "Tertiary companions to close spectroscopic binaries," *A&A* 450, 681, doi: 10.1051/0004-6361:20054427, http://arxiv.org/abs/astro-ph/0601518.

Udalski, A., Jung, Y. K., Han, C., Gould, A., Kozłowski, S., Skowron, J., Poleski, R., Soszyński, I., Pietrukowicz, P., Mróz, P., Szymański, M. K., Wyrzykowski, Ł., Ulaczyk, K., Pietrzyński, G., Shvartzvald, Y., Maoz, D., Kaspi, S., Gaudi, B. S., Hwang, K.-H., Choi, J.-Y., Shin, I.-G., Park, H., & Bozza, V.: 2015, "A Venus-mass Planet Orbiting a Brown Dwarf: A Missing Link between Planets and Moons," *The Astrophysical Journal* 812, 47, doi: 10.1088/0004-637X/812/1/47, http://arxiv.org/abs/1507.02388.

van Dishoeck, E. F., Bergin, E. A., Lis, D. C., & Lunine, J. I.: 2014, Water: "From Clouds to Planets," Protostars and Planets VI 835–858, doi: 10.2458/azu_uapress_9780816531240-ch036, http://arxiv.org/abs/1401.8103.

Winn, J. N. & Fabrycky, D. C.: 2015, "The Occurrence and Architecture of Exoplanetary Systems," *Ann. Rev. Astron. Astrophys.* 53, 409, doi: 10.1146/annurev-astro-082214-122246, http://arxiv.org/abs/1410.4199.

Wood, J. A.: 2005, "The Chondrite Types and Their Origins," in A. N. Krot, E. R. D. Scott, and B. Reipurth (eds.), *Chondrites and the Protoplanetary Disk*, vol. 341 of *Astronomical Society of the Pacific Conference Series*, 953.

Chapter 6: Drifting Through a Planetary System

Armitage, P. J.: 2015, "Physical processes in protoplanetary disks," ArXiv e-prints, http://arxiv.org/abs/1509.06382.

Baruteau, C., Crida, A., Paardekooper, S.-J., Masset, F., Guilet, J., Bitsch, B., Nelson, R., Kley, W., & Papaloizou, J.: 2014, "Planet–Disk Interactions and Early Evolution of Planetary Systems," Protostars and Planets VI 667–89, doi: 10.2458/azu_uapress_9780816531240-ch029, http://arxiv.org/abs/1312.4293.

Batygin, K. & Laughlin, G.: 2015, "Jupiter's decisive role in the inner Solar System's early evolution," *Proceedings of the National Academy of Science* 112, 4214, doi: 10.1073/pnas.1423252112, http://arxiv.org/abs/1503.06945.

Beaugé, C. & Nesvorný, D.: 2012, "Multiple-planet Scattering and the Origin of Hot Jupiters," *The Astrophysical Journal* 751, 119, doi: 10.1088/0004-637X/751/2/119, http://arxiv.org/abs/1110.4392.

Benz, W., Ida, S., Alibert, Y., Lin, D., & Mordasini, C.: 2014, "Planet Population Synthesis," Protostars and Planets VI 691–713, doi: 10.2458/azu_uapress_9780816531240-ch030, http://arxiv.org/abs/1402.7086.

Boley, A. C., Granados Contreras, A. P., & Gladman, B.: 2016, "The In Situ Formation of Giant Planets at Short Orbital Periods," *Astrophysical Journal Letters* 817, L17, doi: 10.3847/2041-8205/817/2/L17, http://arxiv.org/abs/1510.04276.

Coleman, G. A. L. & Nelson, R. P.: 2016, "Giant planet formation in radially structured protoplanetary discs," *Mon. Not. R. Astron. Soc.* 460, 2779, doi: 10.1093/mnras/stw1177, http://arxiv.org/abs/1604.05191.

Davies, M. B., Adams, F. C., Armitage, P., Chambers, J., Ford, E., Morbidelli, A., Raymond, S. N., & Veras, D.: 2014, "The Long-Term Dynamical Evolution of Planetary Systems," Protostars and Planets VI 787–808, doi: 10.2458/azu_uapress_9780816531240-ch034, http://arxiv.org/abs/1311.6816.

Desch, S. J.: 2007, "Mass Distribution and Planet Formation in the Solar Nebula," *The Astrophysical Journal* 671, 878, doi: 10.1086/522825.

Hasegawa, Y.: 2016, "Super-Earths as Failed Cores in Orbital Migration Traps," *The Astrophysical Journal* 832, 83, doi: 10.3847/0004-637X/832/1/83, http://arxiv.org/abs/1609.04798.

Hasegawa, Y. & Pudritz, R. E.: 2013, "Planetary Populations in the Mass-Period Diagram: A Statistical Treatment of Exoplanet Formation and the Role of Planet Traps," *The Astrophysical Journal* 778, 78, doi: 10.1088/0004-637X/778/1/78, http://arxiv.org/abs/1310.2009.

Hayashi, C., Nakazawa, K., & Adachi, I.: 1977, "Long-Term Behavior of Planetesimals and the Formation of the Planets," *Publ. Astron. Soc. Japan* 29, 163.

Helled, R., Bodenheimer, P., Podolak, M., Boley, A., Meru, F., Nayakshin, S., Fortney, J. J., Mayer, L., Alibert, Y., & Boss, A. P.: 2014, "Giant Planet Formation, Evolution, and Internal Structure," Protostars and Planets VI 643–65, doi: 10.2458/azu_uapress_9780816531240-ch028, http://arxiv.org/abs/1311.1142.

Kley, W. & Nelson, R. P.: 2012, "Planet–Disk Interaction and Orbital Evolution," *Ann. Rev. Astron. Astrophys.* 50, 211, doi: 10.1146/annurev-astro-081811-125523, http://arxiv.org/abs/1203.1184.

Lin, D. N. C., Bodenheimer, P., & Richardson, D. C.: 1996, "Orbital migration of the planetary companion of 51 Pegasi to its present location," *Nature* 380, 606, doi: 10.1038/380606a0.

Lin, D. N. C. & Papaloizou, J.: 1980, "On the structure and evolution of the primordial solar nebula," *Mon. Not. R. Astron. Soc.* 191, 37, doi: 10.1093/mnras/191.1.37.

Lin, D. N. C. & Papaloizou, J.: 1986, "On the tidal interaction between protoplanets and the protoplanetary disk. III—Orbital migration of protoplanets," *The Astrophysical Journal* 309, 846, doi: 10.1086/164653.

Lykawka, P. S. & Ito, T.: 2017, "Terrestrial planet formation: constraining the formation of Mercury," *The Astrophysical Journal* 838, 106.

Marois, C., Zuckerman, B., Konopacky, Q. M., Macintosh, B., & Barman, T.: 2010, "Images of a fourth planet orbiting HR 8799," *Nature* 468, 1080, doi: 10.1038/nature09684, http://arxiv.org/abs/1011.4918.

Morbidelli, A., Brasser, R., Gomes, R., Levison, H. F., & Tsiganis, K.: 2010, "Evidence from the Asteroid Belt for a Violent Past Evolution of Jupiter's Orbit," *Astron. J.* 140, 1391, doi: 10.1088/0004-6256/140/5/1391, http://arxiv.org/abs/1009.1521.

Morbidelli, A., Lunine, J. I., O'Brien, D. P., Raymond, S. N., & Walsh, K. J.: 2012, "Building Terrestrial Planets," *Annual Review of Earth and Planetary Sciences* 40, 251, doi: 10.1146/annurev-earth-042711-105319, http://arxiv.org/abs/1208.4694.

Morbidelli, A. & Raymond, S. N.: 2016, "Challenges in planet formation," *Journal of Geophysical Research (Planets)* 121, 1962, doi: 10.1002/2016JE005088, http://arxiv.org/abs/1610.07202.

Morbidelli, A., Tsiganis, K., Crida, A., Levison, H. F., & Gomes, R.: 2007, "Dynamics of the Giant Planets of the Solar System in the Gaseous Protoplanetary Disk and Their Relationship to the Current Orbital Architecture", *Astron. J.* 134, 1790, doi: 10.1086/521705, http://arxiv.org/abs/0706.1713.

Muto, T., Grady, C. A., Hashimoto, J., Fukagawa, M., Hornbeck, J. B., Sitko, M., Russell, R., Werren, C., Curé, M., Currie, T., Ohashi, N., Okamoto, Y., Momose, M., Honda, M., Inutsuka, S., Takeuchi, T., Dong, R., Abe, L., Brandner, W., Brandt, T., Carson, J., Egner, S., Feldt, M., Fukue, T., Goto, M., Guyon, O., Hayano, Y., Hayashi, M., Hayashi, S., Henning, T., Hodapp, K. W., Ishii, M., Iye, M., Janson,

Further Reading

M., Kandori, R., Knapp, G. R., Kudo, T., Kusakabe, N., Kuzuhara, M., Matsuo, T., Mayama, S., McElwain, M. W., Miyama, S., Morino, J.-I., Moro-Martin, A., Nishimura, T., Pyo, T.-S., Serabyn, E., Suto, H., Suzuki, R., Takami, M., Takato, N., Terada, H., Thalmann, C., Tomono, D., Turner, E. L., Watanabe, M., Wisniewski, J. P., Yamada, T., Takami, H., Usuda, T., & Tamura, M.: 2012, "Discovery of Small-scale Spiral Structures in the Disk of SAO 206462 (HD 135344B): Implications for the Physical State of the Disk from Spiral Density Wave Theory," *Astrophysical Journal Letters* 748, L22, doi: 10.1088/2041-8205/748/2/L22, http://arxiv.org/abs/1202.6139.

Nesvorný, D., Vokrouhlický, D., & Morbidelli, A.: 2007, "Capture of Irregular Satellites during Planetary Encounters," *Astron. J.* 133, 1962, doi: 10.1086/512850.

Ramirez, J. Peña, Olvera, L. A., Nijmeijer, H., & Alvarez, J.: 2015, "The sympathy of two pendulum clocks: beyond Huygens observations," *Scientific Reports* 6, 23580, doi: 10.1038/srep23580.

Raymond, S. N. & Izidoro, A.: 2017, "Origin of water in the inner Solar System: Planetesimals scattered inward during Jupiter and Saturn's rapid gas accretion," *Icarus* 297, 134, doi: 10.1016/j.icarus.2017.06.030, http://arxiv.org/abs/1707.01234.

Rubie, D. C., Jacobson, S. A., Morbidelli, A., O'Brien, D. P., Young, E. D., de Vries, J., Nimmo, F., Palme, H., & Frost, D. J.: 2015, "Accretion and differentiation of the terrestrial planets with implications for the compositions of early-formed Solar System bodies and accretion of water," *Icarus* 248, 89, doi: 10.1016/j.icarus.2014.10.015, http://arxiv.org/abs/1410.3509.

Teyssandier, J. & Terquem, C.: 2014, "Evolution of eccentricity and orbital inclination of migrating planets in 2:1 mean motion resonance," *Mon. Not. R. Astron. Soc.* 443, 568, doi: 10.1093/mnras/stu1137, http://arxiv.org/abs/1406.2189.

Walsh, K. J., Morbidelli, A., Raymond, S. N., O'Brien, D. P., & Mandell, A. M.: 2011, "A low mass for Mars from Jupiter's early gas-driven migration," *Nature* 475, 206, doi: 10.1038/nature10201, http://arxiv.org/abs/1201.5177.

Wilder, R. L.: 1967, "The Role of Intuition", *Science* 156, doi: 10.1126/science.156.3775.605.

Zhang, H. & Zhou, J.-L.: 2010, "On the Orbital Evolution of a Giant Planet Pair Embedded in a Gaseous Disk. I. Jupiter-Saturn Configuration," *The Astrophysical Journal* 714, 532, doi: 10.1088/0004-637X/714/1/532, http://arxiv.org/abs/1002.2201.

Chapter 7: Lone Rovers

Amelin, Y., Krot, A. N., Hutcheon, I. D., & Ulyanov, A. A.: 2002, "Lead Isotopic Ages of Chondrules and Calcium-Aluminum-Rich Inclusions," *Science* 297, 1678, doi: 10.1126/science.1073950.

Apai, D.: 2013, "Protoplanetary disks and planet formation around brown dwarfs and very low-mass stars," *Astronomische Nachrichten* 334, 57, doi: 10.1002/asna.201211780.

Barclay, T., Quintana, E. V., Raymond, S. N., & Penny, M. T.: 2017, "The Demographics of Rocky Free-Floating Planets and their Detectability by WFIRST," *The Astrophysical Journal* 841, 86, http://arxiv.org/abs/1704.08749.

Beaugé, C. & Nesvorný, D.: 2012, "Multiple-planet Scattering and the Origin of Hot Jupiters," *The Astrophysical Journal* 751, 119, doi: 10.1088/0004-637X/751/2/119, http://arxiv.org/abs/1110.4392.

Boss, A. P.: 2017, "Triggering Collapse of the Presolar Dense Cloud Core and Injecting Short-Lived Radioisotopes with a Shock Wave. V. Nonisothermal Collapse Regime," *The Astrophysical Journal* 833, 113, http://arxiv.org/abs/1706.09840.

Clanton, C. & Gaudi, B. S.: 2017, "Constraining the Frequency of Free-floating Planets from a Synthesis of Microlensing, Radial Velocity, and Direct Imaging Survey Results," *The Astrophysical Journal* 834(1), 46.

Coles, P.: 2001, "Einstein, Eddington and the 1919 Eclipse," in V. J. Martínez, V. Trimble, and M. J. Pons-Bordería (eds.), *Historical Development of Modern Cosmology*, vol. 252 of *Astronomical Society of the Pacific Conference Series*, 21.

Davies, M. B., Adams, F. C., Armitage, P., Chambers, J., Ford, E., Morbidelli, A., Raymond, S. N., & Veras, D.: 2014, "The Long-Term Dynamical Evolution of Planetary Systems," Protostars and Planets VI 787–808, doi: 10.2458/azu_uapress_9780816531240-ch034, http://arxiv.org/abs/1311.6816.

Desch, S. J.: 2007, "Mass Distribution and Planet Formation in the Solar Nebula," *The Astrophysical Journal* 671, 878, doi: 10.1086/522825.

Drass, H., Haas, M., Chini, R., Bayo, A., Hackstein, M., Hoffmeister, V., Godoy, N., & Vogt, N.: 2016, "The bimodal initial mass function in the Orion nebula cloud," *Mon. Not. R. Astron. Soc.* 461, 1734, doi: 10.1093/mnras/stw1094, http://arxiv.org/abs/1605.03600.

Gorti, U., Liseau, R., Sándor, Z., & Clarke, C.: 2016, "Disk Dispersal: Theoretical Understanding and Observational Constraints," *Space Sci. Rev.* 205, 125, doi: 10.1007/s11214-015-0228-x, http://arxiv.org/abs/1512.04622.

Goswami, J. N. & Vanhala, H. A. T.: 2000, "Extinct Radionuclides and the Origin of the Solar System," Protostars and Planets IV 963.

Hester, J. J., Desch, S. J., Healy, K. R., & Leshin, L. A.: 2004, "The Cradle of the Solar System," Science 304, doi: 10.1126/science.1096808.

Johnstone, D., Hollenbach, D., & Bally, J.: 1998, "Photoevaporation of Disks and Clumps by Nearby Massive Stars: Application to Disk Destruction in the Orion Nebula", The Astrophysical Journal 499, 758, doi: 10.1086/305658.

Lamers, H. J. G. L. M., Gieles, M., Bastian, N., Baumgardt, H., Kharchenko, N. V., & Portegies Zwart, S.: 2005, "An analytical description of the disruption of star clusters in tidal fields with an application to Galactic open clusters," A&A 441, 117, doi: 10.1051/0004-6361:20042241, http://arxiv.org/abs/astro-ph/0505558.

Lee, T., Papanastassiou, D. A., & Wasserburg, G. J.: 1976, "Demonstration of Mg-26 excess in Allende and evidence for Al-26," Geophys. Res. Lett. 3, 41, doi: 10.1029/GL003i001p00041.

Looney, L. W., Tobin, J. J., & Fields, B. D.: 2006, "Radioactive Probes of the Supernova-contaminated Solar Nebula: Evidence that the Sun Was Born in a Cluster," The Astrophysical Journal 652, 1755, doi: 10.1086/508407, http://arxiv.org/abs/astro-ph/0608411.

Mao, S. & Paczynski, B.: 1991, "Gravitational microlensing by double stars and planetary systems," Astrophysical Journal Letters 374, L37, doi: 10.1086/186066.

McCrea, W. H.: 1979, "Einstein: Relations with the Royal Astronomical Society," Quarterly J. R. Astron. Soc. 20, 251.

Murray, N.: 2011, "Star Formation Efficiencies and Lifetimes of Giant Molecular Clouds in the Milky Way," The Astrophysical Journal 729, 133, doi: 10.1088/0004-637X/729/2/133, http://arxiv.org/abs/1007.3270.

Pfalzner, S., Davies, M. B., Gounelle, M., Johansen, A., Münker, C., Lacerda, P., Portegies Zwart, S., Testi, L., Trieloff, M., & Veras, D.: 2015, "The formation of the solar system," Physica Scripta 90(6), 068001, doi: 10.1088/0031-8949/90/6/068001, http://arxiv.org/abs/1501.03101.

Portegies Zwart, S. F.: 2009, "The Lost Siblings of the Sun," Astrophysical Journal Letters 696, L13, doi: 10.1088/0004-637X/696/1/L13, http://arxiv.org/abs/0903.0237.

Russell, S. S., Gounelle, M., & Hutchison, R.: 2001, "Origin of short-lived radionuclides," Philosophical Transactions of the Royal Society of London Series A 359, 1991, doi: 10.1098/rsta.2001.0893.

Scholz, A., Jayawardhana, R., Muzic, K., Geers, V., Tamura, M., & Tanaka, I.: 2012, "Substellar Objects in Nearby Young Clusters (SONYC). VI. The Planetary-mass Domain of NGC 1333," *The Astrophysical Journal* 756, 24, doi: 10.1088/0004-637X/756/1/24, http://arxiv.org/abs/1207.1449.

Schrijver, C. J.: 2009, "On a Transition from Solar-Like Coronae to Rotation-Dominated Jovian-Like Magnetospheres in Ultracool Main-Sequence Stars," *Astrophysical Journal Letters* 699, L148, doi: 10.1088/0004-637X/699/2/L148, http://arxiv.org/abs/0905.1354.

Schrijver, K. & Schrijver, I.: 2015, *Living with the Stars: How the Human Body Is Connected to the Life Cycles of the Earth, the Planets, and the Stars*, Oxford University Press, Oxford.

Strigari, L. E., Barnabè, M., Marshall, P. J., & Blandford, R. D.: 2012, "Nomads of the Galaxy," *Mon. Not. R. Astron. Soc.* 423, 1856, doi: 10.1111/j.1365-2966.2012.21009.x, http://arxiv.org/abs/1201.2687.

Sumi, T., Kamiya, K., Bennett, D. P., Bond, I. A., Abe, F., Botzler, C. S., Fukui, A., Furusawa, K., Hearnshaw, J. B., Itow, Y., Kilmartin, P. M., Korpela, A., Lin, W., Ling, C. H., Masuda, K., Matsubara, Y., Miyake, N., Motomura, M., Muraki, Y., Nagaya, M., Nakamura, S., Ohnishi, K., Okumura, T., Perrott, Y. C., Rattenbury, N., Saito, T., Sako, T., Sullivan, D. J., Sweatman, W. L., Tristram, P. J., Udalski, A., Szymański, M. K., Kubiak, M., Pietrzyński, G., Poleski, R., Soszyński, I., Wyrzykowski, Ł., Ulaczyk, K., & Microlensing Observations in Astrophysics (MOA) Collaboration: 2011, "Unbound or distant planetary mass population detected by gravitational microlensing," *Nature* 473, 349, doi: 10.1038/nature10092, http://arxiv.org/abs/1105.3544.

Thies, I., Kroupa, P., Goodwin, S. P., Stamatellos, D., & Whitworth, A. P.: 2010, "Tidally Induced Brown Dwarf and Planet Formation in Circumstellar Disks," *The Astrophysical Journal* 717, 577, doi: 10.1088/0004-637X/717/1/577, http://arxiv.org/abs/1005.3017.

Varvoglis, H., Sgardeli, V., & Tsiganis, K.: 2012, "Interaction of free-floating planets with a star–planet pair," *Celestial Mechanics and Dynamical Astronomy* 113, 387, doi: 10.1007/s10569-012-9429-8, http://arxiv.org/abs/1201.1385.

Wadhwa, M., Amelin, Y., Bizzarro, M., Kita, N., Kleine, T., Lugmair, G. W., & Yin, Q.: 2007, "Comparison of Short-lived and Long-lived Chronometers: Towards a Consistent Chronology of the Early Solar System," in *Chronology of Meteorites and the Early Solar System*, vol. 1374, p. 173.

Wood, J. A.: 2005, "The Chondrite Types and Their Origins," in A. N. Krot, E. R. D. Scott, and B. Reipurth (eds.), *Chondrites and the Protoplanetary Disk*, vol. 341 of *Astronomical Society of the Pacific Conference Series*, 953.

Chapter 8: Aged Stars and Disrupted Exosystems

Alcock, C., Fristrom, C. C., & Siegelman, R.: 1986, "On the number of comets around other single stars," *The Astrophysical Journal* 302, 462, doi: 10.1086/164005.

Farihi, J.: 2016, "Circumstellar debris and pollution at white dwarf stars," *New Astron. Rev.* 71, 9, doi: 10.1016/j.newar.2016.03.001, http://arxiv.org/abs/1604.03092.

Guo, J., Lin, L., Bai, C., & Liu, J.: 2016, "The effects of solar Reimers η on the final destinies of Venus, the Earth, and Mars," *Astrophysics and Space Science* 361, 122, doi: 10.1007/s10509-016-2684-5.

Hansen, B. M. S. & Liebert, J.: 2003, "Cool White Dwarfs," *Ann. Rev. Astron. Astrophys.* 41, 465, doi: 10.1146/annurev.astro.41.081401.155117.

Jura, M. & Young, E. D.: 2014, "Extrasolar Cosmochemistry," *Annual Review of Earth and Planetary Sciences* 42, 45, doi: 10.1146/annurev-earth-060313-054740.

Kirchhoff, G. & Bunsen, R.: 1860, "Chemische Analyse durch Spectralbeobachtungen," *Annalen der Physik* 186, 161, doi: 10.1002/andp.18601860602.

Koester, D. & Chanmugam, G.: 1990, "Physics of white dwarf stars," *Reports on Progress in Physics* 53, 837, doi: 10.1088/0034-4885/53/7/001.

Kunitomo, M., Ikoma, M., Sato, B., Katsuta, Y., & Ida, S.: 2011, "Planet Engulfment by ~1.5–3 M$_{sun}$ Red Giants," *The Astrophysical Journal* 737, 66, doi: 10.1088/0004-637X/737/2/66, http://arxiv.org/abs/1106.2251.

Napiwotzki, R.: 2009, "The galactic population of white dwarfs", in *Journal of Physics Conference Series*, vol. 172, 012004.

Nordhaus, J. & Spiegel, D. S.: 2013, "On the orbits of low-mass companions to white dwarfs and the fates of the known exoplanets," *Mon. Not. R. Astron. Soc.* 432, 500, doi: 10.1093/mnras/stt569, http://arxiv.org/abs/1211.1013.

Rasio, F. A., Tout, C. A., Lubow, S. H., & Livio, M.: 1996, "Tidal Decay of Close Planetary Orbits," *The Astrophysical Journal* 470, 1187, doi: 10.1086/177941, http://arxiv.org/abs/astro-ph/9605059.

Schrijver, C. J., Bagenal, F., & Sojka, J. J. (eds.): 2016, *Heliophysics: Active Stars, their Astrospheres, and Impacts on Planetary Environments,* Cambridge University Press, Cambridge.

Schrijver, C. J. & Siscoe, G. L. (eds.): 2012, *Heliophysics: Evolving Solar Activity and the Climates of Space and Earth,* Cambridge University Press, Cambridge.

van Maanen, A.: 1920, *No. 182.* "The photographic determination of stellar parallaxes with the 60-inch reflector." Fourth series, Contributions from the Mount Wilson Observatory / Carnegie Institution of Washington 182, 1.

Vanderburg, A., Johnson, J. A., Rappaport, S., Bieryla, A., Irwin, J., Lewis, J. A., Kipping, D., Brown, W. R., Dufour, P., Ciardi, D. R., Angus, R., Schaefer, L., Latham, D. W., Charbonneau, D., Beichman, C., Eastman, J., McCrady, N., Wittenmyer, R. A., & Wright, J. T.: 2015, "A disintegrating minor planet transiting a white dwarf," *Nature* 526, 546, doi: 10.1038/nature15527, http://arxiv.org/abs/1510.06387.

Veras, D.: 2016, "Post-main-sequence planetary system evolution," *Royal Society Open Science* 3, 150571, doi: 10.1098/rsos.150571, http://arxiv.org/abs/1601.05419.

Veras, D., Mustill, A. J., Gänsicke, B. T., Redfield, S., Georgakarakos, N., Bowler, A. B., & Lloyd, M. J. S.: 2016, "Full-lifetime simulations of multiple unequal-mass planets across all phases of stellar evolution," *Mon. Not. R. Astron. Soc.* 458, 3942, doi: 10.1093/mnras/stw476, http://arxiv.org/abs/1603.00025.

Villaver, E., Livio, M., Mustill, A. J., & Siess, L.: 2014, "Hot Jupiters and Cool Stars," *The Astrophysical Journal* 794, 3, doi: 10.1088/0004-637X/794/1/3, http://arxiv.org/abs/1407.7879.

Zuckerman, B.: 2015, "Recognition of the First Observational Evidence of an Extrasolar Planetary System," in Dufour, Bergeron, and Fontaine (eds.), *19th European Workshop on White Dwarfs ASP Conference Series,* vol. 493, ASP, 291.

Chapter 9: The Worlds of Exoplanets

Baines, E. K., McAlister, H. A., ten Brummelaar, T. A., Turner, N. H., Sturmann, J., Sturmann, L., Goldfinger, P. J., & Ridgway, S. T.: 2008, "CHARA Array Measurements of the Angular Diameters of Exoplanet Host Stars," *The Astrophysical Journal* 680, 728, doi: 10.1086/588009, http://arxiv.org/abs/0803.1411.

Baraffe, I., Chabrier, G., & Barman, T.: 2010, "The physical properties of extra-solar planets," *Reports on Progress in Physics* 73(1), 016901, doi: 10.1088/0034-4885/73/1/016901, http://arxiv.org/abs/1001.3577.

Baraffe, I., Chabrier, G., Fortney, J., & Sotin, C.: 2014, "Planetary Internal Structures," Protostars and Planets VI 763–86, doi: 10.2458/azu_uapress_9780816531240-ch033, http://arxiv.org/abs/1401.4738.

Bedell, M., Bean, J. L., Meléndez, J., Mills, S. M., Fabrycky, D. C., Freitas, F. C., Ramírez, I., Asplund, M., Liu, F., & Yong, D.: 2017, "Kepler-11 is a Solar Twin: Revising the Masses and Radii of Benchmark Planets via Precise Stellar Characterization," *The Astrophysical Journal* 839(2), 94.

Birkby, J. L., de Kok, R. J., Brogi, M., Schwarz, H., & Snellen, I. A. G.: 2017, "Discovery of Water at High Spectral Resolution in the Atmosphere of 51 Peg b," *Astron. J.* 153, 138, doi: 10.3847/1538-3881/aa5c87, http://arxiv.org/abs/1701.07257.

Bolmont, E., Raymond, S. N., von Paris, P., Selsis, F., Hersant, F., Quintana, E. V., & Barclay, T.: 2014, "Formation, Tidal Evolution, and Habitability of the Kepler-186 System," *The Astrophysical Journal* 793, 3, doi: 10.1088/0004-637X/793/1/3, http://arxiv.org/abs/1404.4368.

Boutle, I. A., Mayne, N. J., Drummond, B., Manners, J., Goyal, J., Hugo Lambert, F., Acreman, D. M., & Earnshaw, P. D.: 2017, "Exploring the climate of Proxima B with the Met Office Unified Model," *A&A* 601, A120, doi: 10.1051/0004-6361/201630020, http://arxiv.org/abs/1702.08463.

Burgasser, A. J. & Mamajek, E. E.: 2017, "On the Age of the TRAPPIST-1 System," *The Astrophysical Journal* 845, 2, doi: 10.3847/1538-4357/aa7fea, http://arxiv.org/ abs/1706.02018.

Campante, T. L., Barclay, T., Swift, J. J., Huber, D., Adibekyan, V. Z., Cochran, W., Burke, C. J., Isaacson, H., Quintana, E. V., Davies, G. R., Silva Aguirre, V., Ragozzine, D., Riddle, R., Baranec, C., Basu, S., Chaplin, W. J., Christensen-Dalsgaard, J., Metcalfe, T. S., Bedding, T. R., Handberg, R., Stello, D., Brewer, J. M., Hekker, S., Karoff, C., Kolbl, R., Law, N. M., Lundkvist, M., Miglio, A., Rowe, J. F., Santos, N. C., Van Laerhoven, C., Arentoft, T., Elsworth, Y. P., Fischer, D. A., Kawaler, S. D., Kjeldsen, H., Lund, M. N., Marcy, G. W., Sousa, S. G., Sozzetti, A., & White, T. R.: 2015, "An Ancient Extrasolar System with Five Sub-Earth-size Planets," *The Astrophysical Journal* 799, 170, doi: 10.1088/0004-637X/799/2/170, http://arxiv.org/abs/1501.06227.

D'Angelo, G. & Bodenheimer, P.: 2016, "In Situ and Ex Situ Formation Models of Kepler 11 Planets," *The Astrophysical Journal* 828, 33, doi: 10.3847/0004-637X/828/1/33, http://arxiv.org/abs/1606.08088.

Dorn, C., Hinkel, N. R., & Venturini, J.: 2017, "Bayesian analysis of interiors of HD 219134b, Kepler-10b, Kepler-93b, CoRoT-7b, 55 Cnc e, and HD 97658b using stellar abundance proxies," A&A 597, A38, doi: 10.1051/0004-6361/201628749, http://arxiv.org/abs/1609.03909.

Doyle, L. R., Carter, J. A., Fabrycky, D. C., Slawson, R. W., Howell, S. B., Winn, J. N., Orosz, J. A., Prsa, A., Welsh, W. F., Quinn, S. N., Latham, D., Torres, G., Buchhave, L. A., Marcy, G. W., Fortney, J. J., Shporer, A., Ford, E. B., Lissauer, J. J., Ragozzine, D., Rucker, M., Batalha, N., Jenkins, J. M., Borucki, W. J., Koch, D., Middour, C. K., Hall, J. R., McCauliff, S., Fanelli, M. N., Quintana, E. V., Holman, M. J., Caldwell, D. A., Still, M., Stefanik, R. P., Brown, W. R., Esquerdo, G. A., Tang, S., Furesz, G., Geary, J. C., Berlind, P., Calkins, M. L., Short, D. R., Steffen, J. H., Sasselov, D., Dunham, E. W., Cochran, W. D., Boss, A., Haas, M. R., Buzasi, D., & Fischer, D.: 2011, "Kepler-16: A Transiting Circumbinary Planet," Science 333, 1602, doi: 10.1126/science.1210923, http://arxiv.org/abs/1109.3432.

Dunhill, A. C. & Alexander, R. D.: 2013, "The curiously circular orbit of Kepler-16b," Mon. Not. R. Astron. Soc. 435, 2328, doi: 10.1093/mnras/stt1456, http://arxiv.org/abs/1308.0596.

Dupuy, T. J., Kratter, K. M., Kraus, A. L., Isaacson, H., Mann, A. W., Ireland, M. J., Howard, A. W., & Huber, D.: 2016, "Orbital Architectures of Planet-hosting Binaries. I. Forming Five Small Planets in the Truncated Disk of Kepler-444A," The Astrophysical Journal 817, 80, doi: 10.3847/0004-637X/817/1/80, http://arxiv.org/abs/1512.03428.

Figueira, P., Santerne, A., Suárez Mascareño, A., Gomes da Silva, J., Abe, L., Adibekyan, V. Z., Bendjoya, P., Correia, A. C. M., Delgado-Mena, E., Faria, J. P., Hebrard, G., Lovis, C., Oshagh, M., Rivet, J.-P., Santos, N. C., Suarez, O., & Vidotto, A. A.: 2016, "Is the activity level of HD 80606 influenced by its eccentric planet?," A&A 592, A143, doi: 10.1051/0004-6361/201628981, http://arxiv.org/abs/1606.05549.

Fortney, J. J. & Nettelmann, N.: 2010, "The Interior Structure, Composition, and Evolution of Giant Planets," Space Sci. Rev. 152, 423, doi: 10.1007/s11214-009-9582-x, http://arxiv.org/abs/0912.0533.

Gillon, M., Triaud, A. H. M. J., Demory, B., Jehin, E., Agol, E., Deck, K. M., Lederer, S. M., de Wit, J., Burdanov, A., Ingalls, J. G., Bolmont, E, Leconte, J., Raymond, S. N., Selsis, F., Turbet, M., Barkaoui, K., Burgasser, A., Burleigh, M. R., Carey, S. J., Chaushev, A., Copperwheat, C. M., Delrez, L., Fernandes, C. S., Holdsworth, D. L., Kotze, E. J., Van Grootel, V., Almleaky, Y., Benkhaldoun, Z., Magain, P., & Queloz, D.: 2017, "Seven temperate terrestrial planets around the

nearby ultracool dwarf star TRAPPIST-1," *Nature* 542, 456, doi: 10.1038/nature21360.

Hatzes, A. P., Guenther, E. W., Endl, M., Cochran, W. D., Döllinger, M. P., & Bedalov, A.: 2005, "A giant planet around the massive giant star HD 13189," *A&A* 437, 743, doi: 10.1051/0004-6361:20052850.

Hausoel, A., Karolak, M., ŞaşÉ©oğlu, E., Lichtenstein, A., Held, K., Katanin, A., Toschi, A., & Sangiovanni, G.: 2017, "Local magnetic moments in iron and nickel at ambient and Earth's core conditions," *Nat. Commun.* 8, 16062, doi: 10.1038/ncomms16062.

Helled, R., Bodenheimer, P., Podolak, M., Boley, A., Meru, F., Nayakshin, S., Fortney, J. J., Mayer, L., Alibert, Y., & Boss, A. P.: 2014, "Giant Planet Formation, Evolution, and Internal Structure," Protostars and Planets VI 643–665, doi: 10.2458/azu_uapress_9780816531240-ch028, http://arxiv.org/abs/1311.1142.

Kane, S. R., Hill, M. L., Kasting, J. F., Kopparapu, R. K., Quintana, E. V., Barclay, T., Batalha, N. M., Borucki, W. J., Ciardi, D. R., Haghighipour, N., Hinkel, N. R., Kaltenegger, L., Selsis, F., & Torres, G.: 2016, "A Catalog of Kepler Habitable Zone Exoplanet Candidates," *The Astrophysical Journal* 830, 1, doi: 10.3847/0004-637X/830/1/1, http://arxiv.org/abs/1608.00620.

Laughlin, G., Deming, D., Langton, J., Kasen, D., Vogt, S., Butler, P., Rivera, E., & Meschiari, S.: 2009, "Rapid heating of the atmosphere of an extrasolar planet," *Nature* 457, 562, doi: 10.1038/nature07649.

Léger, A., Grasset, O., Fegley, B., Codron, F., Albarede, A. F., Barge, P., Barnes, R., Cance, P., Carpy, S., Catalano, F., Cavarroc, C., Demangeon, O., Ferraz-Mello, S., Gabor, P., Grießmeier, J.-M., Leibacher, J., Libourel, G., Maurin, A.-S., Raymond, S. N., Rouan, D., Samuel, B., Schaefer, L., Schneider, J., Schuller, P. A., Selsis, F., & Sotin, C.: 2011, "The extreme physical properties of the CoRoT-7b super-Earth," *Icarus* 213, 1, doi: 10.1016/j.icarus.2011.02.004, http://arxiv.org/abs/1102.1629.

Lissauer, J. J., Dawson, R. I., & Tremaine, S.: 2014, "Advances in exoplanet science from Kepler," *Nature* 513, 336, doi: 10.1038/nature13781, http://arxiv.org/abs/1409.1595.

Lopez, E. D., Fortney, J. J., & Miller, N.: 2012, "How Thermal Evolution and Mass-loss Sculpt Populations of Super-Earths and Sub-Neptunes: Application to the Kepler-11 System and Beyond," *The Astrophysical Journal* 761, 59, doi: 10.1088/0004-637X/761/1/59, http://arxiv.org/abs/1205.0010.

Marcy, G. W., Butler, R. P., Williams, E., Bildsten, L., Graham, J. R., Ghez, A. M., & Jernigan, J. G.: 1997, "The Planet around 51 Pegasi," *The Astrophysical Journal* 481, 926, doi: 10.1086/304088.

Margalit, B. & Metzger, B. D.: 2017, "Merger of a white dwarf-neutron star binary to 10 29 carat diamonds: origin of the pulsar planets," *Mon. Not. R. Astron. Soc.* 465, 2790, doi: 10.1093/mnras/stw2640, http://arxiv.org/abs/1608.08636.

Marois, C., Zuckerman, B., Konopacky, Q. M., Macintosh, B., & Barman, T.: 2010, "Images of a fourth planet orbiting HR 8799," *Nature* 468, 1080, doi: 10.1038/nature09684, http://arxiv.org/abs/1011.4918.

Mortier, A., Santos, N. C., Sousa, S. G., Adibekyan, V. Z., Delgado Mena, E., Tsantaki, M., Israelian, G., & Mayor, M.: 2013, "New and updated stellar parameters for 71 evolved planet hosts. On the metallicity–giant planet connection," *A&A* 557, A70, doi: 10.1051/0004-6361/201321641, http://arxiv.org/abs/1307.7870.

Ramirez, R. M. & Kaltenegger, L.: 2014, "The Habitable Zones of Pre-main-sequence Stars," *Astrophysical Journal Letters* 797, L25, doi: 10.1088/2041-8205/797/2/L25, http://arxiv.org/abs/1412.1764.

Ribas, I., Bolmont, E., Selsis, F., Reiners, A., Leconte, J., Raymond, S. N., Engle, S. G., Guinan, E. F., Morin, J., Turbet, M., Forget, F., & Anglada-Escudé, G.: 2016, "The habitability of Proxima Centauri b. I. Irradiation, rotation and volatile inventory from formation to the present," *A&A* 596, A111, doi: 10.1051/0004-6361/201629576, http://arxiv.org/abs/1608.06813.

Sanchis-Ojeda, R., Rappaport, S., Pallè, E., Delrez, L., DeVore, J., Gandolfi, D., Fukui, A., Ribas, I., Stassun, K. G., Albrecht, S., Dai, F., Gaidos, E., Gillon, M., Hirano, T., Holman, M., Howard, A. W., Isaacson, H., Jehin, E., Kuzuhara, M., Mann, A. W., Marcy, G. W., Miles-Páez, P. A., Montañés-Rodríguez, P., Murgas, F., Narita, N., Nowak, G., Onitsuka, M., Paegert, M., Van Eylen, V., Winn, J. N., & Yu, L.: 2015, "The K2-ESPRINT Project I: Discovery of the Disintegrating Rocky Planet K2-22b with a Cometary Head and Leading Tail," *The Astrophysical Journal* 812, 112, doi: 10.1088/0004-637X/812/2/112, http://arxiv.org/abs/1504.04379.

Schrijver, C. J., Bagenal, F., & Sojka, J. J. (eds.): 2016, *Heliophysics: Active Stars, their Astrospheres, and Impacts on Planetary Environments*, Cambridge University Press, Cambridge.

Schrijver, C. J. & Siscoe, G. L. (eds.): 2011, *Heliophysics: Plasma Physics of the Local Cosmos*, Cambridge University Press, Cambridge.

Schuler, S. C., Kim, J. H., Tinker, Jr., M. C., King, J. R., Hatzes, A. P., & Guenther, E. W.: 2005, "High-Resolution Spectroscopy of the Planetary Host HD 13189: Highly Evolved and Metal-poor," *Astrophysical Journal Letters* 632, L131, doi: 10.1086/497988, http://arxiv.org/abs/astro-ph/0509270.

Showman, A. P. & Guillot, T.: 2002, "Atmospheric circulation and tides of "51 Pegasus b-like" planets," *A&A* 385, 166, doi: 10.1051/0004-6361:20020101, http://arxiv.org/abs/astro-ph/0202236.

Silva Aguirre, V., Davies, G. R., Basu, S., Christensen-Dalsgaard, J., Creevey, O., Metcalfe, T. S., Bedding, T. R., Casagrande, L., Handberg, R., Lund, M. N., Nissen, P. E., Chaplin, W. J., Huber, D., Serenelli, A. M., Stello, D., Van Eylen, V., Campante, T. L., Elsworth, Y., Gilliland, R. L., Hekker, S., Karoff, C., Kawaler, S. D., Kjeldsen, H., & Lundkvist, M. S.: 2015, "Ages and fundamental properties of Kepler exoplanet host stars from asteroseismology," *Mon. Not. R. Astron. Soc.* 452, 2127, doi: 10.1093/mnras/stv1388, http://arxiv.org/abs/1504.07992.

Silvotti, R., Schuh, S., Janulis, R., Solheim, J.-E., Bernabei, S., Østensen, R., Oswalt, T. D., Bruni, I., Gualandi, R., Bonanno, A., Vauclair, G., Reed, M., Chen, C.-W., Leibowitz, E., Paparo, M., Baran, A., Charpinet, S., Dolez, N., Kawaler, S., Kurtz, D., Moskalik, P., Riddle, R., & Zola, S.: 2007, "A giant planet orbiting the 'extreme horizontal branch' star V391 Pegasi," *Nature* 449, 189, doi: 10.1038/nature06143.

Turbet, M., Bolmont, E., Leconte, J., Forget, F., Selsis, F., Tobie, G., Caldas, A., Naar, J., & Gillon, M.: 2017, "Climate diversity on cool planets around cool stars with a versatile 3-D Global Climate Model: the case of TRAPPIST-1 planets," ArXiv e-prints, http://arxiv.org/abs/1707.06927.

Turbet, M., Leconte, J., Selsis, F., Bolmont, E., Forget, F., Ribas, I., Raymond, S. N., & Anglada-Escudé, G.: 2016, "The habitability of Proxima Centauri b. II. Possible climates and observability," *A&A* 596, A112, doi: 10.1051/0004-6361/201629577, http://arxiv.org/abs/1608.06827.

Vida, K., Kővári, Z., Pál, A., Oláh, K., & Kriskovics, L.: 2017, "Frequent Flaring in the TRAPPIST-1 System—Unsuited for Life?," *The Astrophysical Journal* 841, 124, doi: 10.3847/1538-4357/aa6f05, http://arxiv.org/abs/1703.10130.

Winn, J. N., Howard, A. W., Johnson, J. A., Marcy, G. W., Gazak, J. Z., Starkey, D., Ford, E. B., Colón, K. D., Reyes, F., Nortmann,

L., Dreizler, S., Odewahn, S., Welsh, W. F., Kadakia, S., Vanderbei, R. J., Adams, E. R., Lockhart, M., Crossfield, I. J., Valenti, J. A., Dantowitz, R., & Carter, J. A.: 2009, "The Transit Ingress and the Tilted Orbit of the Extraordinarily Eccentric Exoplanet HD 80606b," *The Astrophysical Journal* 703, 2091, doi: 10.1088/0004-637X/703/2/2091, http://arxiv.org/abs/0907.5205.

Wolszczan, A.: 2008, "Fifteen years of the neutron star planet research," *Physica Scripta Volume T* 130(1), 014005, doi: 10.1088/0031-8949/2008/T130/014005.

Chapter 10: Habitability of Planets and Moons

Domagal-Goldman, S. D., Wright, K. E., Adamala, K., Arina de la Rubia, L., Bond, J., Dartnell, L. R., Goldman, A. D., Lynch, K., Naud, M.-E., Paulino-Lima, I. G., Singer, K., Walter-Antonio, M., Abrevaya, X. C., Anderson, R., Arney, G., Atri, D., Azúa-Bustos, A., Bowman, J. S., Brazelton, W. J., Brennecka, G. A., Carns, R., Chopra, A., Colangelo-Lillis, J., Crockett, C. J., DeMarines, J., Frank, E. A., Frantz, C., de la Fuente, E., Galante, D., Glass, J., Gleeson, D., Glein, C. R., Goldblatt, C., Horak, R., Horodyskyj, L., Kaçar, B., Kereszturi, A., Knowles, E., Mayeur, P., McGlynn, S., Miguel, Y., Montgomery, M., Neish, C., Noack, L., Rugheimer, S., Stüeken, E. E., Tamez-Hidalgo, P., Walker, S. I., & Wong, T.: 2016, "The Astrobiology Primer v2.0," *Astrobiology* 16, 561, doi: 10.1089/ast.2015.1460.

Güdel, M.: 2007, "The Sun in Time: Activity and Environment," *Living Reviews in Solar Physics* 4, 3, doi: 10.12942/lrsp-2007-3, http://arxiv.org/abs/0712.1763.

Hatzes, A. P.: 2016, "Periodic Hα variations in GL 581: Further evidence for an activity origin to GL 581d," *A&A* 585, A144, doi: 10.1051/0004-6361/201527135, http://arxiv.org/abs/1512.00878.

Hörst, S. M.: 2017, "Titan's atmosphere and climate," *Journal of Geophysical Research (Planets)* 122, 432, doi: 10.1002/2016JE005240, http://arxiv.org/abs/1702.08611.

Jakosky, B. M., Slipski, M., Benna, M., Mahaffy, P., Elrod, M., Yelle, R., Stone, S., & Alsaeed, N.: 2017, "Mars' atmospheric history derived from upper-atmosphere measurements of $^{38}Ar/^{36}Ar$," *Science* 355, 1408, doi: 10.1126/science.aai7721.

Joshi, M.: 2003, "Climate Model Studies of Synchronously Rotating Planets," *Astrobiology* 3, 415, doi: 10.1089/153110703769016488.

Lammer, H., Blanc, M., Benz, W., Fridlund, M., Foresto, V. C. d., Güdel, M., Rauer, H., Udry, S., Bonnet, R.-M., Falanga, M., Charbonneau, D., Helled, R., Kley, W., Linsky, J., Elkins-Tanton, L. T., Alibert, Y., Chassefière, E., Encrenaz, T., Hatzes, A. P., Lin, D., Liseau, R., Lorenzen, W., & Raymond, S. N.: 2013, "The Science of Exoplanets and Their Systems," *Astrobiology* 13, 793, doi: 10.1089/ast.2013.0997.

Lammer, H., Güdel, M., Kulikov, Y., Ribas, I., Zaqarashvili, T. V., Khodachenko, M. L., Kislyakova, K. G., Gröller, H., Odert, P., Leitzinger, M., Fichtinger, B., Krauss, S., Hausleitner, W., Holmström, M., Sanz-Forcada, J., Lichtenegger, H. I. M., Hanslmeier, A., Shematovich, V. I., Bisikalo, D., Rauer, H., & Fridlund, M.: 2012, "Variability of solar/stellar activity and magnetic field and its influence on planetary atmosphere evolution," *Earth, Planets, and Space* 64, 179, doi: 10.5047/eps.2011.04.002.

Lopes, R. M. C., Kirk, R. L., Mitchell, K. L., Legall, A., Barnes, J. W., Hayes, A., Kargel, J., Wye, L., Radebaugh, J., Stofan, E. R., Janssen, M. A., Neish, C. D., Wall, S. D., Wood, C. A., Lunine, J. I., & Malaska, M. J.: 2013, "Cryovolcanism on Titan: New results from Cassini RADAR and VIMS," *Journal of Geophysical Research (Planets)* 118, 416, doi: 10.1002/jgre.20062.

Oremland, R. S.: 1989, "Present-day biogeochemical activities of anaerobic bacteria and their relevance to future exobiological investigations," *Advances in Space Research* 9, 127, doi: 10.1016/0273-1177(89)90218-4.

Owen, T.: 1982, "The atmosphere of Titan," *J. Mol. Evol.* 18, 150, doi: 10.1007/BF01733040.

Raulin, F.: 2008, "Astrobiology and habitability of Titan," *Space Sci. Rev.* 135, 37, doi: 10.1007/s11214-006-9133-7.

Robert, F.: 2006, "Solar System Deuterium/Hydrogen Ratio," in Lauretta, D. S. & McSween, H. Y. (eds.) *Meteorites and the Early Solar System II*, University of Arizona Press, Tucson, 341–51.

Sagan, C.: 1997, "Titan in 2097," *Planet. Space Sci.* 45, 887, doi: 10.1016/S0032-0633(97)80259-0.

Schrijver, C. J. & Beer, J.: 2014, "Space Weather From Explosions on the Sun: How Bad Could It Be?," *EOS Transactions* 95, 201, doi: 10.1002/2014EO240001.

Schrijver, C. J., Dobbins, R., Murtagh, W., & Petrinec, S. M.: 2014, "Assessing the impact of space weather on the electric power grid based on insurance claims for industrial electrical equipment,"

Space Weather 12, 487, doi: 10.1002/2014SW001066, http://arxiv.org/abs/1406.7024.

Schrijver, C. J. & Zwaan, C.: 2000, *Solar and Stellar Magnetic Activity*, Cambridge University Press, New York.

Shields, A. I., Ballard, S., & Johnson, J. A.: 2016, "The habitability of planets orbiting M-dwarf stars," *Physics Reports* 663, 1, doi: 10.1016/j.physrep.2016.10.003, http://arxiv.org/abs/1610.05765.

Showman, A. P., Wordsworth, R. D., Merlis, T. M., & Kaspi, Y.: 2013, "Atmospheric Circulation of Terrestrial Exoplanets," in S. J. Mackwell, A. A. Simon-Miller, J. W. Harder, and M. A. Bullock (eds.), *Comparative Climatology of Terrestrial Planets*, p. 277.

von Bloh, W., Bounama, C., Cuntz, M., & Franck, S.: 2007, "The habitability of super-Earths in Gliese 581," *A&A* 476, 1365, doi: 10.1051/0004-6361:20077939, http://arxiv.org/abs/0705.3758.

Way, M. J., Del Genio, A. D., Kiang, N. Y., Sohl, L. E., Grinspoon, D. H., Aleinov, I., Kelley, M., & Clune, T.: 2016, "Was Venus the first habitable world of our solar system?," *Geophys. Res. Lett.* 43, 8376, doi: 10.1002/2016GL069790, http://arxiv.org/abs/1608.00706.

Woese, C. R. & Fox, G. E.: 1977, "Phylogenetic structure of the prokayotic domain: the primary kingdoms," *Proc. Nat. Acad. of Sciences* 74, 5088, doi: 10.1073/pnas.74.11.5088.

Wordsworth, R. D., Forget, F., Selsis, F., Millour, E., Charnay, B., & Madeleine, J.-B.: 2011, "Gliese 581d is the First Discovered Terrestrial-mass Exoplanet in the Habitable Zone," *Astrophysical Journal Letters* 733, L48, doi: 10.1088/2041-8205/733/2/L48, http://arxiv.org/abs/1105.1031.

Chapter 11: The Long View of Planetary Systems

Behroozi, P. & Peeples, M. S.: 2015, "On the history and future of cosmic planet formation," *Mon. Not. R. Astron. Soc.* 454, 1811, doi: 10.1093/mnras/stv1817, http://arxiv.org/abs/1508.01202.

Davies, M. B., Adams, F. C., Armitage, P., Chambers, J., Ford, E., Morbidelli, A., Raymond, S. N., & Veras, D.: 2014, "The Long-Term Dynamical Evolution of Planetary Systems," *Protostars and Planets VI* 787–808, doi: 10.2458/azu_uapress_9780816531240-ch034, http://arxiv.org/abs/1311.6816.

Kenyon, S. J. & Bromley, B. C.: 2017, "Numerical Simulations of Collisional Cascades at the Roche Limits of White Dwarf Stars," *The Astrophysical Journal* 844, 116, http://arxiv.org/abs/1706.08579.

Further Reading

Kirkwood, D.: 1869, "On the Nebular Hypothesis, and the Approximate Commensurability of the Planetary Periods," *Mon. Not. R. Astron. Soc.* 29, 96.

Krauss, L. M. & Scherrer, R. J.: 2007, "The return of a static universe and the end of cosmology," *General Relativity and Gravitation* 39, 1545, doi: 10.1007/s10714-007-0472-9, http://arxiv.org/abs/0704.0221.

Krauss, L. M. & Starkman, G. D.: 2000, "Life, the Universe, and Nothing: Life and Death in an Ever-expanding Universe," *The Astrophysical Journal* 531, 22, doi: 10.1086/308434, http://arxiv.org/abs/astro-ph/9902189.

Madau, P. & Dickinson, M.: 2014, "Cosmic Star-Formation History," *Ann. Rev. Astron. Astrophys.* 52, 415, doi: 10.1146/annurev-astro-081811-125615, http://arxiv.org/abs/1403.0007.

Nagamine, K. & Loeb, A.: 2003, "Future evolution of nearby large-scale structures in a universe dominated by a cosmological constant," *Nature* 8, 439, doi: 10.1016/S1384-1076(02)00234-8, http://arxiv.org/abs/astro-ph/0204249.

Pfalzner, S., Davies, M. B., Gounelle, M., Johansen, A., Münker, C., Lacerda, P., Portegies Zwart, S., Testi, L., Trieloff, M., & Veras, D.: 2015, "The formation of the solar system," *Physica Scripta* 90(6), 068001, doi: 10.1088/0031-8949/90/6/068001, http://arxiv.org/abs/1501.03101.

Sawyer Hogg, H.: 1950, "Out of Old Books (Kirkwood's Gaps in Asteroid Orbits)," *Journal of the Royal Astronomical Society of Canada* 44, 163.

Schrijver, C. J., Bagenal, F., & Sojka, J. J. (eds.): 2016, *Heliophysics: Active Stars, their Astrospheres, and Impacts on Planetary Environments,* Cambridge University Press, Cambridge.

Zackrisson, E., Calissendorff, P., Gonzalez, J., Benson, A., Johansen, A., & Janson, M.: 2016, "Terrestrial planets across space and time," *The Astrophysical Journal* 833(2), 214.

Chapter 12: Living on a Pale Blue Dot

Bromham, L., Dinnage, R., & Hua, X: 2016, "Interdisciplinary research has consistently lower funding success." *Nature.* 2016 Jun 30; 534(7609):684-7. doi: 10.1038/nature18315.

Burke, C. J., Christiansen, J. L., Mullally, F., Seader, S., Huber, D., Rowe, J. F., Coughlin, J. L., Thompson, S. E., Catanzarite, J., Clarke, B. D., Morton, T. D., Caldwell, D. A., Bryson, S. T., Haas, M. R., Batalha, N. M., Jenkins, J. M., Tenenbaum, P., Twicken, J. D., Li, J., Quintana, E., Barclay, T., Henze, C. E., Borucki, W. J., Howell, S. B., & Still, M.:

2015, "Terrestrial Planet Occurrence Rates for the Kepler GK Dwarf Sample," *The Astrophysical Journal* 809, 8, doi: 10.1088/0004-637X/809/1/8, http://arxiv.org/abs/1506.04175.

Firestone, R. B.: 2014, "Observation of 23 Supernovae That Exploded <300 pc from Earth during the past 300 kyr," *The Astrophysical Journal* 789, 29, doi: 10.1088/0004-637X/789/1/29.

Forget, F.: 2013, "On the probability of habitable planets," *International Journal of Astrobiology* 12, 177, doi: 10.1017/S1473550413000128, http://arxiv.org/abs/1212.0113.

Güdel, M., Dvorak, R., Erkaev, N., Kasting, J., Khodachenko, M., Lammer, H., Pilat-Lohinger, E., Rauer, H., Ribas, I., & Wood, B. E.: 2014, "Astrophysical Conditions for Planetary Habitability," Protostars and Planets VI 883–906, doi: 10.2458/azu_uapress_9780816531240-ch038, http://arxiv.org/abs/1407.8174.

Hess, S. L., Henry, R. M., & Tillman, J. E.: 1979, "The seasonal variation of atmospheric pressure on Mars as affected by the south polar CAP," *J. Geophys. Research* 84, 2923, doi: 10.1029/JB084iB06p02923.

Holland, H. D.: 2006, "The oxygenation of the atmosphere and oceans," *Phil. Trans. Royal Soc. B* 361, 903, doi: 10.1098/rstb.2006.1838.

Kenyon, S. J., Najita, J. R., & Bromley, B. C.: 2016, "Rocky Planet Formation: Quick and Neat," *The Astrophysical Journal* 831, 8, doi: 10.3847/0004-637X/831/1/8.

Lammer, H., Blanc, M., Benz, W., Fridlund, M., Foresto, V. C. d., Güdel, M., Rauer, H., Udry, S., Bonnet, R.-M., Falanga, M., Charbonneau, D., Helled, R., Kley, W., Linsky, J., Elkins-Tanton, L. T., Alibert, Y., Chassefière, E., Encrenaz, T., Hatzes, A. P., Lin, D., Liseau, R., Lorenzen, W., & Raymond, S. N.: 2013, "The Science of Exoplanets and Their Systems," *Astrobiology* 13, 793, doi: 10.1089/ast.2013.0997.

Lissauer, J. J., Dawson, R. I., & Tremaine, S.: 2014, "Advances in exoplanet science from Kepler," *Nature* 513, 336, doi: 10.1038/nature13781, http://arxiv.org/abs/1409.1595.

Marcy, G. W., Weiss, L. M., Petigura, E. A., Isaacson, H., Howard, A. W., & Buchhave, L. A.: 2014, "Occurrence and core–envelope structure of 1–4 times Earth-size planets around Sun-like stars," *Proceedings of the National Academy of Science* 111, 12655, doi: 10.1073/pnas.1304197111, http://arxiv.org/abs/1404.2960.

Morbidelli, A. & Raymond, S. N.: 2016, "Challenges in planet formation," *Journal of Geophysical Research (Planets)* 121, 1962, doi: 10.1002/2016JE005088, http://arxiv.org/abs/1610.07202.

Pavlov, A. A., Toon, O. B., Pavlov, A. K., Bally, J., & Pollard, D.: 2005, "Passing through a giant molecular cloud: 'Snowball' glaciations produced by interstellar dust," *Geophys. Res. Lett.* 32, L03705, doi: 10.1029/2004GL021890.

Petigura, E. A., Howard, A. W., & Marcy, G. W.: 2013, "Prevalence of Earth-size planets orbiting Sun-like stars," *Proceedings of the National Academy of Science* 110, 19273, doi: 10.1073/pnas.1319909110, http://arxiv.org/abs/1311.6806.

Pierrehumbert, R. T., Abbot, D. S., Voigt, A., & Koll, D.: 2011, "Climate of the Neoproterozoic," *Annual Review of Earth and Planetary Sciences* 39, 417, doi: 10.1146/annurev-earth-040809-152447.

Schrijver, C. J., Bagenal, F., & Sojka, J. J. (eds.): 2016, *Heliophysics: Active Stars, their Astrospheres, and Impacts on Planetary Environments*, Cambridge University Press, Cambridge.

Schrijver, C. J. & Siscoe, G. L. (eds.): 2012, *Heliophysics: Evolving Solar Activity and the Climates of Space and Earth*, Cambridge University Press, Cambridge.

Sleep, N. H.: 2010, "The Hadean–Archaean environment," *Cold Spring Harb. Perspect. Biol.* 2, a00257, doi: 10.1101/cshperspect.a002527.

Walker, J. C. G., Hays, P. B., & Kasting, J. F.: 1981, "A negative feedback mechanism for the long-term stabilization of the earth's surface temperature," *J. Geophys. Research* 86, 9776, doi: 10.1029/JC086iC10p09776.

Ward, P. & Brownlee, D.: 2000, "Rare Earth: Why Complex Life is Uncommon in the Universe," Copernicus, Springer-Verlag, New York.

Wilson, L.: 2009, "Volcanism in the Solar System," *Nature Geoscience* 2, 389, doi: 10.1038/ngeo529.

Winn, J. N. & Fabrycky, D. C.: 2015, "The Occurrence and Architecture of Exoplanetary Systems," *Ann. Rev. Astron. Astrophys.* 53, 409, doi: 10.1146/annurev-astro-082214-122246, http://arxiv.org/abs/1410.4199.

Wolf, E. T., Shields, A. L., Kopparapu, R. K., Haqq-Misra, J., & Toon, O. B.: 2017, "Constraints on Climate and Habitability for Earth-like Exoplanets Determined from a General Circulation Model," *The Astrophysical Journal* 837, 107, doi: 10.3847/1538-4357/aa5ffc.

Zackrisson, E., Calissendorff, P., Gonzalez, J., Benson, A., Johansen, A., & Janson, M.: 2016, "Terrestrial planets across space and time," *The Astrophysical Journal* 833(2), 214.

Acknowledgments

Working as a stellar astrophysicist with a focus on the Sun and stars like it, I have witnessed the discoveries in planetary and exoplanetary sciences in parallel to my own work: I have seen the many exciting stories in the media, attended presentations at astrophysical conferences, and engaged in discussions with colleagues around the world. I have savored the excitement of the fast-moving field of exoplanets and wanted to digest what was being learned about planetary systems, including our own. But the time to adequately study the literature simply has not been there, until recently.

The foundation of this book was laid over a decade ago, however, when Madhulika Guhathakurta at NASA convinced me to partner with George Siscoe in creating a summer school for recent postgraduates on the multitude of physical connections between the Sun and its planets. In the first years of that project George guided me into the world of planetary sciences, while Fran Bagenal and Jan Sojka did so in subsequent cycles. I am grateful to all of them for introducing me to worlds adjacent to mine. Over the decade that I was involved in that project, we invited colleagues from around the world to teach at the school and to write chapters on many of the topics that are touched upon in this book; I am grateful to them for sharing their knowledge and enthusiasm.

I am grateful to the efficient and thoughtful support of my editors at Oxford University Press: Sonke Adlung and Ania Wronski, in particular, for their guidance on content and layout, and Henry MacKeith for his meticulous copy-editing.

As in everything, I thank my fabulous partner on this planet, Iris, for her interest in this project, for her encouragement and

patience throughout the years during which this project took shape and I appeared as a result often absent-minded, and for her thoughtful, conscientious comments on the first full version of the manuscript, pointing out where my enthusiasm or impatience had caused me to leap over an intelligible formulation.

List of Illustrations

Web Resources

Astrobiology web:
http://astrobiology.com
Astrophysics Data System to search the professional literature:
http://adsabs.harvard.edu/abstract_service.html
European Space Agency space science and satellite missions:
http://www.esa.int/Our_Activities/Space_Science/
Exoplanet archive of the NASA Exoplanet Science Institute:
https://exoplanetarchive.ipac.caltech.edu
Exoplanet exploration at NASA:
https://exoplanets.nasa.gov
Exoplanet Science Institute at NASA:
http://nexsci.caltech.edu
Extrasolar planets encyclopaedia:
http://exoplanet.eu
Exoplanet Data Explorer:
http://exoplanets.org
Hubble Heritage site at NASA:
http://heritage.stsci.edu;
see also the gallery index:
http://heritage.stsci.edu/gallery/gallery_category.html
Kepler and K2 satellite mission home page:
https://www.nasa.gov/mission_pages/kepler/
NASA space missions access page:
https://www.nasa.gov/missions
SETI Institute home page:
http://www.seti-inst.edu

Index

ONE OF TEN BILLION EARTHS